NUMERICAL METHODS FOR ORDINARY DIFFERENTIAL SYSTEMS

NUMERICAL METHODS FOR ORDINARY DIFFERENTIAL SYSTEMS

The Initial Value Problem

J. D. Lambert
Professor of Numerical Analysis,
University of Dundee, Scotland

JOHN WILEY & SONS
Chichester · New York · Brisbane · Toronto · Singapore

Reprinted June 1997, February 1999, April 2000

Other Wiley Editorial Offices

John Wiley & Sons, Inc., 605 Third Avenue, New York, NY 10158-0012, USA

Jacaranda Wiley Ltd, G.P.O. Box 859, Brisbane, Queensland 4001, Australia

John Wiley & Sons (Canada) Ltd, 22 Worcester Road, Rexdale, Ontario M9W 1L1, Canada

John Wiley & Sons (SEA) Pte Ltd, 37 Jalan Pemimpin 05-04, Block B, Union Industrial Building, Singapore 2057

Library of Congress Cataloging-in-Publication Data:

Lambert, J. D. (John Denholm), 1932–
Numerical methods for ordinary differential systems: the initial value problem / J. D. Lambert.
p. cm.
Includes bibliographical references and index.
ISBN 0 471 92990 5
1. Initial value problems—Numerical solutions. I. Title.
QA328.L36 1991
515'.35—dc20 90-28513
CIP

A catalogue record for this book is available from the British Library

Typeset in 10/12 pt Times by Thomson Press (India) Ltd
Printed and bound in Great Britain by Antony Rowe Ltd, Eastbourne

To my extended family

Heather

Jenny, Roger, Ian, Douglas and Rachel

Ian

Rod and Elaine

Contents

Preface

In the late 1960s and early 1970s, I wrote a book on the numerical analysis of ordinary differential equations entitled *Computational Methods in Ordinary Differential Equations*, published in 1973; to my considerable surprise, it is still in print. That book was largely based on a course of lectures I had given to M.Sc. students of numerical analysis in the University of Dundee, a course which I have continued to give ever since. As the years have passed, the role of Lambert (1973) has changed from being virtually the content of that course, through a phase when parts of it were dropped and replaced by newer material, to the current situation, where it is relegated to the status of a background reference. There has never been a revised edition of Lambert (1973). I have always felt lukewarm about the idea of revised editions; too often the end-product seems to resemble the results of altering a house by chopping bits off and throwing up extensions—it fulfils its new purpose, but it is not what you would have designed if you had had a free hand. The present book is intended as a replacement for Lambert (1973), and is by no means a revision of it. Although the general topic remains the same, the overlap between the two books has turned out to be very small indeed. The intended readership is precisely the same as for Lambert (1973), namely postgraduate and advanced undergraduate students and users of numerical methods.

Emphasis in the subject of numerical methods for initial value problems in ordinary differential systems has changed substantially since Lambert (1973) was written. At that time, new methods were constantly being proposed (frequently with scant consideration of the problems of efficient implementation); on the theoretical side, convergence was of course well understood, but stability relied exclusively on a linear theory based on an over-restrictive linear constant coefficient test equation. In contrast, the major codes in use today are based on only a handful of methods (Adams–Bashforth–Moulton, Runge–Kutta and the Backward Differentiation Formulae), but embedded in very sophisticated and well-analysed implementations; moreover, there now exists a much more satisfactory nonlinear stability theory. (It is a little ironic to note that all of the names mentioned in the preceding sentence are from the nineteenth century.)

In this book I have tried to reflect those changes. From the outset, systems of differential equations rather than scalar equations are considered, and the basic topics of consistency, zero-stability and convergence are set in a context of a general class of methods. Linear multistep and predictor–corrector methods are studied, at first in general, but with increasing emphasis on Adams methods. Problems of implementation are considered in much more detail than in Lambert (1973). Many of the results on Runge–Kutta methods in Lambert (1973) are valid only for the scalar problem, and I therefore felt it necessary to include a non-rigorous account of the Butcher theory for Runge–Kutta methods for systems. Of course, this theory existed when Lambert (1973) was written, but I felt at that time that it was too demanding for a text at this level; I no longer believe this, and have found that students can not only assimilate this material, but rather enjoy it. (I have also included a brief account of the alternative approach of Albrecht, which appears not to be as well known as one would expect.) There is much more emphasis on embedded explicit Runge–Kutta methods and on implicit Runge–Kutta methods. The topic of stiffness is treated in much more detail than in Lambert (1973), and includes an account of nonlinear stability theory. Numerical experiments are interspersed with the text, and exercises are inserted at the ends of appropriate sections. In the main, the latter are straightforward and are intended only to illustrate and, occasionally, to extend the text; those which are longer or more demanding are marked ★.

This change of emphasis towards implementation has presented me with two problems. Firstly, I had to decide to leave out a number of methods (such as cyclic methods) which, though of intrinsic interest, do not appear to be competitive. The hardest decision in this respect was to omit extrapolation methods, for which a case for competitiveness can be made; in the end, I concluded that the interests of the intended readership would be better served by not sacrificing any of the material on the major classes of methods to make room for them. I have not included general linear methods on the grounds that they are not yet competitive. The second problem is that studying a smaller number of methods in greater depth tends to raise the level of difficulty. I believe that I have managed to avoid this by giving non-rigorous accounts where appropriate; where rigour is essential, I have quoted theorems, but supplied proofs only where these are constructive (in a numerical analytical sense). My evidence for this belief is that most of the material in this book has been tried out on students at the same level as those I taught in 1973 (and I see no overwhelming indications that 1990 students are any brighter or better prepared!). I do not believe that the change of emphasis has produced a duller book (but the reader will have to judge that for himself). One of the magical properties of mathematics is its ability to keep producing fascinating ideas even when it is attempting to answer practical and technical questions; there can be no better example of this than the emergence of the elegant order star theory of Hairer, Wanner and Nørsett, briefly covered in Chapter 6.

I am grateful to my friend and colleague Des Higham for his careful reading of the manuscript and for his many useful suggestions.

J. D. LAMBERT
Dundee, Scotland
November 1990

1 Background Material

1.1 INTRODUCTION

The level of mathematical background needed for this book is not particularly high; in general, a knowledge of the calculus and of some aspects of numerical linear algebra (vector and matrix norms, LU-decomposition) together with some familiarity with complex numbers will suffice. Inevitably, there will be occasions when we need to employ some additional concepts and techniques, not perhaps part of every reader's experience. In some situations, such as the development of Butcher's algebraic theory of Runge–Kutta methods (Chapter 5), it seems appropriate to develop the necessary tools *in situ*, but in others, where the use of the tools is more incidental, such an approach can be disruptive. Accordingly, in this chapter we collect together a number of these additional concepts and techniques. No attempt is made to treat them in a comprehensive manner, and we settle for taking them just as far as is necessary for an understanding of their use in the remainder of the text.

This chapter is, of necessity, a bit of a rag-bag, and readers who are familiar with its contents are urged to proceed at once to Chapter 2.

1.2 NOTATION

This book is concerned with the numerical solution of *systems* of ordinary differential equations, which means that we shall regularly be dealing with vectors. So widespread will be the use of vectors that any attempt to differentiate between scalars and vectors by setting the latter in bold fount would result in rather an ugly text. For most of the time it will be obvious from the context whether a particular symbol represents a scalar or a vector, but when there is doubt we shall insert statements such as $x \in \mathbb{R}$ to indicate that x is a real scalar and $y \in \mathbb{R}^m$ to indicate that y is a real m-dimensional vector. Alternatively, we may write statements such as

$$y' = f(x, y), \quad f: \mathbb{R} \times \mathbb{R}^m \to \mathbb{R}^m$$

to indicate that x is a scalar and y and f are m-dimensional vectors. Similarly, $\zeta \in \mathbb{C}$, $z \in \mathbb{C}^m$ will indicate that ζ is a complex scalar and z a complex m-dimensional vector.

The tth component of $y \in \mathbb{R}^m$ will be denoted by ${}^t y$, so that we may write

$$y = [{}^1 y, {}^2 y, \dots, {}^m y]^\mathrm{T},$$

where the superscript T denotes transpose (vectors will always be column vectors). The slightly unusual notation of labelling a component of a vector by a left superscript is rather forced upon us; the more conventional positions for such labels are pre-booked for other purposes.

Any norm of a vector y or of a matrix A will be denoted by $\|y\|$ and $\|A\|$ respectively, and when vector and matrix norms appear in the same context it is assumed that the matrix norm is subordinate to the vector norm. When we need to use a specific norm, it will usually be the L_2-norm, $\|\cdot\|_2$, defined by

$$\|y\|_2 = \left[\sum_{t=1}^{m} |{}^t y|^2\right]^{1/2}, \qquad \|A\|_2 = (\text{maximum eigenvalue of } \bar{A}^{\mathrm{T}} A)^{1/2},$$

where $\bar{A}$ is the conjugate of the matrix A.

We shall frequently use the notation

$$F(h) = 0(h^p) \qquad \text{as } h \to 0,$$

where h is a scalar and F may be either a scalar or a vector. It means that there exists a positive constant K such that $\|F(h)\| \leqslant Kh^p$ for h sufficiently close to zero. Normally, we do not bother to add the phrase 'as $h \to 0$' and merely write $F(h) = 0(h^p)$, but 'as $h \to 0$' is still implied. This is of some importance, since $F(h) = 0(h^p)$ is an *asymptotic* statement concerning what happens for sufficiently small h; its interpretation must not be debased into implying that $F(h)$ is roughly the same size as h^p, no matter the size of h!

We shall occasionally use the notations $x \in [a, b]$, $x \in (a, b)$, $x \in (a, b]$ and $x \in [a, b)$ meaning that the scalar x satisfies $a \leqslant x \leqslant b, a < x < b, a < x \leqslant b$ and $a \leqslant x < b$ respectively. We shall also write $y(x) \in C^m[a, b]$ to mean that $y(x)$ possesses m continuous derivatives for $x \in [a, b]$. The rth total derivative of y with respect to x will be denoted by $y^{(r)}(x), r = 0, 1, 2, \ldots,$ where $y^{(0)}(x) \equiv y(x)$ and $y^{(1)}(x) \equiv y'(x)$. We shall also use the notations $A := B$ and $B =: A$ to mean 'A is defined to be B'.

The use of the finite difference operators E, Δ and ∇ will frequently prove helpful. These are defined as follows. Let $\{(x_n, F_n), x_n \in \mathbb{R}, F_n \in \mathbb{R}^m, n = 0, 1, 2, \ldots\}$ be a set of *equally spaced* data points in $\mathbb{R}^{m+1}$; that is, we have that $x_n = x_0 + nh$, where h is a constant. Then the *forward shift operator* E is defined by

$$EF_n = F_{n+1}, E^2 F_n = E(EF_n) = F_{n+2}, \quad \text{etc.}$$

Note that if π is a polynomial of degree k, $\pi(r) = \sum_{j=0}^{k} \gamma_j r^j$, then we may write $\sum_{j=0}^{k} \gamma_j F_{n+j}$ as $\pi(E)F_n$. Negative exponents are also allowed so that, for example, $E^{-3}F_n = F_{n-3}$. The *forward difference operator* Δ is defined by $\Delta = E - 1$, and the *backward difference operator* ∇ by $\nabla = 1 - E^{-1}$, so that

$$\Delta F_n = F_{n+1} - F_n, \quad \Delta^2 F_n = \Delta(F_{n+1} - F_n) = F_{n+2} - 2F_{n+1} + F_n,$$

$$\nabla F_n = F_n - F_{n-1}, \quad \nabla^2 F_n = \nabla(F_n - F_{n-1}) = F_n - 2F_{n-1} + F_{n-2}, \quad \text{etc.}$$

When the equally spaced data arise from the evaluation of a continuous function, that is when $F_n = F(x_n)$, then, provided $F(x) \in C^k$, we have the useful results that

$$\Delta^k F(x_n) = h^k F^{(k)}(x_n) + 0(h^{k+1}), \qquad \nabla^k F(x_n) = h^k F^{(k)}(x_n) + 0(h^{k+1}).$$

1.3 MEAN VALUE THEOREMS

On several occasions later in this book we shall make use of the standard mean value theorem. Although this result is very familiar in a scalar context, care must be taken when applying it in a vector context. As far as our applications of the theorem will be concerned, the difficulties introduced by the vector context are merely notational, and virtually constitute only nuisance value. In this section we state mean value theorems for the function $F(z) \in C^1$, where each of F and z can be either scalar or vector, and introduce some simplifying notation which will be useful in avoiding the nuisance factor later.

Case 1 $F: \mathbb{R}^1 \to \mathbb{R}^1$ $F(z)$ is a scalar function of the scalar argument z and the mean value theorem takes the familiar form

$$F(z) - F(z^*) = (z - z^*)F'(\zeta), \tag{1.1}$$

where the mean value ζ lies in the open interval with end points z and z^*.

Case 2 $F: \mathbb{R}^1 \to \mathbb{R}^m$ $F(z) := [{}^1F, {}^2F, \ldots, {}^mF]^\mathrm{T}$ is a vector function of the scalar argument z. We can apply (1.1) to each component of F to get

$${}^tF(z) - {}^tF(z^*) = (z - z^*){}^tF'(\zeta_t), \qquad t = 1, 2, \ldots, m, \tag{1.2}$$

but it is not in general true that the *same* mean value will apply for each component, hence the presence of the subscript in ζ_t. A vector form of (1.2) would be much more convenient, so we force this by writing

$$F(z) - F(z^*) = (z - z^*)\bar{F}'(\zeta), \tag{1.3}$$

where F' is the vector with components ${}^tF'$, $t = 1, 2, \ldots, m$, and the bar indicates that each component of F' is evaluated at a different mean value; ζ now merely symbolizes a typical mean value. That is,

$$\bar{F}'(\zeta) := [{}^1F'(\zeta_1), {}^2F'(\zeta_2), \ldots, {}^mF'(\zeta_m)]^\mathrm{T},$$

where each of the mean values ζ_t, $t = 1, 2, \ldots, m$ lies in the open interval with end points z and z^*.

Case 3 $F: \mathbb{R}^m \to \mathbb{R}^1$ $F(z)$ is a scalar function of the vector argument $z := [{}^1z, {}^2z, \ldots, {}^mz]^\mathrm{T}$, and the mean value theorem takes the form

$$F(z) - F(z^*) = \sum_{t=1}^{m} ({}^tz - {}^tz^*)\frac{\partial F}{\partial {}^tz}(\zeta), \tag{1.4}$$

where ζ is an internal point of the line segment in $\mathbb{R}^m$ joining z to z^*.

Case 4 $F: \mathbb{R}^m \to \mathbb{R}^m$ $F(z) := [{}^1F, {}^2F, \ldots, {}^mF]^\mathrm{T}$ is a vector function of the vector argument $z := [{}^1z, {}^2z, \ldots, {}^mz]^\mathrm{T}$. We can apply (1.4) to each component of F to obtain

$${}^sF(z) - {}^sF(z^*) = \sum_{t=1}^{m} ({}^tz - {}^tz^*)\frac{\partial {}^sF}{\partial {}^tz}(\zeta_s), \qquad s = 1, 2, \ldots, m \tag{1.5}$$

where, as in Case 2, we do not have the same mean value ζ_s for each component. Using the notation

$$F_{st}(z) := \frac{\partial^s F}{\partial^t z}(z)$$

we can write (1.5) in the more convenient form

$$F(z) - F(z^*) = \bar{J}(\zeta)(z - z^*), \tag{1.6}$$

where J is the Jacobian matrix of F with respect to z, and the bar indicates that each row of J is evaluated at a different mean value; that is

$$\bar{J}(\zeta) = \begin{bmatrix} F_{11}(\zeta_1) & F_{12}(\zeta_1) & \cdots & F_{1m}(\zeta_1) \\ F_{21}(\zeta_2) & F_{22}(\zeta_2) & \cdots & F_{2m}(\zeta_2) \\ \vdots & & & \\ F_{m1}(\zeta_m) & F_{m2}(\zeta_m) & \cdots & F_{mm}(\zeta_m) \end{bmatrix}.$$

Each of the mean values ζ_t, $t = 1, 2, \ldots, m$ is an internal point of the line segment in $\mathbb{R}^m$ joining z to z^*.

There is another mean value theorem, the *generalized mean value theorem for integrals*, which we shall have occasion to use. In its scalar form it states that if $\varphi(x)$ and $g(x)$ are scalar functions of the scalar variable x where, in an interval $[c, d]$ of x, $\varphi(x)$ is continuous and $g(x)$ is integrable and of constant sign, then there exists a mean value $\xi \in (c, d)$ such that

$$\int_c^d \varphi(x) g(x)\,\mathrm{d}x = \varphi(\xi) \int_c^d g(x)\,\mathrm{d}x. \tag{1.7}$$

In the situation we shall meet, $\varphi(x)$ is a vector function of x but $g(x)$ remains a scalar function, and clearly (1.7) can be applied to each component of φ (provided, of course, that the stated conditions on φ and g hold), giving the result

$$\int_c^d \varphi(x) g(x)\,\mathrm{d}x = \bar{\varphi}(\xi) \int_c^d g(x)\,\mathrm{d}x, \tag{1.8}$$

where the notation $\bar{\varphi}(\xi)$ implies that each component of φ is evaluated at a different mean value in the interval (c, d).

1.4 FIRST-ORDER SYSTEMS OF ORDINARY DIFFERENTIAL EQUATIONS

Throughout this book we shall be concerned with a *first-order system of ordinary differential equations* of the form

$$\left.\begin{aligned} {}^1y' &= {}^1f(x, {}^1y, {}^2y, \ldots, {}^my) \\ {}^2y' &= {}^2f(x, {}^1y, {}^2y, \ldots, {}^my) \\ \vdots &\quad \vdots \\ {}^my' &= {}^mf(x, {}^1y, {}^2y, \ldots, {}^my) \end{aligned}\right\}, \tag{1.9}$$

where ${}^t y' = (\mathrm{d}/\mathrm{d}x)^t y(x)$. We immediately abbreviate the phrase italicized above to 'first-order system', or just 'system'. The system (1.9) can be written in vector form as

$$y' = f(x, y), \tag{1.10}$$

where $y = [{}^1y, {}^2y, \ldots, {}^my]^{\mathrm{T}}$ and $f = [{}^1f, {}^2f, \ldots, {}^mf]^{\mathrm{T}}$, so that $f: \mathbb{R} \times \mathbb{R}^m \to \mathbb{R}^m$.

We note that each ${}^tf$ depends on ${}^1y, {}^2y, \ldots, {}^my$, that is, the system is *coupled*. Were it the case that each ${}^tf$ depended on ${}^ty$ alone, the system would be *uncoupled*, and each equation in it could be handled independently of the rest. It is this coupling that is the essence of a system; an uncoupled system is not essentially different from a scalar differential equation.

The general solution of a first-order system of dimension m contains, in general, m arbitrary constants; thus, for example, it is easily checked by substitution that the two-dimensional system

$$\left.\begin{aligned} {}^1y' &= {}^1y/x + {}^2yx \\ {}^2y' &= x[({}^2y)^2 - 1]/{}^1y \end{aligned}\right\} \tag{1.11}$$

is satisfied by

$$\begin{aligned} {}^1y(x) &= x[\cos(C_1x + C_2)]/C_1 \\ {}^2y(x) &= -\sin(C_1x + C_2) \end{aligned}$$

for any values (with the exception of $C_1 = 0$) of the arbitrary constants C_1, C_2. For the general m-dimensional system, the m arbitrary constants can be fixed by imposing m side conditions. If these m conditions take the form of demanding that the ${}^ty$, $t = 1, 2, \ldots, m$ all take given values *at the same initial point*, then the system together with the conditions constitute an *initial value problem*. Writing the system in the vector form (1.10), the general initial value problem thus takes the form

$$y' = f(x, y), \quad y(a) = \eta, \qquad f: \mathbb{R} \times \mathbb{R}^m \to \mathbb{R}^m, \tag{1.12}$$

where

$$\eta = [{}^1\eta, {}^2\eta, \ldots, {}^m\eta]^{\mathrm{T}}.$$

Once again, we abbreviate the nomenclature, and henceforth refer to (1.12) as a 'problem'. We regard (1.12) as the standard problem; this book is concerned entirely with numerical processes for solving (1.12).

Not all problems possess a unique solution, or indeed any solution at all. The following standard theorem lays down sufficient conditions for a unique solution to exist; we shall always assume that the hypotheses of this theorem are satisfied:

Theorem 1.1 *Let $f(x, y)$, where $f: \mathbb{R} \times \mathbb{R}^m \to \mathbb{R}^m$, be defined and continuous for all (x, y) in the region D defined by $a \leqslant x \leqslant b$, $-\infty < {}^ty < \infty, t = 1, 2, \ldots, m$, where a and b are finite, and let there exist a constant L such that*

$$\| f(x, y) - f(x, y^*) \| \leqslant L \| y - y^* \| \tag{1.13}$$

holds for every (x, y), $(x, y^*) \in D$. Then for any $\eta \in \mathbb{R}^m$, there exists a unique solution $y(x)$ of the problem (1.12), where $y(x)$ is continuous and differentiable for all $(x, y) \in D$.

The requirement (1.13) is known as a *Lipschitz condition*, and the constant L as a *Lipschitz constant*. Since

$$\begin{aligned} & f(x, y) \text{ continuously differentiable wrt } y \text{ for all } (x, y) \in D \\ \Rightarrow\ & f(x, y) \text{ satisfies a Lipschitz condition wrt } y \text{ for all } (x, y) \in D \\ \Rightarrow\ & f(x, y) \text{ continuous wrt } y \text{ for all } (x, y) \in D, \end{aligned}$$

the condition can be thought of as requiring a little more than continuity but a little less than differentiability. If $f(x, y)$ is differentiable wrt y, then from the mean value theorem (1.6) we have that

$$f(x, y) - f(x, y^*) = \bar{J}(x, \zeta)(y - y^*),$$

where the notation implies that each row of the Jacobian $J = \partial f(x, y)/\partial y$ is evaluated at different mean values, all of which are internal points on the line segment in $\mathbb{R}^{m+1}$ from (x, y) to (x, y^*), that is, all of which are points in D. It follows that the condition (1.13) can be satisfied by choosing the Lipschitz constant to be

$$L = \sup_{(x, y) \in D} \| \partial f(x, y)/\partial y \|. \tag{1.14}$$

If in (1.12) f is independent of x, the problem (and the system it involves) is said to be *autonomous*, and to be *non-autonomous* otherwise. It is always possible, at the cost of raising the dimension by 1, to write a non-autonomous problem in autonomous form. All one need do is add an extra scalar equation ${}^{m+1}y' = 1$ with initial condition ${}^{m+1}y(a) = a$, which implies that ${}^{m+1}y \equiv x$, so that the new $(m+1)$-dimensional system is clearly autonomous. For example, if we add to (1.11) the initial conditions ${}^{1}y(1) = 1$, ${}^{2}y(1) = 0$, the resulting 2-dimensional non-autonomous problem can be rewritten as

$$\begin{aligned} {}^{1}y' &= {}^{1}y/{}^{3}y + {}^{2}y\,{}^{3}y \qquad & {}^{1}y(1) &= 1 \\ {}^{2}y' &= {}^{3}y[({}^{2}y)^2 - 1]/{}^{1}y \qquad & {}^{2}y(1) &= 0 \\ {}^{3}y' &= 1 \qquad & {}^{3}y(1) &= 1, \end{aligned}$$

a 3-dimensional autonomous problem, with solution

$${}^{1}y(x) = x\cos(x - 1), \quad {}^{2}y(x) = -\sin(x - 1), \quad {}^{3}y = x.$$

Since we regard the dimension of the problem as being arbitrary, there is clearly no loss of generality in assuming that the general m-dimensional problem is autonomous. (In fact, we will not generally make that assumption, although it will prove useful to do so in the development of Runge–Kutta theory in Chapter 5.) However, there *is* a loss of generality in assuming that a *scalar* problem is autonomous, since the conversion to autonomous form would raise the dimension by 1, and the problem would no longer be scalar; thus the general scalar problem remains as $y' = f(x, y)$, $y(a) = \eta$.

1.5 HIGHER-ORDER SYSTEMS

The qth-order m-dimensional system of ordinary differential equations of the form

$$y^{(q)} = \varphi(x, y^{(0)}, y^{(1)}, \ldots, y^{(q-1)}), \tag{1.15}$$

where

$$\varphi: \mathbb{R} \times \underbrace{\mathbb{R}^m \times \mathbb{R}^m \times \cdots \times \mathbb{R}^m}_{q\text{-times}} \rightarrow \mathbb{R}^m$$

can be rewritten as a first-order system of dimension qm, by the following device:

Define $Y_r \in \mathbb{R}^m$, $r = 1, 2, \ldots, q$ by

$$\begin{aligned}
Y_1 &:= y & &(\equiv y^{(0)})\\
Y_2 &:= Y'_1 & &(\equiv y^{(1)})\\
Y_3 &:= Y'_2 & &(\equiv y^{(2)})\\
\vdots & \quad \vdots & & \quad \vdots\\
Y_q &:= Y'_{q-1} & &(\equiv y^{(q-1)}).
\end{aligned}$$

The last $q - 1$ of the above equations together with (1.15) gives

$$\begin{aligned}
Y'_1 &= Y_2\\
Y'_2 &= Y_3\\
\vdots & \quad \vdots\\
Y'_{q-1} &= Y_q\\
Y'_q &= \varphi(x, Y_1, Y_2, \ldots, Y_q),
\end{aligned}$$

which is a first-order system of dimension qm. It can be written in more compact form as

$$Y' = F(x, Y),$$

where

$$\begin{aligned}
Y &:= [Y_1^{\mathrm{T}}, Y_2^{\mathrm{T}}, \ldots, Y_q^{\mathrm{T}}]^{\mathrm{T}} \in \mathbb{R}^{qm},\\
F &:= [Y_2^{\mathrm{T}}, Y_3^{\mathrm{T}}, \ldots, Y_q^{\mathrm{T}}, \varphi^{\mathrm{T}}(x, Y_1, Y_2, \ldots, Y_q)]^{\mathrm{T}} \in \mathbb{R}^{qm}.
\end{aligned}$$

The initial value problem consisting of (1.15) together with the initial conditions $y^{(r)}(a) = \eta_{r+1}$, $r = 0, 1, \ldots, q - 1$, can thus be written in the form

$$Y' = F(x, Y), \quad Y(a) = \chi,$$

where $\chi := [\eta_1, \eta_2, \ldots, \eta_q]^{\mathrm{T}}$.

When seeking numerical solutions of initial value problems, it is standard practice first to reduce a qth-order system to a first-order system. The only exception is when the system is second order (and in particular, when such a system does not involve the first derivatives), for which special numerical methods have been devised. Even then, whether or not it is better to make the reduction is an unresolved question the investigation of which leads us into the no man's land of trying to compare norms over different spaces. In any event, the availability of sophisticated software for the numerical solution of first-order systems is a strong incentive always to make the reduction.

1.6 LINEAR SYSTEMS WITH CONSTANT COEFFICIENTS

The first-order system $y' = f(x, y)$, $f\colon \mathbb{R} \times \mathbb{R}^m \to \mathbb{R}^m$ is said to be *linear* if $f(x, y)$ takes the form $f(x, y) = A(x)y + \varphi(x)$, where $A(x)$ is an $m \times m$ matrix and $\varphi \in \mathbb{R}^m$. Further, if $A(x) = A$, independent of x, the system is said to be *linear with constant coefficients.* Associated with such a system

$$y' = Ay + \varphi(x) \tag{1.16}$$

is the *homogeneous* form

$$y' = Ay. \tag{1.17}$$

If $\tilde{y}(x)$ is the general solution of (1.17) and $\psi(x)$ is any particular solution of (1.16), then $y(x) = \tilde{y}(x) + \psi(x)$ is the general solution of (1.16). ($\tilde{y}(x)$ is the *complementary function* of (1.16), and $\psi(x)$ is a *particular integral.*)

A set of M solutions $\{y_t(x), t = 1, 2, \ldots, M\}$ of (1.17) is said to be *linearly independent* if

$$\sum_{t=1}^{M} C_t y_t(x) \equiv 0 \Rightarrow C_t = 0, \qquad t = 1, 2, \ldots, M.$$

A set of m linearly independent solutions $\{\tilde{y}_t(x), t = 1, 2, \ldots, m\}$ of (1.17) is said to form a *fundamental system* of (1.17), and the general solution of (1.17) is then a linear combination of the solutions which form the fundamental system. It is easily checked by substitution that $\tilde{y}_t(x) = \exp(\lambda_t x)c_t$, where λ_t is an eigenvalue of A and c_t the corresponding eigenvector, satisfies (1.17). In the case when A has distinct eigenvalues (the only case we shall need) the set of eigenvectors c_t, $t = 1, 2, \ldots, m$ are indeed linearly independent and thus the solutions $\{\exp(\lambda_t x)c_t, t = 1, 2, \ldots, m\}$ form a fundamental system. We then have that the general solution of (1.17) is

$$\tilde{y}(x) = \sum_{t=1}^{m} \varkappa_t \exp(\lambda_t x)c_t, \tag{1.18}$$

where the $\varkappa_t$ are arbitrary constants, and that the general solution of (1.16) is

$$y(x) = \sum_{t=1}^{m} \varkappa_t \exp(\lambda_t x)c_t + \psi(x), \tag{1.19}$$

where $\psi(x)$ is a particular solution of (1.16).

The $\varkappa_t$ are of course uniquely specified if an initial condition $y(a) = \eta$ is added to (1.16); note that $\{c_t, t = 1, 2, \ldots, m\}$ forms a basis of m-dimensional vector space.

The eigenvalues and eigenvectors of A are, in general, complex, as will be the constants $\varkappa_t$; but, due to the presence of complex conjugates, the solution (1.19) will be real—as indeed it must be. For example, consider the 2-dimensional initial value problem

$$y' = Ay + \varphi(x), \qquad y(0) = \eta,$$

where

$$A = \begin{bmatrix} 1 & 1 \\ -1 & 1 \end{bmatrix}, \quad \varphi(x) = \begin{bmatrix} 1 \\ x \end{bmatrix}, \quad \eta = \begin{bmatrix} 2 \\ \frac{1}{2} \end{bmatrix}.$$

The eigenvalues of A are $1+\mathrm{i}$ and $1-\mathrm{i}$, and the corresponding eigenvectors are $[1,\mathrm{i}]^{\mathrm{T}}$ and $[\mathrm{i},1]^{\mathrm{T}}$ respectively. Note that there is no need to normalize the eigenvectors, since they are going to be multiplied by the *arbitrary* constants $\varkappa_1$ and $\varkappa_2$. By trying a particular integral of the form $[ax+b, cx+d]^{\mathrm{T}}$, we establish that $\psi=[x/2, -(1+x)/2]^{\mathrm{T}}$ is a particular integral. The general solution is therefore

$$y(x)=\varkappa_1\exp[(1+\mathrm{i})x]\begin{bmatrix}1\\ \mathrm{i}\end{bmatrix}+\varkappa_2\exp[(1-\mathrm{i})x]\begin{bmatrix}\mathrm{i}\\ 1\end{bmatrix}+\begin{bmatrix}x/2\\ -(1+x)/2\end{bmatrix}.$$

The initial conditions are seen to be satisfied when $\varkappa_1=1-\mathrm{i}/2$, $\varkappa_2=1/2-\mathrm{i}$. On substituting these values into the general solution and simplifying, we find the solution of the problem is given by

$$^1y(x)=(2\cos x+\sin x)\exp(x)+x/2$$

$$^2y(x)=(\cos x-2\sin x)\exp(x)-(1+x)/2.$$

Exercises

1.6.1. Solve the initial value problem $y'=Ay$, $y(0)=[1,0,-1]^{\mathrm{T}}$, where

$$A=\begin{bmatrix}-21 & 19 & -20\\ 19 & -21 & 20\\ 40 & -40 & -40\end{bmatrix}.$$

1.6.2. Write the scalar differential equation $y^{(3)}=ay^{(2)}+by^{(1)}+cy+\varphi(x)$ as a first-order system $y'=Ay+\Phi(x)$. Show that the eigenvalues of A are the roots of the polynomial r^3-ar^2-br-c. Show also that if $\psi(x)$ is a particular integral of the given scalar differential equation, then $\Psi:=[\psi(x),\psi^{(1)}(x),\psi^{(2)}(x)]^{\mathrm{T}}$ is a particular integral of the equivalent first-order system.

1.6.3. The differential equation $y^{(3)}+y=x^2+\exp(-2x)$ has a particular integral $x^2-[\exp(-2x)]/7$. Find the equivalent first-order system and, using the results of the preceding exercise, find its general solution.

1.7 SYSTEMS OF LINEAR DIFFERENCE EQUATIONS WITH CONSTANT COEFFICIENTS

Let $\{y_n, n=n_0, n_0+1, n_0+2,\ldots\}$ be a sequence of vectors in $\mathbb{R}^m$. Then the system of difference equations

$$\sum_{j=0}^{k}\gamma_j y_{n+j}=\varphi_n,\qquad n=n_0, n_0+1, n_0+2,\ldots,\tag{1.20}$$

where the γ_j are scalar constants (that is, are independent of n) and $\varphi_n\in\mathbb{R}^m$, constitutes a kth-order *system of linear difference equations with constant coefficients.* Note that the solution of such a difference system is a *sequence* $\{y_n\}$ of vectors. The technique for establishing the general solution of (1.20) is a direct analogue of that for the system (1.16) of linear differential equations with constant coefficients. Let $\{\hat{y}_n\}$ be the general solution of the *homogeneous* form

$$\sum_{j=0}^{k} \gamma_j y_{n+j} = 0, \qquad n = n_0, n_0 + 1, n_0 + 2, \ldots, \tag{1.21}$$

and let $\{\psi_n\}$ be a particular solution of (1.20); then the general solution of (1.20) is $\{y_n\}$, where $y_n = \tilde{y}_n + \psi_n$.

A set of K solutions $\{\{y_{n,t}\}, t = 1, 2, \ldots, K\}$ of (1.21) is said to be *linearly independent* if

$$\sum_{t=1}^{K} C_t y_{n,t} = 0, \qquad n = n_0, \quad n_0 + 1, n_0 + 2, \ldots \Rightarrow C_t = 0, \quad t = 1, 2, \ldots, K.$$

A set of k linearly independent solutions $\{\{\tilde{y}_{n,t}\}, t = 1, 2, \ldots, k\}$ of (1.21) is said to form a *fundamental system* of (1.21), and the general solution of (1.21) is then a linear combination of the solutions which form the fundamental system. Let us attempt to find a solution of (1.21) of the form $y_{n,t} = r_t^n$. By substitution, we find that this is indeed a solution provided that r_t is a root of the *characteristic polynomial*

$$\pi(r) := \sum_{j=0}^{k} \gamma_j r^j.$$

If $\pi(r)$ has k distinct roots then it can be shown that the set of solutions $\{r_t^n\}$, $t = 1, 2, \ldots, k$ forms a fundamental system of (1.21) and the general solution of (1.20) is then $\{y_n\}$ where

$$y_n = \sum_{t=1}^{k} r^n d_t + \psi_n,$$

where the d_t are arbitrary vectors, which will be specified if k initial values or *starting values* are given. If r_1 is a root of $\pi(r)$ of multiplicity μ and the remaining $k - \mu$ roots are distinct then the set of solutions $\{r_1^n\}$, $\{nr_1^n\}, \ldots, \{n^{\mu-1}r_1^n\}$ and $\{r_t^n\}$, $t = \mu + 1$, $\mu + 2, \ldots, k$ form a fundamental system, and the general solution of (1.20) becomes $\{y_n\}$, where

$$y_n = \sum_{j=1}^{\mu} n^{j-1} r_1^n d_{1j} + \sum_{t=\mu+1}^{k} r_t^n d_t + \psi_n,$$

where the d_{1j} and the d_t are arbitrary vectors. When the roots of π are complex then the corresponding vectors d_t are likewise complex; the presence of complex conjugates ensures that the solution will be real. For example, consider the 2-dimensional 4th-order difference system

$$y_{n+4} - 6y_{n+3} + 14y_{n+2} - 16y_{n+1} + 8y_n = [n, 1]^T, \quad y_n \in \mathbb{R}^2,$$

with starting values $y_0 = [1, 0]^T$, $y_1 = [2, 1]^T$, $y_2 = [3, 2]^T$, $y_3 = [4, 3]^T$. By trying a particular solution of the form $\psi_n = na + b$, $a, b, \in \mathbb{R}^2$, we find that $\psi_n = [n + 2, 1]^T$ is such a solution. The characteristic polynomial is

$$\pi(r) = r^4 - 6r^3 + 14r^2 - 16r + 8 = (r - 2)^2(r^2 - 2r + 2),$$

which has roots $2, 2, 1 + \mathrm{i}, 1 - \mathrm{i}$. The general solution thus has the form

$$y_n = 2^n d_{11} + n2^n d_{12} + (1 + \mathrm{i})^n d_3 + (1 - \mathrm{i})^n d_4 + [n + 2, 1]^T.$$

Using the given starting values, we obtain a set of four equations for the four arbitrary vectors d_{11}, d_{12}, d_3, d_4, whose solution gives

$$d_{11} = \begin{bmatrix} -1 \\ -\frac{3}{2} \end{bmatrix} \quad d_{12} = \begin{bmatrix} \frac{1}{4} \\ \frac{1}{2} \end{bmatrix} \quad d_3 = \begin{bmatrix} -\mathrm{i}/4 \\ (1-3\mathrm{i})/4 \end{bmatrix} \quad d_4 = \begin{bmatrix} \mathrm{i}/4 \\ (1+3\mathrm{i})/4 \end{bmatrix}$$

whence

$$y_n = 2^n \begin{bmatrix} n/4 - 1 \\ (n-3)/2 \end{bmatrix} + (1+\mathrm{i})^n \begin{bmatrix} -\mathrm{i}/4 \\ (1-3\mathrm{i})/4 \end{bmatrix} + (1-\mathrm{i})^n \begin{bmatrix} \mathrm{i}/4 \\ (1+3\mathrm{i})/4 \end{bmatrix} + \begin{bmatrix} n+2 \\ 1 \end{bmatrix}$$

On writing $1 \pm \mathrm{i} = \sqrt{2}(\cos \pi/4 \pm \mathrm{i} \sin \pi/4)$ we obtain the solution in real form:

$$y_n = 2^n \begin{bmatrix} n/4 - 1 \\ (n-3)/2 \end{bmatrix} + 2^{(n-2)/2} \begin{bmatrix} \sin n\pi/4 \\ \cos n\pi/4 + 3 \sin n\pi/4 \end{bmatrix} + \begin{bmatrix} n+2 \\ 1 \end{bmatrix}.$$

Exercises

1.7.1. For the example at the end of the above section, calculate y_4, y_5 and y_6 directly from the difference system and the given starting values and show that the values so found coincide with those given by the general solution.

1.7.2. If $y_{n+2} - 2\mu y_{n+1} + \mu y_n = c$, $n = 0, 1, 2, \ldots$, where y_n, $c \in \mathbb{R}^m$, $\mu \in \mathbb{R}$, c is constant and $0 < \mu < 1$, show that $y_n \to c/(1-\mu)$ as $n \to \infty$.

1.7.3. Let r_1 and r_2 be the roots, assumed distinct, of the quadratic $r^2 - ar - b$. Show that the solution of the inhomogeneous linear constant coefficient difference system $y_{n+2} - ay_{n+1} - by_n = T_n$, satisfying the initial conditions $y_0 = \delta_0$, $y_1 = \delta_1$, where y_n, T_n, δ_0, $\delta_1 \in \mathbb{R}^m$, is given by

$$y_n = \frac{1}{r_1 - r_2}\left[(r_1^n - r_2^n)\delta_1 - (r_1^{n-1} - r_2^{n-1})r_1 r_2 \delta_0 + \sum_{j=0}^{n-2} (r_1^{n-j-1} - r_2^{n-j-1})T_j\right] \qquad n = 0, 1, \ldots,$$

where the summation term is taken to be zero when the upper limit of summation is negative.

1.7.4. The Fibonacci numbers are a sequence of integers $\{\varphi_n | n = 0, 1, \ldots\}$ such that each member of the sequence is the sum of the two preceding it, the first two being 0 and 1. Construct the first eleven Fibonacci numbers and compute the ratio φ_{n+1}/φ_n, $n = 1, 2, \ldots, 9$. Do you see any signs of this ratio converging as n increases? By solving the appropriate scalar difference equation, prove that

$$\lim_{n \to \infty} \varphi_{n+1}/\varphi_n = \tfrac{1}{2}(1 + \sqrt{5}).$$

1.8 ITERATIVE METHODS FOR NONLINEAR SYSTEMS OF ALGEBRAIC EQUATIONS

We shall frequently need to find numerical solutions of systems of nonlinear algebraic equations of the form

$$y = \varphi(y), \quad \varphi: \mathbb{R}^m \to \mathbb{R}^m. \tag{1.22}$$

This is done iteratively by one of two different methods. The first is *fixed point iteration*, which consists of constructing a sequence $\{y^{[\nu]}\}$ defined by

$$y^{[\nu+1]} = \varphi(y^{[\nu]}), \qquad \nu = 0, 1, 2, \ldots, y^{[0]} \text{ arbitrary}. \tag{1.23}$$

The following theorem states conditions under which (1.22) possesses a unique solution to which the iteration (1.23) will converge:

Theorem 1.2 Let $\varphi(y)$ satisfy a Lipschitz condition

$$\| \varphi(y) - \varphi(y^*) \| \leqslant M \| y - y^* \|$$

for all y, y^*, where the Lipschitz constant M satisfies $0 \leqslant M < 1$. Then there exists a unique solution $y = \alpha$ of (1.22), and if $\{y^{[\nu]}\}$ is defined by (1.23), then $y^{[\nu]} \to \alpha$ as $\nu \to \infty$.

Occasions will arise where we are unable to satisfy the hypotheses of Theorem 1.2, and the iteration (1.23) diverges. In such circumstances we turn to another form of iteration, *Newton iteration*, and usually just hope that a unique solution of (1.22) exists. Newton iteration (or the *Newton–Raphson process*) is most familiar when applied to the scalar problem $F(y) = 0$, $F: \mathbb{R} \to \mathbb{R}$, when it takes the form

$$y^{[\nu+1]} = y^{[\nu]} - F(y^{[\nu]})/F'(y^{[\nu]}), \qquad \nu = 0, 1, 2, \ldots. \tag{1.24}$$

The interpretation of (1.24) in terms of drawing tangents to the curve $z = F(y)$ and determining where they cut the y-axis will be familiar to most readers. Such an interpretation is enough to indicate that Newton iteration, unlike fixed point iteration, has only local and not global convergence; that is, it will not converge for arbitrary $y^{[0]}$ but only for $y^{[0]}$ sufficiently close to the solution. There exist theorems telling us how close to the solution $y^{[0]}$ has to be, but these are seldom of value in applications, and the usual practice is simply to guess $y^{[0]}$; if the iteration fails to converge, we abort it and seek a better first guess. If convergence is achieved, then it is quadratic, that is, the error in the current iterate is asymptotically proportional to the square of the error in the previous iterate.

Newton iteration applied to the system $F(y) = 0$, $F: \mathbb{R}^m \to \mathbb{R}^m$, takes the analogous form

$$y^{[\nu+1]} = y^{[\nu]} - J^{-1}(y^{[\nu]})F(y^{[\nu]}), \qquad \nu = 0, 1, 2, \ldots, \tag{1.25}$$

where $J(y) = (\partial F/\partial y)(y)$, the Jacobian matrix of F with respect to y. Applied to (1.22) (putting $F(y) = y - \varphi(y)$) we clearly get

$$y^{[\nu+1]} = y^{[\nu]} - \left[I - \frac{\partial \varphi}{\partial y}(y^{[\nu]}) \right]^{-1} [y^{[\nu]} - \varphi(y^{[\nu]})], \qquad \nu = 0, 1, 2, \ldots.$$

In practice, it is more efficient not to invert the matrix but instead use LU decomposition (see, for example, Atkinson and Harley (1983)), to solve, at each step of the iteration, the linear algebraic system

$$\left[I - \frac{\partial \varphi}{\partial y}(y^{[\nu]}) \right] \tilde{\Delta} y^{[\nu]} = -y^{[\nu]} + \varphi(y^{[\nu]}), \qquad \nu = 0, 1, 2, \ldots, \tag{1.26}$$

where

$$\tilde{\Delta} y^{[\nu]} := y^{[\nu+1]} - y^{[\nu]}$$

is the increment that must be added to the old iterate to obtain the new one. (We have added the tilde to avoid later confusion with Δ, defined in §1.2; $\tilde{\Delta}$ operates on the iteration superscript, whilst Δ operates on the subscripts in a set of discrete values $\{y_n\}$.) Newton iteration is considerably more expensive on computing time than is fixed point iteration. Each step of the latter costs just one function evaluation, whereas each step of the former calls for the updating of the Jacobian and a new LU decomposition and back substitution. In order to cut down on this computational effort, one can decline to update the Jacobian, so that (1.26) is replaced by

$$\left[I - \frac{\partial \varphi}{\partial y}(y^{[0]})\right]\tilde{\Delta} y^{[\nu]} = -y^{[\nu]} + \varphi(y^{[\nu]}), \qquad \nu = 0, 1, 2, \ldots. \tag{1.27}$$

This means that the same LU decomposition can be used for every step of the iteration, and only new back substitutions need be performed. The iteration (1.27) is known as *modified Newton iteration* (and sometimes as *quasi-Newton* or *pseudo-Newton iteration*). Note that it is the analogue of drawing all tangents parallel to the first one in the interpretation of Newton iteration for a scalar problem.

Exercises

1.8.1. Show that there exists a unique solution of the scalar equation $y = \varphi(y)$, where $\varphi(y) = \cos y$. Fixed point iteration for this equation can be nicely demonstrated on a hand calculator. Set any number on the calculator (set to compute in radians, of course) and repeatedly press the cosine key. Hence demonstrate that the iteration converges to the solution, $y = 0.739085\ldots$, no matter what starting value is used.

1.8.2. Using a microcomputer (or a programmable calculator) show that, for the problem in Exercise 1.8.1, modified Newton iteration sometimes converges and sometimes diverges, depending on the starting value. Show also that if the starting value is reasonably close to the solution, then it converges considerably faster than does fixed-point iteration.

1.9 SCHUR POLYNOMIALS

We shall frequently be concerned with the question of whether the roots of a polynomial with real coefficients lie within the unit circle. There is a handy phrase for describing such polynomials:

Definition A polynomial $\pi(r)$ of degree k is said to be **Schur** *if its roots r_t satisfy $|r_t| < 1$, $t = 1, 2, \ldots, k$.*

It is mildly surprising that the conditions for a polynomial to be Schur turn out not to be particularly easy or natural. There exist several criteria each of which throws up a set of inequalities that must be satisfied by the coefficients of the polynomial. In the author's experience, the criterion which usually produces the most easily solved set of inequalities is the *Routh–Hurwitz criterion*. This is in fact a criterion for the roots of a polynomial to lie in the left half-plane, so it is necessary first to make the transformation $r \to z$, $r, z \in \mathbb{C}$, where

$$r = (1 + z)/(1 - z).$$

This transformation maps the boundary of the circle $|r| = 1$ onto the imaginary axis $\operatorname{Re} z = 0$, and the interior of the circle $|r| = 1$ onto the left half-plane $\operatorname{Re} z < 0$. Define

$$P(z) := (1 - z)^k \pi[(1 + z)/(1 - z)] = a_0 z^k + a_1 z^{k-1} + \cdots + a_k, \tag{1.28}$$

where we may assume, without loss of generality, that $a_0 > 0$. The necessary and sufficient conditions for the roots of $P(z)$ to lie in the half-plane $\operatorname{Re} z < 0$, that is, for $\pi(r)$ to be Schur, are that all leading principal minors of Q be positive, where Q is the $k \times k$ matrix defined by

$$Q = \begin{bmatrix} a_1 & a_3 & a_5 & \cdots & a_{2k-1} \\ a_0 & a_2 & a_4 & \cdots & a_{2k-2} \\ 0 & a_1 & a_3 & \cdots & a_{2k-3} \\ 0 & a_0 & a_2 & \cdots & a_{2k-4} \\ \vdots & \vdots & \vdots & & \vdots \\ 0 & 0 & 0 & \cdots & a_k \end{bmatrix},$$

(where a_j is to be taken to be zero if $j > k$). It can be shown that these conditions imply that $a_j > 0$, $j = 0, 1, \ldots, k$, so that the positivity of the coefficients a_j in (1.28) is a necessary but not sufficient condition for $\pi(r)$ to be Schur. For $k = 2, 3, 4$, the Routh–Hurwitz conditions turn out to be:

$$\begin{aligned} k = 2;\quad & a_j > 0,\ j = 0, 1, 2 \\ k = 3;\quad & a_j > 0,\ j = 0, 1, 2, 3,\ a_1 a_2 - a_3 a_0 > 0 \\ k = 4;\quad & a_j > 0,\ j = 0, 1, 2, 3, 4,\ a_1 a_2 a_3 - a_0 a_3^2 - a_4 a_1^2 > 0. \end{aligned}$$

We illustrate by an example, which is itself a useful result. Consider the quadratic

$$\pi(r) = r^2 + \alpha r + \beta. \tag{1.29}$$

Applying (1.28), we have that

$$P(z) = (1 + z)^2 + \alpha(1 - z^2) + \beta(1 - z)^2 = a_0 z^2 + a_1 z + a_2,$$

where

$$a_0 = 1 - \alpha + \beta, \quad a_1 = 2(1 - \beta), \quad a_2 = 1 + \alpha + \beta.$$

The necessary and sufficient conditions for the quadratic (1.29) to be Schur are therefore that the point (α, β) lies in the interior of the triangle in the α, β plane bounded by the lines

$$\beta = 1, \quad \beta = \alpha - 1, \quad \beta = -\alpha - 1$$

(see Figure 1.1).

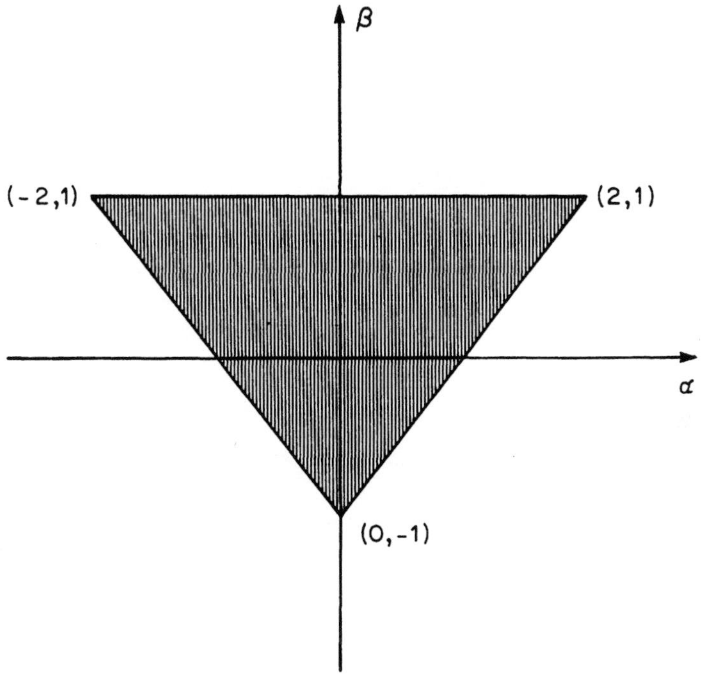

Figure 1.1

Exercises

1.9.1. Use the Routh–Hurwitz criterion to find the interval of α for which the polynomial $2r^3 + (2\alpha - 1)r^2 + (5\alpha - 2)r + 1 - 3\alpha$ is Schur. Check your result by finding a linear factor of the polynomial and using Figure 1.1.

1.9.2* $P^p_q(z, w)$, $z, w \in \mathbb{C}$, is a polynomial of exact degree p in z, whose coefficients are themselves polynomials in w of degree at most q, where at least one of these polynomial coefficients has exact degree q. Specifically,

$$P^p_q(z,w) = \sum_{i=0}^{p} \gamma_i(w) z^i, \quad \gamma_p(w) \not\equiv 0, \quad \gamma_i(w) = \sum_{j=0}^{q} \gamma_{ij} w^j, \qquad i = 0, 1, \ldots, p$$

where there exists at least one $i^* \in \{0, 1, \ldots, p\}$ such that $\gamma_{i^*q} \neq 0$.

We shall say that $P^p_q(z, w)$ is *ultimately Schur* if there exists a positive constant K such that $P^p_q(z, w)$, regarded as a polynomial in z, is Schur for all w satisfying $\operatorname{Re} w < -K < 0$. Prove that if $P^p_q(z, w)$ is ultimately Schur, then $\gamma_{pq} \neq 0$.

1.10 INTERPOLATION FORMULAE

Consistent with our policy for this chapter, as stated in §1.1, we shall gather together in this section only those results from interpolation theory which will be needed later in this book. The reader who wishes to see a full account, including proofs, is referred to Isaacson and Keller (1966). In most accounts of interpolation theory, the points to be interpolated are taken to be (x_{n+j}, F_{n+j}), $j = 0, 1, \ldots, q$, but for our purposes it will be more natural to taken them as (x_{n-j}, F_{n-j}), $j = 0, 1, \ldots, q$.

There exists a unique vector polynomial (that is, a polynomial with scalar argument and vector coefficients) of degree at most q which interpolates (that is, passes through) the $q+1$ distinct data points (x_{n-j}, F_{n-j}) $j = 0, 1, \ldots, q$, where $x_{n-j} \in \mathbb{R}$ and $F_{n-j} \in \mathbb{R}^m$. We shall denote this polynomial by $I_q(x)$, and we shall be particularly interested in representations of $I_q(x)$ which enable us readily to compute $I_{q+1}(x)$ from $I_q(x)$.

The interpolant takes a particularly simple form in the case when the data points are equally spaced, that is, when $x_{n-j} = x_n - jh$, $j = 0, 1, \ldots, q$, and h is a constant. In such circumstances it is advantageous to make use of the backward difference operator ∇, defined in §1.2.

The Newton–Gregory backward interpolation formula

When the data are evenly spaced, $I_q(x)$ may be written in terms of the backward differences of F as

$$I_q(x) = P_q(r) = \sum_{i=0}^{q} (-1)^i \binom{-r}{i} \nabla^i F_n, \tag{1.30}$$

where $x = x_n + rh$, and $\binom{-r}{i}$ is the binomial coefficient. Illustrating in the case $q = 2$ we have

$$P_2(r) = F_n + r\nabla F_n + \tfrac{1}{2}r(r+1)\nabla^2 F_n$$

$$x = x_n \quad \Leftrightarrow \quad r = 0 \quad \Rightarrow \quad P_2(r) = F_n$$

$$x = x_{n-1} \quad \Leftrightarrow \quad r = -1 \quad \Rightarrow \quad P_2(r) = F_n - \nabla F_n = F_{n-1}$$

$$x = x_{n-2} \quad \Leftrightarrow \quad r = -2 \quad \Rightarrow \quad P_2(r) = F_n - 2\nabla F_n + \nabla^2 F_n = F_{n-2}.$$

If the data points have arisen from evaluating a function $F(x) \in C^{q+1}$, that is if $F_{n-j} = F(x_{n-j})$, $j = 0, 1, \ldots, q$, then the error in the interpolation can be written in the form

$$F(x_n + rh) - I_q(x_n + rh) = (-1)^{q+1} \binom{-r}{q+1} h^{q+1} \bar{F}^{(q+1)}(\xi) \tag{1.31}$$

where, using the notation of §1.3, the bar indicates that each component ${}^tF^{(q+1)}$ is evaluated at different mean values ξ_t, each of which is an interior point of the smallest interval containing $x_n, x_{n+1}, \ldots, x_{n+q}$ and $x_n + rh$.

Note that from (1.30) it is straightforward to generate $I_{q+1}(x)$ in terms of $I_q(x)$.

The Lagrange interpolation formula

In the case when the data are unevenly spaced, the easiest interpolation formula is that due to Lagrange. Define

$$L_{q,j}(x) := \prod_{\substack{i=0\\ i\neq j}}^{q} \frac{x - x_{n-i}}{x_{n-j} - x_{n-i}}.$$

It is obvious that $L_{q,j}(x)$ is a polynomial in x of degree q, and moreover that

$$L_{q,j}(x_{n-i}) = \begin{Bmatrix} 1 & \text{if } i = j \\ 0 & \text{if } i \neq j \end{Bmatrix}, \qquad i = 1, 2, \ldots, q.$$

It follows that $I_q(x)$ can be written in the form

$$I_q(x) = \sum_{j=0}^{q} L_{q,j}(x) F_{n-j}. \tag{1.32}$$

Although the Lagrange formula is conceptually simple, it suffers a serious disadvantage. If we wish to add a further data point (x_{n-q-1}, F_{n-q-1}) then, using (1.32) with q replaced by $q+1$, we can obtain an expression for $I_{q+1}(x)$, but there is no easy way in which we can generate $I_{q+1}(x)$ directly from $I_q(x)$. An alternative form overcomes this difficulty.

The Newton divided difference interpolation formula

Given the $q+1$ (unequally spaced) data points (x_{n-j}, F_{n-j}), $j = 0, 1, \ldots, q$, we define the *divided differences* $F[n, n-1, \ldots, n-i]$, $i = 0, 1, \ldots, q$ recursively by

$$\begin{aligned} F[n] &:= F_n \\ F[n, n-1, \ldots, n-i] &= \frac{F[n, n-1, \ldots, n-i+1] - F[n-1, n-2, \ldots, n-i]}{x_n - x_{n-i}}. \end{aligned} \tag{1.33}$$

The interpolating polynomial $I_q(x)$ can be written in the form

$$\begin{aligned} I_q(x) = F[n] &+ (x - x_n)F[n, n-1] + \cdots \\ &+ (x - x_n)(x - x_{n-1}) \ldots (x - x_{n-q+1})F[n, n-1, \ldots, n-q]. \end{aligned} \tag{1.34}$$

Illustrating this in the case $q = 2$, we have

$$I_2(x) = F[n] + (x - x_n)F[n, n-1] + (x - x_n)(x - x_{n-1})F[n, n-1, n-2],$$

whence

$$\begin{aligned} I_2(x_n) &= F[n] = F_n \\ I_2(x_{n-1}) &= F[n] - (F[n] - F[n-1]) = F[n-1] = F_{n-1} \\ I_2(x_{n-2}) &= F[n] + (x_{n-2} - x_n)F[n, n-1] - (x_{n-2} - x_{n-1})(F[n, n-1] - F[n-1, n-2]) \\ &= F[n] + (x_{n-1} - x_n)F[n, n-1] + (x_{n-2} - x_{n-1})F[n-1, n-2] \\ &= F[n] - (F[n] - F[n-1]) - (F[n-1] - F[n-2]) \\ &= F[n-2] = F_{n-2}. \end{aligned}$$

Now (1.34) and (1.32) are different representations of the *unique* polynomial of degree at most q interpolating the data (x_{n-j}, F_{n-j}). We can therefore equate the coefficients of x^q in the two polynomials to get

$$F[n, n-1, \ldots, n-q] = \sum_{j=0}^{q} \left[F_{n-j} \Big/ \prod_{\substack{i=0 \\ i \neq j}}^{q} (x_{n-j} - x_{n-i}) \right], \qquad q = 0, 1, \ldots.$$

It follows that the value of $F[n, n-1, \ldots, n-q]$ is independent of the order of the integers $n, n-1, \ldots, n-q$. More precisely,

$$F[n, n-1, \ldots, n-q] = F[p_0, p_1, \ldots, p_q], \qquad q = 0, 1, \ldots, \tag{1.35}$$

where $\{p_0, p_1, \ldots, p_q\}$ is any permutation of the integers $\{n, n-1, \ldots, n-q\}$.

Now let us suppose that we wish to add an extra point (x_{n-q-1}, F_{n-q-1}) to the data set. Then it follows from (1.34) that the unique polynomial $I_{q+1}(x)$ of degree at most $q+1$ interpolating the data (x_{n-j}, F_{n-j}), $j = 0, 1, \ldots, q+1$ is given by

$$I_{q+1}(x) = I_q(x) + (x - x_n)(x - x_{n-1}) \ldots (x - x_{n-q}) F[n, n-1, \ldots, n-q-1] \tag{1.36}$$

and, in contrast with the Lagrange form, we have an easy way of generating $I_{q+1}(x)$ from $I_q(x)$.

What would happen if we wished to add to the original data set an extra point (x_{n+1}, F_{n+1}) at the 'other end'? The answer is that nothing new happens at all, since there is no such thing as the 'other end'! The data, being unevenly spaced, can be distributed in any manner, and we certainly have not assumed that $x_n > x_{n-1} > \cdots > x_{n-q}$. There is nothing to stop us labelling the extra point (x_{n+1}, F_{n+1}) rather than (x_{n-q-1}, F_{n-q-1}) and, noting that x_{n-q-1} does not appear in (1.36), we can rewrite that equation in the form

$$I_{q+1}(x) = I_q(x) + (x - x_n)(x - x_{n-1}) \ldots (x - x_{n-q}) F[n, n-1, \ldots, n-q, n+1].$$

In view of (1.35), this can be rewritten as

$$I_{q+1}(x) = I_q(x) + (x - x_n)(x - x_{n-1}) \ldots (x - x_{n-q}) F[n+1, n, \ldots, n-q] \tag{1.37}$$

a representation of the polynomial interpolating (x_{n-j}, F_{n-j}), $j = -1, 0, 1, \ldots, q$ in terms of the polynomial interpolating (x_{n-j}, F_{n-j}), $j = 0, 1, \ldots, q$, which we shall need later in this book. We illustrate this result in the case $q = 1$, when (1.37) reads

$$I_2(x) = I_1(x) + (x - x_n)(x - x_{n-1}) F[n+1, n, n-1],$$

where

$$I_1(x) = F[n] + (x - x_n) F[n, n-1].$$

Since $I_1(x)$ interpolates (x_n, F_n) and (x_{n-1}, F_{n-1}) then so does $I_2(x)$, since the added term is zero at $x = x_n, x_{n-1}$. Further,

$$\begin{aligned} I_2(x_{n+1}) &= F[n] + (x_{n+1} - x_n) F[n, n-1] + (x_{n+1} - x_n)(F[n+1, n] - F[n, n-1]) \\ &= F[n] + (x_{n+1} - x_n) F[n+1, n] = F[n+1] = F_{n+1} \end{aligned}$$

Note that when the data become equally spaced, then the divided differences do *not* revert to backward differences. It is easily seen that when $x_{n-j} = x_n - jh$

$$F[n, n-1, \ldots, n-i] = \frac{1}{i!h^i}\nabla^i F_n, \qquad i = 1, 2, \ldots, q-1. \tag{1.38}$$

However, it is readily checked that on putting $x_{n-j} = x_n - jh$ and using (1.38), the Newton divided difference interpolation formula (1.34) reverts to the Newton–Gregory backward interpolation formula (1.30) for equally spaced data.

Exercises

1.10.1. Find the quadratic polynomial $I_2(x)$ which interpolates the data points $(x, 1/x)$ for $x = 1.0$, 0.9, 0.8 using (i) the Newton–Gregory backward interpolation formula and (ii) the Lagrange interpolation formula.

1.10.2. Find the cubic polynomial $I_3(x)$ which interpolates the data points $(x, 1/x)$ for $x = 1.0, 0.9, 0.8, 0.75$ using (i) the Lagrange interpolation formula and (ii) the result found in Exercise 1.10.1 together with equation (1.36). You should be persuaded of the advantage of the divided difference approach.

1.11 THE DIRECT PRODUCT OF MATRICES

Suppose we were dealing with a scalar differential equation $y' = f(x, y)$, $f: \mathbb{R} \times \mathbb{R} \to \mathbb{R}$; there arise occasions when we need to consider an $s \times s$ matrix whose elements are values of $\partial f/\partial y$. However, in this book we shall be dealing exclusively with the system of differential equations $y' = f(x, y)$, $f: \mathbb{R} \times \mathbb{R}^m \to \mathbb{R}^m$; the corresponding matrices will have dimension $ms \times ms$, the scalar element $\partial f/\partial y$ being replaced by the $m \times m$ Jacobian matrix $\partial f/\partial y$. This leads to somewhat heavy notation, which tends to obscure what is going on. A useful notation which helps overcome this problem is that of the *direct product* of two matrices. In this section we define the direct product and list only those properties which we shall need. A fuller treatment can be found in Lancaster (1969).

Definition *Let $A = [a_{ij}]$ be an $s \times s$ matrix and let B be an $m \times m$ matrix. Then the* **direct product** *of A and B, denoted by $A \otimes B$, is an $ms \times ms$ matrix defined by*

$$A \otimes B = \begin{bmatrix} a_{11}B & a_{12}B & \cdot & \cdot & a_{1s}B \\ a_{21}B & a_{22}B & \cdot & \cdot & a_{2s}B \\ \cdot & \cdot & \cdot & \cdot & \cdot \\ \cdot & \cdot & \cdot & \cdot & \cdot \\ a_{s1}B & a_{s2}B & \cdot & \cdot & a_{ss}B \end{bmatrix}.$$

Properties

(1) $(A \otimes B)(C \otimes D) = AC \otimes BD$, where A and C are $s \times s$, B and D are $m \times m$.
(2) $(A \otimes B)^{-1} = A^{-1} \otimes B^{-1}$.

(3) If the eigenvalues of A are $p_i, i = 1, 2, \ldots, s$, and those of B are $q_j, j = 1, 2, \ldots, m$, then $A \otimes B$ has eigenvalues $p_i q_j$, $i = 1, 2, \ldots, s, j = 1, 2, \ldots, m$.

Exercises

1.11.1. Prove Property (1) and deduce Property (2).

1.11.2. Verify that Property (3) holds for the case

$$A = \begin{pmatrix} 1 & 0 \\ 1 & 2 \end{pmatrix}, \quad B = \begin{pmatrix} 2 & 2 \\ -1 & 5 \end{pmatrix}.$$

2 Introduction to Numerical Methods

2.1 THE ROLE OF NUMERICAL METHODS FOR INITIAL VALUE PROBLEMS

The mathematical modelling of many problems in physics, engineering, chemistry, biology etc. gives rise to systems of ordinary differential equations (henceforth shortened to 'systems'). Yet, the number of instances where an exact solution can be found by analytical means is very limited. Indeed, the only general class of systems for which exact solutions can always be found (subject to being able to find a particular integral) consists of linear constant coefficient systems of the form

$$y' = Ay + F(x), \tag{2.1}$$

where A is a constant matrix. There are of course many examples of particular linear variable coefficient or nonlinear systems for which exact solutions are known, but, in general, for such systems we must resort to either an approximate or a numerical method.

In this context, by 'approximate methods' we mean techniques such as solution in series, solutions which hold only asymptotically for large x, etc. Sometimes the view is taken (wrongly in the author's opinion) that since powerful and well-tested numerical procedures are now commonly available, such approximate methods are obsolete. Such a view ignores the fact that approximate methods frequently (as in the case of linear variable coefficient systems) produce approximate *general* solutions, whereas numerical methods produce *particular* solutions satisfying given initial or boundary conditions; specifically, numerical methods solve initial or boundary value *problems*, not *systems*. Situations can arise where a low accuracy approximate general solution of the system is more revealing than high accuracy numerical solutions of a range of initial or boundary value problems. Further, even when the task in hand is the solution of a specific initial or boundary value problem, the system may contain a number of unspecified parameters; approximate methods can sometimes cast more light on the influence of these parameters than can repeated applications of a numerical method for ranges of the parameters, a procedure which is not only time-consuming but often hard to interpret.

Conversely, some mathematical modellers seem loath to turn to numerical procedures, even in circumstances where they are entirely appropriate, and do so only when all else fails. Just because an approximate—or even an exact—method exists is no reason always to use it, rather than a numerical method, to produce a numerical solution. For example, calculation of the complementary function in the exact solution of the simple system (2.1) involves, as we have seen in §1.6, the computation of all of the eigenvalues and eigenvectors of the matrix A; it is not difficult to construct examples where a numerical solution of an initial value problem involving (2.1) computed via the eigensystem will be considerably less accurate and efficient than one computed by an appropriate numerical method applied directly to the problem. Modern numerical methods, packaged in highly-tuned automatic algorithms are powerful and well-tested procedures which, together with other techniques, should surely find a place in the toolkit of any mathematical modeller.

2.2 NUMERICAL METHODS; NOMENCLATURE AND EXAMPLES

This book is concerned with numerical methods for *initial value* problems only. As we have seen in §1.5, a higher-order differential equation or system of equations can always be rewritten as a first-order system, and we shall always assume that this has been done, so that the standard problem we attempt to solve numerically is

$$y' = f(x, y), \quad y(a) = \eta, \tag{2.2}$$

where $y = [{}^1y, {}^2y, \ldots, {}^my]^T$, $f = [{}^1f, {}^2f, \ldots, {}^mf]^T$ and $\eta = [{}^1\eta, {}^2\eta, \ldots, {}^m\eta]^T$ are m-dimensional (column) vectors, and x and a are scalars. A solution is sought on the interval $[a, b]$ of x, where a and b are finite. It is assumed that the hypotheses of Theorem 1.1 (see §1.4) hold, so that there exists a unique solution $y(x)$ of (2.2).

All of the numerical methods we shall discuss in this book involve the idea of *discretization*; that is, the continuous interval $[a, b]$ of x is replaced by the discrete point set $\{x_n\}$ defined by $x_n = a + nh$, $n = 0, 1, 2, \ldots, N = (b - a)/h$. The parameter h is called the *steplength*; for the time being (in fact for quite some time) we shall regard it as being a constant, though we remark in passing that much of the power of modern algorithms derives from their ability to change h automatically as the computation proceeds. We let y_n denote an approximation to the solution $y(x_n)$ of (2.2) at x_n,

$$y_n \simeq y(x_n), \tag{2.3}$$

and our aim is to find a means of producing a sequence of values $\{y_n\}$ which approximates the solution of (2.2) on the discrete point set $\{x_n\}$; such a sequence constitutes a *numerical solution* of the problem (2.2).

A *numerical method* (henceforth shortened to 'method') is a difference equation involving a number of consecutive approximations $y_{n+j}, j = 0, 1, \ldots, k$, from which it will be possible to compute sequentially the sequençe $\{y_n | n = 0, 1, 2, \ldots, N\}$; naturally, this difference equation will also involve the function f. The integer k is called the *stepnumber* of the method; if $k = 1$, the method is called a *one-step* method, while if $k > 1$, the method is called a *multistep* or *k-step* method.

An *algorithm* or *package* is a computer code which implements a method. In addition to computing the sequence $\{y_n\}$ it may perform other tasks, such as estimating the error in the approximation (2.3), monitoring and updating the value of the steplength h and deciding which of a family of methods to employ at a particular stage in the solution.

Numerical methods can take many forms.

Example 1

$$y_{n+2} + y_{n+1} - 2y_n = \frac{h}{4}[f(x_{n+2}, y_{n+2}) + 8f(x_{n+1}, y_{n+1}) + 3f(x_n, y_n)].$$

Example 2

$$y_{n+2} - y_{n+1} = \frac{h}{3}[3f(x_{n+1}, y_{n+1}) - 2f(x_n, y_n)].$$

Example 3

$$y_{n+3} + \tfrac{1}{4}y_{n+2} - \tfrac{1}{2}y_{n+1} - \tfrac{3}{4}y_n = \frac{h}{8}[19f(x_{n+2}, y_{n+2}) + 5f(x_n, y_n)].$$

Example 4

$$y_{n+2} - y_n = h[f(x_{n+2}, y^*_{n+2}) + f(x_n, y_n)],$$

where

$$y^*_{n+2} - 3y_{n+1} + 2y_n = \frac{h}{2}[f(x_{n+1}, y_{n+1}) - 3f(x_n, y_n)].$$

Example 5

$$y_{n+1} - y_n = \frac{h}{4}(k_1 + 3k_3),$$

where

$$k_1 = f(x_n, y_n)$$
$$k_2 = f(x_n + \tfrac{1}{3}h, y_n + \tfrac{1}{3}hk_1)$$
$$k_3 = f(x_n + \tfrac{2}{3}h, y_n + \tfrac{2}{3}hk_2).$$

Example 6

$$y_{n+1} - y_n = \frac{h}{2}(k_1 + k_2),$$

where

$$k_1 = f(x_n, y_n)$$
$$k_2 = f(x_n + h, y_n + \tfrac{1}{2}hk_1 + \tfrac{1}{2}hk_2).$$

Clearly, Examples 5 and 6 are one-step methods, and on putting $y_0 = \eta$ the sequence $\{y_n\}$ can be computed sequentially by setting $n = 0, 1, 2, \ldots$ in the difference equation. Examples 1, 2 and 4, however, are 2-step methods, and it will be necessary to provide an additional *starting value* y_1 before the sequence $\{y_n\}$ can be computed; in the case of Example 3, a 3-step method, it will be necessary to provide two additional starting values, y_1 and y_2. Finding such additional starting values presents no serious difficulty.

One can always employ a separate one-step method to do this, but in practice all modern algorithms based on multistep methods have a self-starting facility; this will be discussed in Chapter 4.

If the method is such that, given $y_{n+j}, j = 0, 1, \ldots, k-1$, the difference equation yields y_{n+k} explicitly, it is said to be *explicit*; this is clearly the case for Examples 2, 3, 4 and 5. If the value y_{n+k} cannot be computed without solving an implicit system of equations, as is the case for Examples 1 and 6 (note that in the latter k_2 is defined implicitly) then the method is said to be *implicit*. Since the function f is in general nonlinear in y, implicit methods involve the solution of a nonlinear system of equations at every step of the computation, and are thus going to be much more computationally costly than explicit methods. Note that in the case of explicit methods the provision of the necessary starting values essentially converts the method into an algorithm, albeit a rather rudimentary one. In contrast, an implicit method is some way from being an algorithm, since we would have to incorporate in the latter a subroutine which numerically solves the implicit system of equations at each step.

Examples 1, 2 and 3 are examples of *linear multistep* methods, a class in which the difference equation involves only linear combinations of $y_{n+j}, f(x_{n+j}, y_{n+j}), j = 0, 1, \ldots, k$. Example 4 is a *predictor–corrector* method, in which an explicit linear multistep method (the predictor) is combined with an implicit one (the corrector); note that the resulting method is explicit. Examples 5 and 6 fall in the class of *Runge–Kutta* method, a class with a much more complicated structure.

All of the above examples, and indeed (almost) all of the methods covered in this book can be written in the general form

$$\sum_{j=0}^{k} \alpha_j y_{n+j} = h\phi_f(y_{n+k}, y_{n+k-1}, \ldots, y_n, x_n; h), \tag{2.4}$$

where the subscript f on the right-hand side indicates that the dependence of ϕ on y_{n+k}, $y_{n+k-1}, \ldots, y_n, x_n$ is through the function $f(x, y)$. We impose two conditions on (2.4), namely

$$\left.\begin{aligned} &\phi_f \equiv 0(y_{n+k}, y_{n+k-1}, \ldots, y_n, x_n; h) \equiv 0, \\ &\|\phi_f(y_{n+k}, y_{n+k-1}, \ldots, y_n, x_n; h) - \phi_f(y^*_{n+k}, y^*_{n+k-1}, \ldots, y^*_n, x_n; h)\| \\ &\leqslant M \sum_{j=0}^{k} \|y_{n+j} - y^*_{n+j}\| \end{aligned}\right\} \tag{2.5}$$

where M is a constant. These conditions are not at all restrictive; for all methods of the form (2.4) considered in this book, the first is satisfied, while the second is a consequence of the fact that the initial value problem (2.2) is assumed to satisfy a Lipschitz condition (Theorem 1.1 of §1.4). Thus, Example 4 can be re-cast in the form

$$y_{n+2} - y_n = h\phi_f(y_{n+1}, y_n, x_n; h),$$

where

$$\phi_f = f\left(x_{n+2}, 3y_{n+1} - 2y_n + \frac{h}{2}f(x_{n+1}, y_{n+1}) - \frac{3h}{2}f(x_n, y_n)\right) + f(x_n, y_n)$$

and it is clear that the first of the conditions (2.5) is satisfied. By repeatedly applying the Lipschitz condition

$$\|f(x,y)-f(x,y^*)\| \leqslant L\|y-y^*\|,$$

it is straightforward to show that

$$\begin{aligned}&\|\phi_f(y_{n+1},y_n,x_n;h)-\phi_f(y^*_{n+1},y^*_n,x_n;h)\|\\ &\leqslant\left(3+\frac{h}{2}L\right)L\|y_{n+1}-y^*_{n+1}\|+\left(3+\frac{3h}{2}L\right)L\|y_n-y^*_n\|,\end{aligned}$$

whence the second of (2.5) is satisfied with

$$M=(3+\tfrac{3}{2}hL)L.$$

Each of the major classes of methods will be studied in detail in separate chapters later in this book, but certain fundamental properties, common to all these classes, can be developed for the general class (2.4), and this will be done in the following sections.

Exercise

2.2.1. Show that the conditions (2.5) are satisfied for all of the six examples given in this section.

2.3 CONVERGENCE

Let us consider the numerical solution of the initial value problem (2.2) given by the general method (2.4) with appropriate starting values, that is the solution given by

$$\left.\begin{aligned}\sum_{j=0}^{k}\alpha_j y_{n+j}&=h\phi_f(y_{n+k},y_{n+k-1},\dots,y_n,x_n;h)\\ y_\mu&=\eta_\mu(h),\mu=0,1,\dots,k-1.\end{aligned}\right\}\qquad(2.6)$$

In the limit as the steplength $h\to 0$, the discrete point set $\{x_n|x_n=a+nh, n=0,1,\dots,N=(b-a)/h\}$ becomes the continuous interval $[a.b]$. An obvious property to require of any numerical method is that, in this limit, the numerical solution $\{y_n, n=0,1,\dots,N=(b-a)/h\}$, where $\{y_n\}$ is defined by (2.6), becomes the exact solution $y(x)$, $x\in[a,b]$. This, in loose terms, is what is meant by *convergence*; the concept is straightforward, but there are notational difficulties to be considered.

Consider a sequence of numerical solutions given by (2.6) obtained by repeatedly halving the steplength, that is by taking $h=h_0,h_1,h_2,\dots$, where $h_i=h_0/2^i$, and let us temporarily adopt the notation $y_n(h_i)$ to denote the value y_n given by (2.6) when the steplength is h_i.

Figure 2.1 typifies the sort of behaviour we envisage for a convergent method; here the solid line represents the exact solution for a component ${}^t y(x)$ of $y(x)$, and the points marked □, ◇, + represent the numerical solutions $\{{}^t y_n(h_0)\}$, $\{{}^t y_n(h_1)\}$, $\{{}^t y_n(h_2)\}$

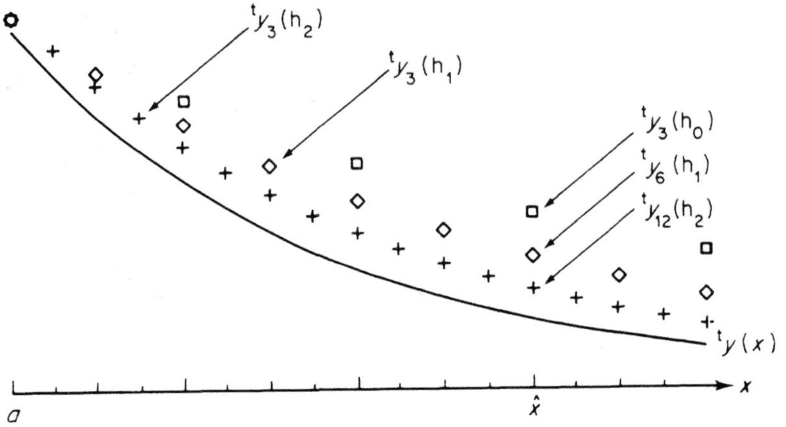

Figure 2.1

respectively. It would be quite inappropriate to consider the convergence of the sequence ${}^t y_n(h_0), {}^t y_n(h_1), {}^t y_n(h_2), \ldots$, *for fixed n*; for any fixed n such a sequence would clearly tend to the initial value ${}^t\eta$. What is appropriate is to consider a fixed value, say $\hat{x} = a + 3h_0$, of x and, noting that

$$\hat{x} = a + 3h_0 = a + 6h_1 = a + 12h_2 = \ldots,$$

consider the convergence of the sequence ${}^t y_3(h_0)$, ${}^t y_6(h_1)$, ${}^t y_{12}(h_2), \ldots$, to ${}^t y(\hat{x})$; moreover, we clearly want this to happen for all $x \in [a, b]$ and for $t = 1, 2, \ldots, m$. Of course, we need to consider not only the case where the steplength is repeatedly halved, but more general sequences of steplengths tending to zero. We are thus led to the idea of a limiting process in which $h \to 0$ and $n \to \infty$ simultaneously in such a way that $x = a + nh$ remains fixed. Such a limit is denoted by

$$\lim_{\substack{h \to 0 \\ x = a + nh}} F(h, n)$$

and is called a *fixed station limit*. It is nothing more than an ordinary limit in which we must substitute $(x - a)/h$ for n in $F(h, n)$ before letting $h \to 0$ (or, alternatively, substitute $(x - a)/n$ for h in $F(h, n)$ before letting $n \to \infty$). For example,

$$\lim_{\substack{h \to 0 \\ x = a + nh}} (1 + h)^n = \lim_{\substack{h \to 0 \\ x = a + nh}} \exp[n \ln(1 + h)] = \lim_{h \to 0} \exp\left[\frac{x - a}{h} \ln(1 + h)\right] = \exp(x - a).$$

Alternatively,

$$\lim_{\substack{h \to 0 \\ x = a + nh}} \exp[n \ln(1 + h)] = \lim_{n \to \infty} \exp\left[n \ln\left(1 + \frac{x - a}{n}\right)\right] = \exp(x - a).$$

There remain two other points to be considered before we attempt to frame a definition of convergence. Firstly, it is clear that we have to impose some restrictions on the starting values $\eta_\mu(h)$, $\mu = 0, 1, \ldots, k-1$; referring to Figure 2.1 again, the obvious restriction is that as $h \to 0$, $\eta_\mu(h) \to \eta$, $\mu = 0, 1, \ldots, k-1$. Secondly, if convergence is to be a property of the method, then convergence must take place for all initial value problems.

Definition *The method defined by (2.6) is said to be* **convergent** *if, for all initial value problems satisfying the hypotheses of Theorem 1.1, we have that*

$$\lim_{\substack{h \to 0 \\ x = a + h}} y_n = y(x)$$

holds for all $x \in [a, b]$ and for all solutions $\{y_n\}$ of the difference equation in (2.6) satisfying starting conditions $y_\mu = \eta_\mu(h)$ for which $\lim \eta_\mu(h) = \eta$, $\mu = 0, 1, \ldots, k-1$. A method which is not convergent is said to be **divergent**.

An alternative equivalent definition is possible:

Alternative Definition *The method defined by (2.6) is said to be* **convergent** *if, for all initial value problems satisfying the hypotheses of Theorem 1.1, we have that*

$$\max_{0 \leqslant n \leqslant N} \| y(x_n) - y_n \| \to 0 \qquad \text{as } h \to 0.$$

Note that the starting values in (2.6) as well as the solution $\{y_n | n = k, k+1, \ldots, N\}$ are thus required to converge. Although this definition is much simpler, it does not alert us to the notational difficulties that will be encountered when we attempt to use the definition in any analysis.

2.4 CONSISTENCY

We now turn to the question of what conditions a numerical method must satisfy if it is to be convergent. We would expect that one such condition would be that it has to be a sufficiently accurate representation of the differential system. It would be an infinitely accurate representation if the difference equation (2.4) were satisfied exactly when we replaced the numerical solution y_{n+j} at x_{n+j}, by the exact solution $y(x_{n+j})$, for $j = 0, 1, 2, \ldots, k$. We therefore take as a measure of accuracy the value of the *residual* R_{n+k} which results on making this substitution. We thus define R_{n+k} by

$$R_{n+k} := \sum_{j=0}^{k} \alpha_j y(x_{n+j}) - h\phi_f(y(x_{n+k}), y(x_{n+k-1}), \ldots, y(x_n), x_n; h). \tag{2.7}$$

R_{n+k} is essentially the *local truncation error*, which we shall discuss in detail for the various classes of methods later in this book. There are, however, several variants of the definition of local truncation error; sometimes it is taken to be R_{n+k}/h, and sometimes it is further scaled by a constant multiplier independent of h.

A first thought on the appropriate level of accuracy that might be needed for convergence is that we should ask that $R_{n+k} \to 0$ as $h \to 0$. Further thought shows that this is not going to be enough. If we let $h \to 0$ in (2.7) then (assuming that ϕ_f does not tend to ∞ as $h \to 0$, as is certainly always the case) we have that

$$R_{n+k} \to \sum_{j=0}^{k} \alpha_j y(x_{n+j})$$

and the condition $R_{n+k} \to 0$ as $h \to 0$ can put a constraint only on the coefficients α_j in the method (2.4). It does not constrain the function ϕ_f in any way, and we cannot believe that convergence can be obtained with arbitrary ϕ_f, since that would be equivalent to choosing the function f arbitrarily; in short, the method would not even know which differential system it was dealing with. As we shall see presently, the appropriate level of accuracy is to demand that $R_{n+k}/h \to 0$ as $h \to 0$.

Definition *The method* (2.4) *is said to be* **consistent** *if, for all initial value problems satisfying the hypotheses of Theorem 1.1, the residual* R_{n+k} *defined by* (2.7) *satisfies*

$$\lim_{\substack{h\to 0\\ x=a+nh}} \frac{1}{h} R_{n+k} = 0. \tag{2.8}$$

(The word 'consistent' is shorthand for the phrase 'consistent with the differential system'.)

We now establish the conditions which (2.4) must satisfy if it is to be consistent. From Theorem 1.1 of §1.4 we can assume the existence of $y'(x)$ in $[a, b]$, but not necessarily that of higher derivatives. This means that, in investigating the limit (2.8), we must use the mean value theorem rather than any expansion in powers of h. Referring to §1.3, Case 2, we may write

$$y(x_{n+j}) - y(x_n) = jh\bar{y}'(\xi_j), \qquad j = 0, 1, \ldots, k,$$

where

$$\bar{y}'(\xi_j) = [{}^1y'(\xi_{1j}), {}^2y'(\xi_{2j}), \ldots, {}^my'(\xi_{mj})]^{\mathrm{T}},$$

and $\xi_{tj} \in (x_n, x_{n+j})$, $t = 1, 2, \ldots, m$. It then follows from (2.7) that

$$\begin{aligned}\frac{1}{h} R_{n+k} &= \frac{1}{h} \sum_{j=0}^{k} \alpha_j [y(x_n) + jh\bar{y}'(\xi_j)] - \phi_f(y(x_{n+k}), y(x_{n+k-1}), \ldots, y(x_n), x_n; h)\\ &= \frac{1}{h}\left(\sum_{j=0}^{k} \alpha_j\right) y(x_n) + \sum_{j=0}^{k} j\alpha_j \bar{y}'(\xi_j) - \phi_f(y(x_{n+k}), y(x_{n+k-1}), \ldots, y(x_n), x_n; h).\end{aligned}$$

In the limit as $h \to 0$, $x_n = a + nh$,

$$\bar{y}'(\xi_j) = y'(x_n), \qquad j = 0, 1, \ldots, k$$

and

$$\phi_f(y(x_{n+k}), y(x_{n+k-1}), \ldots, y(x_n), x_n; h) = \phi_f(y(x_n), y(x_n), \ldots, y(x_n), x_n; 0)$$

since, as $h \to 0$, $x_{n+j} := x_n + jh \to x_n, j = 0, 1, \ldots, k$. It follows that (2.8) holds, for all initial value problems, if

$$\sum_{j=0}^{k} \alpha_j = 0 \tag{2.9i}$$

and

$$\left(\sum_{j=0}^{k} j\alpha_j\right) y'(x_n) = \phi_f(y(x_n), y(x_n), \dots, y(x_n), x_n; 0),$$

whence, since $y(x)$ is a solution of the differential system $y' = f(x, y)$,

$$\phi_f(y(x_n), y(x_n), \dots, y(x_n), x_n; 0) \Big/ \left(\sum_{j=0}^{k} j\alpha_j\right) = f(x_n, y(x_n)). \tag{2.9ii}$$

The method (2.4) is thus consistent if it satisfies the two conditions (2.9).

The role of each of these conditions can be illustrated as follows. Assume that the sequence $\{y_n\}$ converges, in the sense of the preceding section, to *some* function $z(x) \in C^1[a, b]$, where $z(x) \not\equiv 0$. Then

$$y_{n+j} \to z(x_n) \qquad \text{as } h \to 0, \quad x_n = a + nh, \qquad j = 0, 1, \dots, k,$$

and from (2.4) we obtain in the limit

$$\left(\sum_{j=0}^{k} \alpha_j\right) z(x_n) = 0,$$

whence the condition (2.9i) must hold. Thus convergence of $\{y_n\}$ to any non-trivial function, not necessarily related to the solution of the initial value problem, is enough to imply (2.9i). This is in agreement with our earlier remark that requiring that $R_{n+k} \to 0$ as $h \to 0$ cannot be enough. Now assume that (2.9i) holds. Then, since

$$\sum_{j=0}^{k} \alpha_j (y_{n+j} - y_n) = \sum_{j=0}^{k} \alpha_j y_{n+j} - \left(\sum_{j=0}^{k} \alpha_j\right) y_n = \sum_{j=0}^{k} \alpha_j y_{n+j}$$

we may write (2.4) in the form

$$\sum_{j=0}^{k} j\alpha_j (y_{n+j} - y_n)/jh = \phi_f(y_{n+k}, y_{n+k-1}, \dots, y_n, x_n; h). \tag{2.10}$$

In the limit as $h \to 0$, $x_n = a + nh$,

$$(y_{n+j} - y_n)/jh = z'(x_n)$$

and

$$\phi_f(y_{n+k}, y_{n+k-1}, \dots, y_n, x_n; h) = \phi_f(z(x_n), z(x_n), \dots, z(x_n), x_n; 0)$$

so that (2.10) gives

$$\left(\sum_{j=0}^{k} j\alpha_j\right) z'(x_n) = \phi_f(z(x_n), z(x_n), \dots, z(x_n), x_n; 0).$$

Thus, if $z(x)$ satisfies the differential system $z' = f(x, z)$, then the second of the consistency conditions (2.9ii) is satisfied. We have thus shown that if $\{y_n\}$ converges to *any* non-trivial function then (2.9i) is satisfied, and if that function is a solution of the given differential system, then (2.9ii) is satisfied.

Note that if the method (2.4) satisfies (2.9i) but, instead of satisfying (2.9ii) it satisfies

$$\phi_f(y(x_n), y(x_n), \dots, y(x_n), x_n; 0) \Big/ \left(\sum_{j=0}^{k} j\alpha_j \right) = \mathrm{K} f(x_n, y(x_n)),$$

where K is a constant, then it will attempt to solve the initial value problem for the differential system $y' = \mathrm{K} f(x, y)$.

It is appropriate at this stage to introduce the *first characteristic polynomial* ρ associated with the general method (2.4), defined by

$$\rho(\zeta) := \sum_{j=0}^{k} \alpha_j \zeta^j, \tag{2.11}$$

where $\zeta \in \mathbb{C}$ is a dummy variable. It is then possible to write the necessary and sufficient conditions (2.9) for the method (2.4) to be consistent in the following alternative form:

$$\rho(1) = 0 \qquad \text{(2.12i)}$$

$$\phi_f(y(x_n), y(x_n), \dots, y(x_n), x_n; 0)/\rho'(1) = f(x_n, y(x_n)). \qquad \text{(2.12ii)}$$

We conclude this section by applying the conditions (2.12) to each of the Examples 1–6 of §2.2. It is easily seen that (2.12i) is satisfied for all six examples, and it is straightforward to see that (2.12ii) holds for Examples 1 and 3. For Example 2, $\rho(r) = r^2 - r$, whence $\rho'(1) = 1$ and

$$\phi_f(y(x_n), y(x_n), \dots, y(x_n), x_n; 0)/\rho'(1) = \tfrac{1}{3} f(x_n, y(x_n))$$

and the method is inconsistent; if it is applied to the initial value problem (2.2) it will attempt to solve instead the problem $y' = \frac{1}{3} f(x, y)$, $y(a) = \eta$. For Example 4, $\rho(r) = r^2 - 1, \rho'(1) = 2$ and

$$\phi_f(y_{n+1}, y_n, x_n; h) = f\left(x_{n+2}, 3y_{n+1} - 2y_n + \frac{h}{2} f(x_{n+1}, y_{n+1}) - \frac{3h}{2} f(x_n, y_n) \right) + f(x_n, y_n),$$

whence

$$\phi_f(y(x_n), y(x_n), x_n; 0) = 2f(x_n, y(x_n))$$

and (2.12ii) is clearly satisfied. For Examples 5 and 6, it is clear that when $h = 0$ and y_n is replaced by $y(x_n)$, each of the k_i reduces to $f(x_n, y(x_n))$, and (2.12ii) is satisfied. Thus, all of the Examples except Example 2 are consistent.

Exercises

2.4.1. Apply the method

$$y_{n+2} - y_{n+1} = \frac{h}{12} [4f(x_{n+2}, y_{n+2}) + 8f(x_{n+1}, y_{n+1}) - f(x_n, y_n)]$$

to the scalar initial value problem $y' = x$, $y(0) = 0$ to get a one-step difference equation of the form $y_{n+2} - y_{n+1} = \varphi(n, h)$. By trying a particular solution of the form $y_n = An^2 + Bn$, find the exact solution of this difference equation satisfying the initial condition $y_1 = \frac{1}{2}h^2$ (which coincides with the exact solution of the problem at $x = h$). Hence show that as $h \to 0$, $n \to \infty$, $x = nh$, the sequence $\{y_n\}$ so obtained does converge, but not to the solution of the initial value problem. Why is this?

2.4.2. Use the method of Exercise 2.4.1 to compute numerical solutions of the scalar initial value problem $y' = 4xy^{1/2}$, $y(0) = 1$ for $0 \leqslant x \leqslant 2$, using the steplengths $h = 0.1$, 0.05 and 0.025. Compare the results with the exact solution $y(x) = (1 + x^2)^2$ and deduce that the numerical solutions are not converging to the exact solution as $h \to 0$.

2.5 ZERO-STABILITY

Although, as we have seen in the preceding section, convergence implies consistency, the converse is not true. It can happen that the difference system produced by applying a numerical method to a given initial value problem suffers an in-built instability which persists even in the limit as $h \to 0$ and prevents convergence. Various forms of stability will be discussed later in this book; the form to be considered here is called *zero-stability*, since it is concerned with the stability of the difference system in the limit as h tends to zero.

We start by considering a stability property of the initial value problem (2.2). Suppose that in the problem (2.2) we perturb both the function f and the initial value η and ask how sensitive the solution is to such perturbations. The perturbation $(\delta(x), \delta)$ and the perturbed solution $z(x)$ are defined by the perturbed initial value problem

$$z' = f(x, z) + \delta(x), \quad z(a) = \eta + \delta, \qquad x \in [a, b].$$

Definition (Hahn, 1967; Stetter, 1971) Let $(\delta(x), \delta)$ and $(\delta^*(x), \delta^*)$ be any two perturbations of (2.2) and let $z(x)$ and $z^*(x)$ be the resulting perturbed solutions. Then if there exists a positive constant S such that, for all $x \in [a, b]$,

$$\| z(x) - z^*(x) \| \leqslant S\varepsilon$$

whenever (2.13)

$$\| \delta(x) - \delta^*(x) \| \leqslant \varepsilon \quad \text{and} \quad \| \delta - \delta^* \| \leqslant \varepsilon,$$

then the initial value problem (2.2) is said to be **totally stable**.

To ask that an initial value problem be totally stable (or, equivalently, *properly-posed*) is not asking for much; note that S can be as large as we please as long as it is a (finite) constant. Indeed, it is straightforward to show that the hypotheses of Theorem 1.1 of §1.4 are sufficient for the initial value problem (2.2) to be totally stable (see, for example, Gear (1971a)).

Any numerical method applied to (2.2) will introduce errors due to discretization and round-off, and these could be interpreted as being equivalent to perturbing the problem; if (2.13) is not satisfied, then no numerical method has any hope of producing an acceptable solution. The same will be true if the difference equation produced by the method is itself over-sensitive to perturbations. We therefore consider the effects of perturbations of the function ϕ_f and the starting values $\eta_\mu(h)$ in (2.6). The perturbation

$\{\delta_n, n = 0, 1, \ldots, N\}$ and the perturbed solution $\{z_n, n = 0, 1, \ldots, N\}$ of (2.6) are defined by the perturbed difference system

$$\left.\begin{aligned} &\sum_{j=0}^{k} \alpha_j z_{n+j} = h[\phi_f(z_{n+k}, z_{n+k-1}, \ldots, z_n, x_n; h) + \delta_{n+k}] \\ &z_\mu = \eta_\mu(h) + \delta_\mu, \qquad \mu = 0, 1, \ldots, k-1 \end{aligned}\right\} \tag{2.14}$$

Definition *Let* $\{\delta_n, n = 0, 1, \ldots, N\}$ *and* $\{\delta_n^*, n = 0, 1, \ldots, N\}$ *be any two perturbations of (2.6), and let* $\{z_n, n = 0, 1, \ldots, N\}$ *and* $\{z_n^*, n = 0, 1, \ldots, N\}$ *be the resulting perturbed solutions. Then if there exist constants* S *and* h_0 *such that, for all* $h \in (0, h_0]$,

$$\left.\begin{aligned} &\|z_n - z_n^*\| \leqslant S\varepsilon, \qquad 0 \leqslant n \leqslant N \\ \textit{whenever} \qquad &\|\delta_n - \delta_n^*\| \leqslant \varepsilon, \qquad 0 \leqslant n \leqslant N, \end{aligned}\right\} \tag{2.15}$$

we say that the method (2.4) is **zero-stable**.

Several comments can be made about this definition:

(a) Zero-stability requires that (2.15) holds for *all* $h \in (0, h_0]$; it is therefore concerned with what happens in the limit as $h \to 0$.
(b) Zero-stability is a property of the method, not of the system. Our assumption that (2.2) satisfies a Lipschitz condition ensures that the *problem* is totally stable and therefore insensitive to perturbations; zero-stability is simply a requirement that the *difference system* which the method generates be likewise insensitive to perturbations. It is equivalent to saying that the difference system is *properly posed.*
(c) A very practical interpretation can be put on the definition. No computer can calculate to infinite precision, so that inevitably round-off errors arise whenever ϕ_f is computed; in (2.14), $\{\delta_n, n = k, k+1, \ldots, N\}$ could be interpreted as these round-off errors. Likewise, the starting values cannot always be represented on the computer to infinite precision, and $\{\delta_n, n = 0, 1, \ldots, k-1\}$ could be interpreted as round-off errors in the starting values. If (2.15) is not satisfied, then the solutions of the difference system generated by the method, using two different rounding procedures—for example, using two different computers—could result in two numerical solutions of the same difference system being infinitely far apart, no matter how fine the precision. In other words, if the method is zero-unstable, then the sequence $\{y_n\}$ is essentially uncomputable.

Before stating the necessary and sufficient conditions for the method (2.4) to be zero-stable, let us consider an example. We shall consider the solution of the scalar initial value problem

$$y' = -y, \quad y(0) = 1,$$

whose exact solution is $y(x) = \exp(-x)$, by the consistent implicit two-step method

$$y_{n+2} - (1+\alpha)y_{n+1} + \alpha y_n = \tfrac{1}{2}h[f(x_{n+2}, y_{n+2}) + (1-\alpha)f(x_{n+1}, y_{n+1}) - \alpha f(x_n, y_n)] \tag{2.16}$$

(where α is a free parameter) satisfying the initial conditions

$$y_0 = 1, \quad y_1 = 1.$$

On substituting $-y$ for $f(x, y)$ in (2.16) we obtain the difference equation

$$(1 + h/2)y_{n+2} - [1 + \alpha - (1 - \alpha)h/2]y_{n+1} + \alpha(1 - h/2)y_n = 0, \quad y_0 = y_1 = 1.$$

We consider the simple perturbation in which $\delta_n = \delta$, a constant, for $n = 0, 1, \ldots, N$. The perturbed difference equation is thus

$$(1 + h/2)z_{n+2} - [1 + \alpha - (1 - \alpha)h/2]z_{n+1} + \alpha(1 - h/2)z_n = h\delta \tag{2.17}$$

$$z_0 = 1 + \delta, \quad z_1 = 1 + \delta. \tag{2.18}$$

Following §1.7, the characteristic polynomial of (2.17) is

$$(1 + h/2)r^2 - [1 + \alpha - (1 - \alpha)h/2]r + \alpha(1 - h/2)$$

with roots α and $(1 - h/2)/(1 + h/2)$. A particular solution is found to be $\delta/(1 - \alpha)$ if $\alpha \neq 1$ and $n\delta$ if $\alpha = 1$. Thus the general solution of (2.17) can be written in the form

$$z_n = P\alpha^n + Q[(1 - h/2)/(1 + h/2)]^n + \begin{cases} \delta/(1 - \alpha) & \text{if } \alpha \neq 1 \\ n\delta & \text{if } \alpha = 1 \end{cases} \tag{2.19}$$

where P and Q are arbitrary constants.

Case $\alpha \neq 1$ After some manipulation, we find that the solution of (2.17) satisfying the starting conditions (2.18) is

$$\left.\begin{aligned} z_n &= \frac{1}{C}\left[A(\delta)\alpha^n + B(\delta)\left(\frac{1 - h/2}{1 + h/2}\right)^n\right] + \frac{\delta}{1 - \alpha} \\ \text{where} \quad A(\delta) &= h[\alpha\delta/(1 - \alpha) - 1] \\ B(\delta) &= (1 - \alpha - \alpha\delta)(1 + h/2) \\ C &= 1 - \alpha - h(1 + \alpha)/2 \end{aligned}\right\} . \tag{2.20}$$

If we replace the constant perturbation δ by another constant perturbation δ^*, then the resulting perturbed solution $\{z_n^*\}$ is obviously given by (2.20) with δ replaced by δ^*. On subtracting we obtain

$$z_n - z_n^* = \left\{\frac{1}{C}\left[\frac{h\alpha}{1 - \alpha}\alpha^n - (1 + h/2)\alpha\left(\frac{1 - h/2}{1 + h/2}\right)^n\right] + \frac{1}{1 - \alpha}\right\}(\delta - \delta^*). \tag{2.21}$$

Now, for all $h \in (0, h_0]$, $\alpha \in [-1, 1)$,

$$|h\alpha^{n+1}/(1 - \alpha)| \leqslant h_0/(1 - \alpha), \quad |(1 + h/2)\alpha| \leqslant 1 + h_0/2, \quad \left|\frac{1 - h/2}{1 + h/2}\right| < 1$$

and

$$\left|\frac{1}{C}\right| \leqslant \frac{1}{|1 - \alpha - h_0(1 + \alpha)/2|}$$

provided that we choose h_0 such that $0 < h_0 < 2(1-\alpha)/(1+\alpha)$. It follows that (2.15) holds with

$$S = \frac{h_0/(1-\alpha) + 1 + h_0/2}{|1 - \alpha - h_0(1+\alpha)/2|} + 1/(1-\alpha).$$

This will not be the case, however, if $|\alpha| > 1$. Consider the term $h\alpha^n$ on the right-hand side of (2.21). Since

$$\lim_{\substack{h\to 0\\ x=nh}} h\alpha^n = x \lim_{n\to\infty} \alpha^n/n = \infty$$

this term becomes unbounded as $h \to 0$, and (2.15) cannot hold.

Case $\alpha = 1$ Using (2.19), the solution of (2.17) satisfying the starting conditions (2.18) turns out to be

$$z_n = 1 + \frac{h-2}{2h}\delta + \frac{h+2}{2h}\delta\left(\frac{1-h/2}{1+h/2}\right)^n + n\delta \tag{2.22}$$

and on replacing z_n by z_n^*, δ by δ^*, and subtracting, we obtain

$$z_n - z_n^* = \left[\frac{h-2}{2h} + \frac{h+2}{2h}\left(\frac{1-h/2}{1+h/2}\right)^n + n\right](\delta - \delta^*).$$

As $h \to 0$, $nh = x$, the term within the square bracket becomes unbounded, and (2.15) cannot hold.

Thus, for this example, the condition (2.15) is satisfied if and only if $-1 \leqslant \alpha < 1$.

It is easily checked that the method used above is consistent, and we can see as follows that we have convergence if $-1 \leqslant \alpha < 1$, but divergence otherwise (thus demonstrating that consistency is not sufficient for convergence). Let $h \to 0$, $nh = x$ and $\delta \to 0$ in (2.20), noting that the latter ensures that the conditions $y_\mu \to \eta$ as $h \to 0$, $\mu = 0, \ldots, k-1$, appearing in the definition of convergence, are satisfied. Then

$$B(\delta)/C \to 1, \quad A(\delta)\alpha^n/C \to \begin{cases} 0 & \text{if } -1 \leqslant \alpha < 1 \\ \infty & \text{if } |\alpha| > 1 \end{cases}$$

and therefore

$$z_n \to \begin{cases} \left(\dfrac{1-h/2}{1+h/2}\right)^n & \text{if } -1 \leqslant \alpha < 1 \\ \infty & \text{if } |\alpha| > 1. \end{cases}$$

On comparing expansions in powers of h, it is easily seen that

$$\frac{1-h/2}{1+h/2} = \exp(-h) + 0(h^3),$$

whence $z_n \to \exp(-nh) = \exp(-x)$, the exact solution of the initial value problem. We thus have convergence if $-1 \leqslant \alpha < 1$, and divergence if $|\alpha| > 1$. On applying the same limiting process to (2.22), it is easy to establish that divergence also occurs when $\alpha = 1$.

The above example shows that zero-stability is concerned with the roots of the characteristic polynomial of the difference equation (2.17). Of course, had we applied the method to a nonlinear problem, then the resulting difference equation would have been nonlinear and there would have been no such thing as a characteristic polynomial. Nevertheless, for general problems, as $h \to 0$, the method (2.4) tends to the linear constant coefficient difference system

$$\sum_{j=0}^{k} \alpha_j y_{n+j} = 0$$

whose characteristic polynomial is $\rho(\zeta)$, the first characteristic polynomial of the method, defined by (2.11); it is no surprise that it is the location of the roots of $\rho(\zeta)$ that controls the zero-stability of the method.

Let the roots of $\rho(\zeta)$ be ζ_i, $i = 1, 2, \ldots, k$. If we assume the method is consistent, then by (2.12i), one of the roots must be $+1$. We call this root *principal root*, and always label it ζ_1 $(= +1)$. The remaining roots ζ_i, $i = 2, 3, \ldots, k$, are the *spurious roots*, and arise because we choose to represent a first-order differential system by a kth-order difference system. Obviously, for a one-step method there are no spurious roots. (Note that in our example, it is the root $(1 - h/2)/(1 + h/2)$ of the characteristic polynomial of (2.17) which produces an approximation to the desired solution, and it is this root which tends to the principal root of ρ as $h \to 0$; the other root tends to the spurious root α of ρ (it so happens that it coincides with α) and this is the root which has the capability of invoking zero-instability.) It turns out that zero-stability is ensured if the roots of ρ satisfy the following condition:

Definition *The method (2.4) is said to satisfy the* **root condition** *if all of the roots of the first characteristic polynomial have modulus less than or equal to unity, and those of modulus unity are simple.*

The roots of a polynomial being complex, an alternative statement of the root condition is that all of the roots of the first characteristic polynomial must lie in or on the unit circle, and there must be no multiple roots on the unit circle. Note that all consistent one-step methods satisfy the root condition. Example 1 of §2.2 clearly fails to satisfy the root condition, since ρ has a spurious root at -2; the remaining five Examples all satisfy the root condition.

Theorem 2.1 *The necessary and sufficient condition for the method given by (2.4) and (2.5) to be zero-stable is that it satisfies the root condition.*

For a proof of this theorem, see, for example, Isaacson and Keller (1966). We note that our example corroborates this result, since both roots 1 and α of ρ lie in or on the unit circle when $-1 \leqslant \alpha < 1$, there is a root outside the circle when $|\alpha| > 1$, and there is a multiple root on the circle when $\alpha = 1$.

Some authors (for example, Lambert (1973)) adopt the following alternative definition of zero-stability:

Alternative Definition *The method (2.4) is said to be* **zero-stable** *if it satisfies the root condition.*

In view of Theorem 2.1 the two definitions are clearly equivalent, but there are two disadvantages in adopting the alternative form. Firstly, it does not have the flavour of a stability definition in the way that our first definition has. Secondly, it does not draw attention to the fact that zero-stability, being nothing more than a requirement that the difference system be properly posed, is a minimal demand.

We are now in a position to state the necessary and sufficient conditions for convergence.

Theorem 2.2 *The necessary and sufficient conditions for the method (2.4) to be convergent are that it be both consistent and zero-stable.*

Theorem 2.2 is the fundamental theorem of this subject. It was first proved for linear multistep methods by Dahlquist (1956)—see also Henrici (1962). A proof for the class (2.4) can be found in Isaacson and Keller (1966). Proofs for yet wider classes of methods can be found in Gear (1965), Butcher (1966), Spijker (1966), Chartres and Stepleman (1972) and Mäkela, Nevanlinna and Sipilä (1974).

Exercises

2.5.1. Find the range of α for which the method

$$y_{n+2} + (\alpha - 1)y_{n+1} - \alpha y_n = \frac{h}{4}[(\alpha + 3)f(x_{n+2}, y_{n+2}) + (3\alpha + 1)f(x_n, y_n)]$$

is zero-stable. Apply the method, with $\alpha = -1$ to the scalar initial value problem $y' = y$, $y(0) = 1$, and solve exactly the resulting difference equation, taking the starting values to be $y_0 = y_1 = 1$. Hence show that the numerical solution diverges as $h \to 0$, $n \to \infty$.

2.5.2. Quade's method is given by

$$y_{n+4} - \frac{8}{19}(y_{n+3} - y_{n+1}) - y_n = \frac{6h}{19}(f_{n+4} + 4f_{n+3} + 4f_{n+1} + f_n),$$

where $f_{n+j} = f(x_{n+j}, y_{n+j})$, $j = 0, 1, \dots, 4$. Show that the method is convergent.

2.5.3. A method is given by

$$y_{n+2} = \hat{y}_{n+2} - 6\alpha(y_{n+1} - y_n) + \alpha h(f_{n+2} - 4f_{n+3/2} + 7f_{n+1} + 2f_n)$$

$$\hat{y}_{n+2} = 2y_{n+1} - y_n + \frac{h}{3}(4f_{n+3/2} - 3f_{n+1} - f_n),$$

where $f_{n+j} = f(x_{n+j}, y_{n+j})$, $j = 0, 1, 2$, $f_{n+3/2} = f(x_{n+3/2}, y_{n+3/2})$ and $y_{n+3/2}$ is given by a formula of the form

$$y_{n+3/2} + \tilde{\alpha}_1 y_{n+1} + \tilde{\alpha}_0 y_n = h(\tilde{\beta}_1 f_{n+1} + \tilde{\beta}_0 f_n).$$

Show that the method satisfies the conditions (2.5) of §2.2 and is consistent. Find the range of α for which it is zero-stable.

2.5,4. Demonstrate the effect of zero-instability by using the method

$$y_{n+2}-(1+\alpha)y_{n+1}+\alpha y_n=\frac{h}{2}[(3-\alpha)f(x_{n+1},y_{n+1})-(1+\alpha)f(x_n,y_n)]$$

with (i) $\alpha=0$, (ii) $\alpha=-5$ to compute numerical solutions of the scalar initial value problem $y'=4xy^{1/2}$, $y(0)=1$ for $0\leqslant x\leqslant 2$, using the steplengths $h=0.1, 0.05, 0.025$.

2.5.5*. The family of methods (2.16) is a sub-family of the two-parameter family of methods

$$y_{n+2}-(1+\alpha)y_{n+1}+\alpha y_n=h[(1+\beta)f_{n+2}-(\alpha+\beta+\alpha\beta)f_{n+1}+\alpha\beta f_n], \tag{1}$$

where $f_{n+j}=f(x_{n+j},y_{n+j})$, $j=0,1,2$. The family (1) is a useful one for illustrative purposes since it has the property that when it is applied to the scalar initial value problem $y'=y$, $y(0)=1$ the resulting difference equation can be solved exactly. Show that this solution, with starting values $y_0=\eta_0$ $[=\eta_0(h)]$, $y_1=\eta_1$ $[=\eta_1(h)]$ is

$$y_n=\left[A\alpha^n+\left(\frac{1-\beta h}{1-(1+\beta)h}\right)^n\right]\Big/C,$$

where

$$A=(-1+\beta h)\eta_0+[1-(1+\beta)h]\eta_1,\quad B=[1-(1+\beta)h](\alpha\eta_0-\eta_1),\quad C=\alpha-1-(\alpha-\beta+\alpha\beta)h.$$

Show further that

$$\left(\frac{1-\beta h}{1-(1+\beta)h}\right)^n=\begin{cases}\exp(x_n)[1+(\frac{1}{2}+\beta)x_nh+0(h^2)] & \text{if } \beta\neq-\frac{1}{2}\\ \exp(x_n)[1+\frac{1}{12}x_nh^2+0(h^3)] & \text{if } \beta=-\frac{1}{2}\end{cases}.$$

(*Hint*: Consider the expansion of the logarithm of the left side.) We assume that the starting values satisfy

$$\lim_{h\to 0}\eta_i(h)=1,\qquad i=0,1. \tag{2}$$

(i) Demonstrate that when $|\alpha|<1$ the method converges, for all starting values satisfying (2).
(ii) Demonstrate that when $|\alpha|>1$ the method diverges for *general* starting values satisfying (2), but that it converges for the specific starting values $\eta_0=1$, $\eta_1(h)=(1-\beta h)/[1-(1+\beta)h]$ (which satisfy (2)). Why would we not be able to demonstrate this numerically? Try doing so.
(iii) Demonstrate that when $\alpha=1$, there exist some starting values satisfying (2) for which the method converges, and some for which it diverges (sometimes in the sense that $\{y_n\}$ converges to the wrong solution, and sometimes in the sense that $y_n\to\infty$ as $h\to 0$, $n\to\infty$).

2.6 THE SYNTAX OF A STABILITY DEFINITION

Zero-stability is not the only form of stability pertinent to the numerical solution of initial value problems, and several other stability definitions will appear later in this book. In this section we shall discuss a general framework for such definitions and introduce a 'syntax diagram', which the reader may (or may not) find helpful.

A stability definition can be broken down into the following components:

1. We impose certain conditions C_P on the problem (2.2) which force the exact solution $y(x)$, $x\in[a,b]$, to display a certain stability property.

2. We apply the method (2.4) to the problem, assumed to satisfy C_p.
3. We ask what conditions C_m must be imposed on the method in order that the numerical solution $\{y_n, n = 0, 1, \ldots, N\}$ displays a stability property analogous to that displayed by the exact solution.

This 'syntax' can be represented by the diagram below.

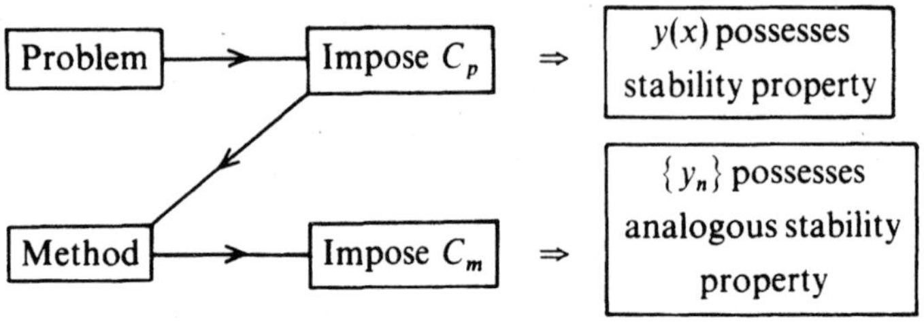

(2.23)

The syntax diagram for zero-stability can therefore be written as shown below.

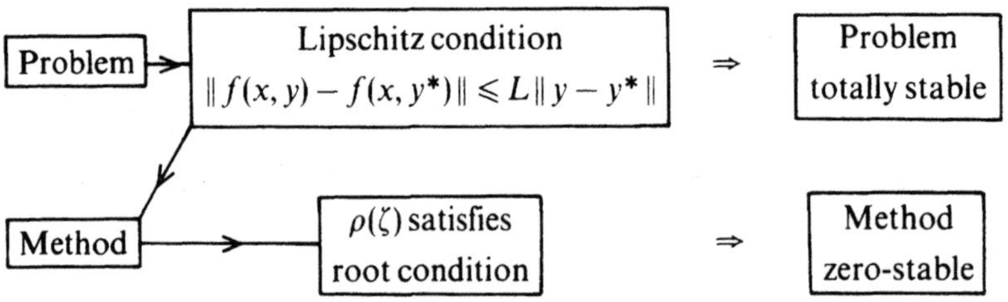

The syntax diagram is not a replacement for a stability definition; thus in the above example, it is still necessary to refer to the formal definitions of total stability and zero-stability. It is more a device for putting stability concepts into context. In the general syntax diagram (2.23), the rightmost lower box normally defines the stability property of the method, and the box to its left defines the conditions for that property to hold; but it can also be the case that the middle box defines the stability property in which case the rightmost box is interpreted as a consequence of the property. Thus, if we adopt the alternative definition of zero-stability given in §2.5, the syntax diagram—now appropriate to convergence rather than zero-stability—becomes as shown below.

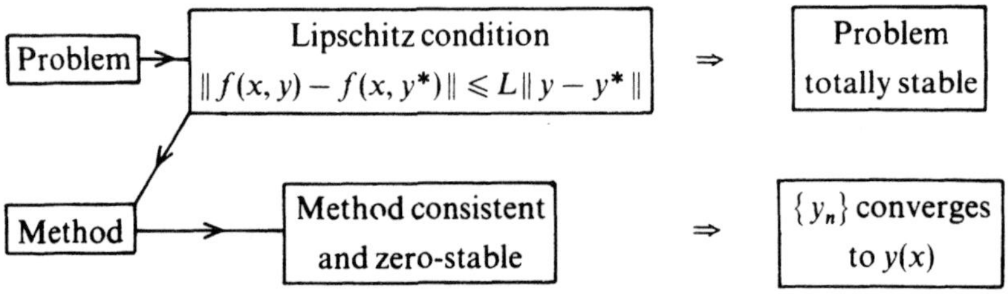

Again, we find the alternative definition of zero-stability less satisfactory than the first definition.

In certain circumstances (which will arise in Chapter 7), it will be appropriate to replace 'problem' by 'system' in the top line of the syntax diagram.

2.7 SOME NUMERICAL EXPERIMENTS

The experiments described in this section consist of applying each of the six Examples of methods given in §2.2 to the same initial value problem, using a range of steplengths. The purpose is two-fold: firstly to demonstrate the effects of the properties defined in the preceding sections, and secondly to persuade the reader that it is by no means guaranteed that a convergent method will always produce acceptable numerical solutions.

The initial value problem to be solved is

$$y' = f(x, y), \quad y(0) = \eta, \qquad x \in [0, 1]$$

where

$$y = [u, v]^{\mathrm{T}}, \tag{2.24}$$

and

$$f(x, y) = [v, v(v-1)/u]^{\mathrm{T}}, \qquad \eta = [1/2, -3]^{\mathrm{T}}.$$

It is easily checked that (2.24) satisfies the hypotheses of Theorem 1.1 and is therefore totally stable. The unique exact solution is

$$u(x) = [1 + 3\exp(-8x)]/8, \quad v(x) = -3\exp(-8x),$$

and we note that the solution decays in the sense that both $|u(x)|$ and $|v(x)|$ decrease monotonically as x increases from 0 to 1. When additional starting values are needed these are taken to coincide with the exact solution. It is impracticable to reproduce the numerical solution at every discretization point, and we present only tables of the error E_n defined by

$$E_n := \| y(x_n) - y_n \|_2$$

at intervals of 0.2 of x, for a range of values of h.

Example 1

$$y_{n+2} + y_{n+1} - 2y_n = \frac{h}{4}[f(x_{n+2}, y_{n+2}) + 8f(x_{n+1}, y_{n+1}) + 3f(x_n, y_n)].$$

This method is consistent but zero-unstable and therefore divergent. See Table 2.1 for numerical results.

Table 2.1

x	$h = 0.1$	$h = 0.05$	$h = 0.025$	$h = 0.0125$
0.2	0.026 53	0.008 23	0.008 98	0.154 05
0.4	0.135 04	0.208 52	4.0807	18 273
0.6	0.902 51	5.890 4	1877.3	2.2×10^{11}
0.8	6.156 8	166.69	863 679	O/F
1.0	42.040	4716.9	4.0×10^{10}	

O/F indicates overflow.

The divergence is clear. For each fixed h, the error increases rapidly as x increases. For $x = 0.2$ the error initially decreases as h decreases, but soon starts to increase; for all other fixed values of x, the error increases as h decreases.

Example 2

$$y_{n+2} - y_{n+1} = \frac{h}{3}[3f(x_{n+1}, y_{n+1}) - 2f(x_n, y_n)].$$

This method is zero-stable but inconsistent and therefore divergent. See Table 2.2 for numerical results.

Table 2.2

x	$h = 0.1$	$h = 0.01$	$h = 0.001$	$h = 0.0001$
0.2	1.273 7	1.110 4	1.157 6	1.162 7
0.4	1.101 9	0.907 84	0.916 16	0.917 15
0.6	0.795 01	0.592 94	0.586 18	0.585 59
0.8	0.553 84	0.365 17	0.354 22	0.353 18
1.0	0.384 25	0.220 80	0.210 18	0.209 17

Divergence due to zero-instability led to an explosion of error in Example 1. Here divergence is caused by inconsistency, and leads to no such explosion, but manifests itself in a persistent error which refuses to decay to zero as h is reduced (even to the excessively small value of 10^{-4}); indeed, for $x = 0.2, 0.4$, the error eventually increases slowly as h is decreased. The sequence $\{y_n\}$ is converging as $h \to 0$, but not to the solution of (2.24). This is exactly what we would expect, since the method satisfies the first of the consistency conditions (2.12) but not the second. Following the discussion towards the end of §2.4, we would expect it to attempt to solve a different initial value problem, namely

$$z' = f(x, z), \qquad z(0) = \eta, \qquad x \in [0, 1],$$

where

$$z = [\tilde{u}, \tilde{v}]^{\mathrm{T}},$$

and

$$f(x, z) = \tfrac{1}{3}[\tilde{v}, \tilde{v}(\tilde{v} - 1)/\tilde{u}]^{\mathrm{T}}, \qquad \eta = [\tfrac{1}{2}, -3]^{\mathrm{T}},$$

whose exact solution is

$$z(x) = [(1 + 3\exp(-8x/3))/8, -3\exp(-8x/3)]^{\mathrm{T}}.$$

The errors E_n^*, where $E_n^* := \| z(x_n) - z_n \|_2$ are given in Table 2.3.

One is persuaded that the method is indeed attempting to solve the above initial value problem.

Table 2.3

x	$h=0.1$	$h=0.01$	$h=0.001$	$h=0.0001$
0.2	0.301 95	0.036 13	0.003 77	0.000 38
0.4	0.396 51	0.043 90	0.004 44	0.000 44
0.6	0.389 50	0.039 37	0.003 91	0.000 39
0.8	0.339 72	0.031 25	0.003 06	0.000 31
1.0	0.277 75	0.023 22	0.002 25	0.000 22

Example 3

$$y_{n+3}+\tfrac{1}{4}y_{n+2}-\tfrac{1}{2}y_{n+1}-\tfrac{3}{4}y_n=\frac{h}{8}[19f(x_{n+2},y_{n+2})+5f(x_n,y_n)].$$

This method is consistent and zero-stable and therefore convergent. See Table 2.4 for numerical results.

Table 2.4

x	$h=0.1$	$h=0.05$	$h=0.025$	$h=0.0125$
0.2	—	0.008 37	9.3×10^{-4}	1.1×10^{-4}
0.4	0.257 76	0.041 05	2.4×10^{-4}	4.5×10^{-5}
0.6	1.497 5	0.119 69	1.6×10^{-4}	1.4×10^{-5}
0.8	8.087 6	0.330 12	2.1×10^{-5}	3.8×10^{-6}
1.0	43.507	0.905 07	6.8×10^{-6}	9.6×10^{-7}

The last two columns demonstrate that the method does indeed converge. However, there appears to exist some value h^* of h between 0.025 and 0.05, such that for fixed $h>h^*$, the error increases as x increases, while for fixed $h<h^*$, the error decreases as x increases. We shall examine this phenomenon in detail later in this book, but for the moment we note that for $h>h^*$ the numerical solution generated by the method is not acceptable. Thus convergent methods do not always give acceptable answers; one has to choose h to be sufficiently small.

Example 4

$$y_{n+2}-y_n=h[f(x_{n+2},y^*_{n+2})+f(x_n,y_n)],$$

where

$$y^*_{n+2}-3y_{n+1}+2y_n=\frac{h}{2}[f(x_{n+1},y_{n+1})-3f(x_n,y_n)].$$

This method is consistent and zero-stable and therefore convergent. See Table 2.5 for numerical results.

Table 2.5

x	$h = 0.1$	$h = 0.01$	$h = 0.001$	$h = 0.0001$
0.2	0.89642	0.00665	2.7×10^{-5}	1.7×10^{-7}
0.4	3.9745	0.08924	1.5×10^{-4}	4.7×10^{-6}
0.6	22.955	1.7131	3.4×10^{-3}	1.2×10^{-4}
0.8	135.02	33.193	0.08209	2.9×10^{-3}
1.0	794.75	643.23	1.9633	0.07147
1.2				1.7489
1.4				42.794

The results here are somewhat unexpected, and appear to be demonstrating divergence rather than convergence. However, on comparing them with those for the divergent Example 1, a significant difference emerges. In Example 1, as h is decreased, the value of x at which the numerical solution parts company with the exact solution—to put it in broad terms—moves *closer* to the initial point $x = 0$; in Example 4, it moves *further away* from the initial point. (In the case of $h = 0.0001$ we have extended the interval of integration to show this.) If we were to keep on reducing h, the same pattern would emerge; the numerical solution would be a reasonable approximation to the exact solution for longer and longer intervals of x, but would always blow up for sufficiently large x. Thus, in the limit as $h \to 0$, the point at which the numerical solution detaches itself from the exact solution tends to infinity, and so the method is convergent; for Example 1, that point tends to the initial point, and the method is divergent. Thus, this example provides a salutary lesson—there exist convergent methods which, no matter how small the steplength, will produce numerical solutions which will blow up for sufficiently long intervals of integration.

Example 5

$$y_{n+1} - y_n = \frac{h}{4}(k_1 + 3k_3)$$

where

$$k_1 = f(x_n, y_n)$$
$$k_2 = f(x_n + \tfrac{1}{3}h, y_n + \tfrac{1}{3}hk_1)$$
$$k_3 = f(x_n + \tfrac{2}{3}h, y_n + \tfrac{2}{3}hk_2).$$

This method is consistent and zero-stable and therefore convergent. See Table 2.6 for numerical results.

The comments made on Example 3 apply equally to this example, except that the value of h^* is considerably larger, lying between 0.2 and 0.4. We conclude that the maximum value of h which gives a solution which does not blow-up depends on the method (and presumably on the problem too).

Table 2.6

x	$h=0.4$	$h=0.2$	$h=0.1$	$h=0.05$
0.2	—	0.618 47	0.039 19	0.003 59
0.4	7.8066	0.123 22	0.015 32	0.001 44
0.6	—	0.024 88	0.004 49	0.000 44
0.8	19.521	0.005 02	0.001 17	0.000 12
1.0	—	0.001 01	0.000 29	0.000 03
1.2	49.622			

Example 6

$$y_{n+1} - y_n = \frac{h}{2}(k_1 + k_2)$$

where

$$k_1 = f(x_n, y_n)$$
$$k_2 = f(x_n + h, y_n + \tfrac{1}{2}hk_1 + \tfrac{1}{2}hk_2).$$

This method is consistent and zero-stable and therefore convergent. See Table 2.7 for numerical results.

Table 2.7

x	$h=0.8$	$h=0.4$	$h=0.2$	$h=0.1$
0.2	—	—	0.274 48	0.055 09
0.4	—	0.820 93	0.085 91	0.021 24
0.6	—	—	0.020 73	0.006 15
0.8	1.5887	0.155 98	0.004 56	0.001 58
1.0	—	—	0.000 96	0.000 38
1.6	0.829 53	0.008 57	0.000 01	0
2.4	0.434 52	0.000 46		

Convergence is again demonstrated. This time, the errors decay as x increases for all the values of h used; indeed this would be the case no matter how large h.

The above examples were chosen to demonstrate various stability phenomena which will be studied in detail later in this book; in particular they illustrate that zero-stability is not the only form of stability that will have to be considered. Such stability phenomena are not related to any particular class of methods, and it would therefore be quite wrong to draw any conclusion, on the basis of these results, about which class of methods performs best.

3 Linear Multistep Methods

3.1 NOTATION AND NOMENCLATURE

In §2.2, we mentioned the class of linear multistep methods, in which the function $\phi_f(y_{n+k}, y_{n+k-1}, \dots, y_n, x_n; h)$ defined by (2.4) takes the form of a linear combination of the values of the function f evaluated at (x_{n+j}, y_{n+j}), $j = 0, 1, \dots, k$. Using the shortened notation

$$f_{n+j} := f(x_{n+j}, y_{n+j}), \qquad j = 0, 1, \dots, k$$

we define a *linear multistep method* or *linear k-step method* in standard form by

$$\sum_{j=0}^{k} \alpha_j y_{n+j} = h \sum_{j=0}^{k} \beta_j f_{n+j}, \tag{3.1}$$

where α_j and β_j are constants subject to the conditions

$$\alpha_k = 1, \qquad |\alpha_0| + |\beta_0| \neq 0. \tag{3.2}$$

The first of these conditions removes, in a convenient manner, the arbitrariness that arises from the fact that we could multiply both sides of (3.1) by the same constant without altering the method. (Other means of removing this arbitrariness are of course possible; thus some authors divide both sides of (3.1) by $\sum_{j=0}^{k} \beta_j$.) The second condition prevents both α_0 and β_0 being zero, and thus precludes methods such as

$$y_{n+2} - y_{n+1} = hf_{n+1},$$

which is essentially a 1-step and not a 2-step method, and is in practice indistinguishable from the 1-step method

$$y_{n+1} - y_n = hf_n. \tag{3.3}$$

Method (3.3) is *Euler's Rule*, the simplest of all numerical methods.

There is an alternative notation for linear multistep methods. In (2.11) we introduced the first characteristic polynomial ρ associated with the general method (2.4), ρ being the polynomial of degree k whose coefficients are α_j. In the case of linear multistep methods it is natural to define a similar polynomial whose coefficients are β_j. We thus define the *first* and *second characteristic polynomials* of (3.1) by

$$\rho(\zeta) := \sum_{j=0}^{k} \alpha_j \zeta^j, \quad \sigma(\zeta) := \sum_{j=0}^{k} \beta_j \zeta^j, \tag{3.4}$$

where $\zeta \in \mathbb{C}$ is a dummy variable. Using the notation of §1.2, the linear multistep method

(3.1) can now be written in the form

$$\rho(E)y_n = h\sigma(E)f_n, \tag{3.5}$$

where E is the forward shift operator. The conditions (3.2) imply that ρ is a *monic* polynomial of degree k and that ρ and σ do not have a common factor ζ. Both notations have advantages, and we shall use whichever is the more convenient for the analysis in hand.

The method (3.1) is clearly explicit if $\beta_k = 0$, and implicit if $\beta_k \neq 0$. Equivalently, we can say that (3.5) is implicit if σ has degree k and explicit if it has degree less than k. For an explicit method, the sequence $\{y_n\}$ can be computed directly, provided the necessary additional starting values have been obtained, whereas for an implicit method it is necessary to solve at each step the nonlinear (in general) system of equations

$$y_{n+k} = h\beta_k f(x_{n+k}, y_{n+k}) + g, \tag{3.6}$$

where g is a known function of previously computed values of y_{n+j}. By Theorem 1.2 of §1.8, this system of equations possesses a unique solution for y_{n+k}, which can be approached arbitrarily closely by the iteration

$$y^{[\nu+1]}_{n+k} = h\beta_k f(x_{n+k}, y^{[\nu]}_{n+k}) + g, \qquad \nu = 0, 1, \ldots, \qquad y^{[0]}_{n+k} \text{ arbitrary} \tag{3.7}$$

provided that $0 \leqslant M < 1$, where M is the Lipschitz constant with respect to y_{n+k} of the right-hand side of (3.6). If the Lipschitz constant of f with respect to y is L, then we can take M to have the value $h|\beta_k|L$ and the iteration (3.7) converges to the unique solution of (3.6) provided

$$h < 1/(|\beta_k|L). \tag{3.8}$$

Except in the case of stiff systems, which will be studied later in this book, condition (3.8) does not present any problems; we find that considerations of accuracy impose restrictions on the steplength h which are far more severe than (3.8). (The situation is very different for stiff systems, for which $L \gg 1$ and the restriction imposed by (3.8) is so severe that an alternative to the iteration (3.7) must be sought.)

Within the general class (3.1) of linear multistep methods, there are several well-known sub-classes. The sub-class of methods of *Adams type* are characterized by

$$\rho(\zeta) = \zeta^k - \zeta^{k-1}.$$

Since the spurious roots of ρ are all situated at the origin of the complex plane, methods of Adams type are clearly zero-stable for all values of k. Methods of Adams type which have the maximum possible accuracy are known as *Adams methods*; if they are explicit they are known as *Adams–Bashforth methods*, and if implicit as *Adams–Moulton methods*. The 1-step Adams–Bashforth method is Euler's Rule (3.3), while the 1-step Adams–Moulton method is the *Trapezoidal Rule*,

$$y_{n+1} - y_n = \frac{h}{2}(f_{n+1} + f_n). \tag{3.9}$$

Adams methods are among the oldest of linear multistep methods, dating back to the nineteenth century; nevertheless, as we shall see later, they continue to play a key role in efficient modern algorithms. Other sub-classes are characterized by

$$\rho(\zeta) = \zeta^k - \zeta^{k-2}$$

and are clearly also zero-stable for all k. Explicit members of this sub-class are known as *Nyström methods*, and implicit members as *Generalized Milne–Simpson methods*. A well-known example of a Nyström method is the *Mid-point Rule*

$$y_{n+2} - y_n = 2hf_{n+1}, \tag{3.10}$$

and of a Generalized Milne–Simpson method is *Simpson's Rule*

$$y_{n+2} - y_n = \frac{h}{3}(f_{n+2} + 4f_{n+1} + f_n). \tag{3.11}$$

A sub-class that is important in dealing with stiffness consists of the *Backward Differentiation Formulae* or *BDF*, which are implicit methods with $\sigma(\zeta) = \beta_k\zeta^k$; as we shall see later, they are zero-stable only for a restricted range of k.

Finally, we observe that although the method (3.1) is linear, in the sense that it equates linear combinations of y_{n+j} and of f_{n+j}, the resulting difference system for $\{y_n\}$ is (in general) nonlinear, since f is (in general) a nonlinear function of y. The analytical study of nonlinear difference systems is much harder than that of the corresponding nonlinear differential systems, and since the major motivation for contemplating numerical methods in the first place is our inability to get very far with the latter study, we cannot be optimistic about our chances of obtaining powerful *analytical* results about the solution of the difference system. Thus, numerical methods involve a trade-off: the price we pay for being able to compute a numerical solution is increased difficulty in analysing that solution.

3.2 THE ASSOCIATED DIFFERENCE OPERATOR; ORDER AND ERROR CONSTANT

In §2.4 we introduced the idea of using the residual, defined by (2.7), as a measure of the accuracy of a method. To be a little more precise, by forming a Taylor expansion about some suitable value of x, we could express the residual as a power series in h; the power of h in the first non-vanishing term is then an indication of accuracy. For example, let us carry out this procedure for Euler's Rule (3.3) and the Trapezoidal Rule (3.9), taking x_n as the origin of the expansions. Using the fact that $y' = f(x, y)$, we obtain

$$R_{n+1} = y(x_{n+1}) - y(x_n) - hy'(x_n) = \frac{h^2}{2}y^{(2)}(x_n) + 0(h^3),$$

and (3.12)

$$R_{n+1} = y(x_{n+1}) - y(x_n) - \frac{h}{2}[y'(x_{n+1}) + y'(x_n)] = -\frac{h^3}{12}y^{(3)}(x_n) + 0(h^4),$$

respectively, from which we conclude that the Trapezoidal Rule is the more accurate by one power of h.

However, there are some difficulties with this approach. Firstly, Theorem 1.1 of §1.4 implies only that $y(x) \in C^1[a, b]$, so that the higher derivatives of $y(x)$ used in the Taylor expansions may not exist. Secondly, it is not immediately clear whether, if we use a different origin for the Taylor expansions, the leading term in the expansion of the residual will have the same power of h and the same numerical coefficient.

The first difficulty is easily overcome. Once we have substituted $y'(x_{n+j})$ for $f(x_{n+j}, y(x_{n+j}))$ in the expression defining the residual, subsequent manipulations involving Taylor expansions make no further use of the fact that $y(x)$ is the exact solution of the initial value problem. The same result would be obtained if we replaced $y(x)$ and $y'(x)$ in (3.12) by $z(x)$ and $z'(x)$ respectively, where $z(x)$ is an arbitrary differentiable function. In other words, the important thing in (3.12) is the difference operation that takes place, not the particular function operated on.

Definition *The* **linear difference operator** $\mathscr{L}$ *associated with the linear multistep method (3.1) is defined by*

$$\mathscr{L}[z(x); h] := \sum_{j=0}^{k} [\alpha_j z(x + jh) - h\beta_j z'(x + jh)], \tag{3.13}$$

where $z(x) \in C^1[a, b]$ *is an arbitrary function.*

We now choose the function $z(x)$ to be differentiable as often as we need, expand $z(x + jh)$ and $z'(x + jh)$ about x, and collect terms in (3.13) to obtain

$$\mathscr{L}[z(x); h] = C_0 z(x) + C_1 h z^{(1)}(x) + \cdots + C_q h^q z^{(q)}(x) + \cdots \tag{3.14}$$

where the C_q are constants.

Definition *The linear multistep method (3.1) and the associated difference operator* $\mathscr{L}$ *defined by (3.13) are said to be of* **order** p *if, in (3.14),* $C_0 = C_1 = \cdots = C_p = 0$, $C_{p+1} \neq 0$.

The following formulae for the constants C_q are easily established:

$$\left.\begin{aligned} C_0 &= \sum_{j=0}^{k} \alpha_j \equiv \rho(1) \\ C_1 &= \sum_{j=0}^{k} (j\alpha_j - \beta_j) \equiv \rho'(1) - \sigma(1) \\ C_q &= \sum_{j=0}^{k} \left[\frac{1}{q!} j^q \alpha_j - \frac{1}{(q-1)!} j^{q-1} \beta_j\right] \qquad q = 2, 3, \ldots \end{aligned}\right\} \tag{3.15}$$

The definition of order will be useless unless we can satisfy ourselves that we get the same result if we choose to expand about a different origin; this is the second difficulty referred to above. Suppose we expand $z(x + jh)$ and $z'(x + jh)$ about $x + th$ rather than about x, where t need not necessarily be an integer. In place of (3.14) we obtain

$$\mathscr{L}[z(x); h] = D_0 z(x + th) + D_1 h z^{(1)}(x + th) + \cdots + D_q h^q z^{(q)}(x + th) + \cdots. \tag{3.16}$$

The functions $z^{(q)}(x+th)$, $q=0,1,2,\ldots$ (where $z^{(0)}(x)\equiv z(x)$) can now each be Taylor expanded about x, thus:

$$z^{(q)}(x+th)=z^{(q)}(x)+thz^{(q+1)}(x)+\cdots+\frac{t^sh^s}{s!}z^{(q+s)}(x)+\cdots \qquad q=0,1,2.\ldots$$

Substituting these expansions into (3.16) yields an expression identical in form with the right-hand side of (3.14). Equating the result term by term with (3.14) yields

$$\begin{aligned}
C_0 &= D_0\\
C_1 &= D_1+tD_0\\
C_2 &= D_2+tD_1+\frac{t^2}{2!}D_0\\
&\vdots\\
C_p &= D_p+tD_{p-1}+\cdots+\frac{t^p}{p!}D_0\\
C_{p+1} &= D_{p+1}+tD_p+\cdots+\frac{t^{p+1}}{(p+1)!}D_0\\
C_{p+2} &= D_{p+2}+tD_{p+1}+\cdots+\frac{t^{p+2}}{(p+2)!}D_0.
\end{aligned}$$

It follows that $C_q=0$, $q=0,1,\ldots,p$, if and only if $D_q=0$, $q=0,1,\ldots,p$. Thus we could equally define the method and the associated difference operator to have order p if the first $p+1$ coefficients in the expansion (3.16) vanish, and this definition is independent of t; that is, the definition of order given above is indeed independent of the origin of the Taylor expansions.

Moreover, if $C_q=0$, $q=0,1,\ldots,p$, $C_{p+1}\neq 0$, then

$$D_{p+1}=C_{p+1},\quad D_{p+2}=C_{p+2}+tC_{p+1},$$

etc. Thus the first non-vanishing coefficient in the expansion (3.16) is independent of t, but subsequent coefficients do depend on t. Clearly, the first non-vanishing coefficient, C_{p+1}, is the only one to have any significance.

Definition A linear multistep method of order p is said to have **error constant** C_{p+1}, *given by (3.15).*

We can obviously use the formulae (3.15) to establish the order and error constant of any given linear multistep method; but we can also use them to construct linear multistep methods of given structure. For example, consider the two-parameter family of linear two-step methods given by

$$y_{n+2}-(1+\alpha)y_{n+1}+\alpha y_n=h[(1+\beta)f_{n+2}-(\alpha+\beta+\alpha\beta)f_{n+1}+\alpha\beta f_n], \qquad (3.17)$$

where $\alpha\ (\neq 0)$ and β are free parameters (see Exercise 2.5.5* of §2.5). Using (3.15) we have

$$C_0 = 1 - (1+\alpha) + \alpha = 0$$

$$C_1 = 2 - (1+\alpha) - [1 + \beta - (\alpha + \beta + \alpha\beta) + \alpha\beta] = 0$$

$$C_2 = \tfrac{1}{2}[4 - (1+\alpha)] - [2(1+\beta) - (\alpha + \beta + \alpha\beta)] = (\alpha - 1)(\beta + \tfrac{1}{2})$$

$$C_3 = \tfrac{1}{6}[8 - (1+\alpha)] - \tfrac{1}{2}[4(1+\beta) - (\alpha + \beta + \alpha\beta)] = \begin{cases} -(\beta + \tfrac{1}{2}) & \text{if } \alpha = 1 \\ \tfrac{1}{12}(\alpha - 1) & \text{if } \beta = -\tfrac{1}{2} \end{cases}$$

$$C_4 = \tfrac{1}{24}[16 - (1+\alpha)] - \tfrac{1}{6}[8(1+\beta) - (\alpha + \beta + \alpha\beta)] = -\tfrac{1}{12} \qquad \text{if } \alpha = 1,\ \beta = -\tfrac{1}{2}.$$

Hence,

if $\alpha \neq 1$, $\beta \neq -\frac{1}{2}$, order $p = 1$, error constant $C_2 = (\alpha - 1)(\beta + \frac{1}{2})$

if $\alpha \neq 1$, $\beta = -\frac{1}{2}$, order $p = 2$, error constant $C_3 = \frac{1}{12}(\alpha - 1)$

if $\alpha = 1$, $\beta \neq -\frac{1}{2}$, order $p = 2$, error constant $C_3 = -(\beta + \frac{1}{2})$

if $\alpha = 1$, $\beta = -\frac{1}{2}$, order $p = 3$, error constant $C_4 = -\frac{1}{12}$.

(Note, however, that the method is zero-unstable for $\alpha = 1$.)

Although the above approach is the standard one for deriving linear multistep methods, it can happen that for methods with large stepnumber it is easier to abandon formulae (3.15), and perform the Taylor expansions *ab initio* about some point other than x, in the hope of utilizing symmetry; see Exercise 3.2.2.

Finally, we observe that (2.7), which defines the residual R_{n+k}, for the general class of methods (2.4), gives in the case of linear multistep methods

$$R_{n+k} = \mathscr{L}[y(x_n); h],$$

where $y(x)$ is the exact solution of the initial value problem. It follows from (3.14) and the discussion in §2.4, that a linear multistep method is consistent if it has order $p \geqslant 1$. It then follows from (3.15) that for a consistent linear multistep method, we have

$$\sum_{j=0}^{k} \alpha_j = 0, \quad \sum_{j=0}^{k} j\alpha_j = \sum_{j=0}^{k} \beta_j,$$

or, equivalently

$$\rho(1) = 0, \quad \rho'(1) = \sigma(1).$$

Note that if $\sigma(1) = 0$, we would have $\rho(1) = \rho'(1) = 0$; ρ would then have a double root at $+1$ and the method would fail to satisfy the root condition. Thus, for all consistent zero-stable linear multistep methods, $\sigma(1) \neq 0$. (This is why some authors normalize linear multistep methods by dividing through by $\sigma(1)$; see §3.1.)

Exercises

3.2.1. Construct a one-parameter family of implicit linear two-step methods of greatest possible order, and find the order and error constant. For which values of the parameter is the method convergent?

3.2.2. Find the order and error constant of Quade's method, given in Exercise 2.5.2. What is the most efficient point about which to take Taylor expansions?

3.2.3. Let $\mathscr{L}$ be the linear difference operator associated with a linear multistep method. Show that the method has order p if and only if

$$\mathscr{L}[x^r; h] \not\equiv 0, \; r = 0, 1, \ldots, p \quad \text{and} \quad \mathscr{L}[x^{p+1}; h] \not\equiv 0,$$

and the error constant C_{p+1} satisfies

$$h^{p+1}(p+1)!C_{p+1} = \mathscr{L}[x^{p+1}; h].$$

3.2.4. A linear multistep method is defined by its first and second characteristic polynomials $\rho(\zeta)$, $\sigma(\zeta)$. Sequences of polynomials $\{\rho_j(\zeta) | j = 1, 2, \ldots\}$, $\{\sigma_j(\zeta) | j = 1, 2, \ldots\}$ are constructed as follows:

$$\begin{aligned} \rho_1(\zeta) &= \rho(\zeta), & \sigma_1(\zeta) &= \sigma(\zeta) \\ \rho_{j+1}(\zeta) &= \zeta\rho_j'(\zeta), & \sigma_{j+1}(\zeta) &= \zeta\sigma_j'(\zeta), \qquad j = 1, 2, \ldots. \end{aligned}$$

Prove that the linear multistep method has order p if and only if

$$\rho_1(1) = 0, \quad \rho_{j+1}(1) = j\sigma_j(1), \quad j = 1, 2, \ldots, p \text{ and } \rho_{p+2}(1) \neq (p+1)\sigma_{p+1}(1).$$

Use this result to verify your answer to Exercise 3.2.2.

3.2.5*. A hybrid method is an extension of a linear multistep method which involves $f(x, y)$ evaluated at an off-step point (x_{n+r}, y_{n+r}), $0 < r < k$, $r \notin \{0, 1, \ldots, k\}$. (The value y_{n+r} is given by a separate formula, but that need not concern us here.) An explicit zero-stable 2-step method of this type has the form

$$y_{n+2} - (1+\alpha)y_{n+1} + \alpha y_n = h[\beta_1 f_{n+1} + \beta_0 f_n + \beta_r f_{n+r}], \quad -1 \leqslant \alpha < 1. \tag{1}$$

The associated linear difference operator, the order and the error constant can be defined in obvious extensions of the corresponding definitions for a linear multistep method.

(i) Show that for any α satisfying $-1 < \alpha < 1$ there exists a value of r for which the method (1) has order 4, and find the relation between r and α which must then hold. Why do we exclude the case $\alpha = -1$?
(ii) Show that there exists a *unique* value for r and for α such that the method (1) has order 5.

3.2.6*. There exists in the literature a family of one-step methods for the numerical solution of our standard problem; these methods are applicable to the general system, but we shall assume here (in order to keep things simple) that the problem is scalar. The methods are given by

$$\left.\begin{aligned} y_{n+1} &= y_n + hf_n + e^{\mathrm{T}}v_n + Kh(f_{n+1} - \hat{f}_n) \\ h\hat{f}_n &= hf_n + \phi^{\mathrm{T}}v_n \\ v_{n+1} &= Uv_n + h(f_{n+1} - \hat{f}_n)c \end{aligned}\right\} \tag{1}$$

where $f_n = f(x_n, y_n)$, U is an $r \times r$ upper triangular matrix, K is a scalar and e, v_n, ϕ, $c \in \mathbb{R}^r$. The vectors e and ϕ are given by

$$e = [1, 1, \ldots, 1]^{\mathrm{T}}, \quad \phi = [2, 3, \ldots, r+1]^{\mathrm{T}},$$

and K, c and U are all constant and can be regarded as the parameters of the family. It is assumed that starting values y_0 and v_0 are available.

(i) By considering the linear combination $\sum_{j=0}^{r}\gamma_j y_{n+j+1}$, with a suitable choice of coefficients γ_j, $j = 0, 1, \ldots, r$, show that the method (1) is equivalent to a linear multistep method (LMM) with stepnumber $r + 1$, and show how the coefficients in the equivalent LMM can be calculated in terms of the parameters appearing in (1). [*Hints*: First find v_{n+j} in terms of v_n, c, f_n, $f_{n+1}, \ldots, f_{n+j}$ and the matrix $M = U - c\phi^{\mathrm{T}}$. The Cayley–Hamilton theorem might come in handy.]

(ii) Illustrate your answer to (i) by finding the LMM equivalent to (1) in the case when

$$r = 2, \quad U = \begin{bmatrix} 1 & 3 \\ 0 & 1 \end{bmatrix}, \quad c = \begin{bmatrix} \frac{3}{4} \\ \frac{1}{6} \end{bmatrix}, \quad K \text{ free.} \tag{2}$$

Hence find the order of the one-parameter family of methods defined by (1) and (2). Find also the values of K for which (i) the method has maximum order, and (ii) the equivalent LMM is a 2-step method; identify the equivalent LMMs in both (i) and (ii).

(iii) Using the results of (i), find, in terms of the eigenvalues of the matrix $M = U - c\phi^{\mathrm{T}}$, the condition for the *general* method (1) to be zero-stable for all values of K.

3.3 THE LINK WITH POLYNOMIAL INTERPOLATION

Linear multistep methods are closely linked with the process of polynomial interpolation. Two distinct such links can be established, the first involving interpolation of the f values, the second involving that of the y values. We shall illustrate by using both approaches to derive Simpson's Rule (3.11).

The method we seek thus has the form

$$y_{n+2} - y_n = h(\beta_2 f_{n+2} + \beta_1 f_{n+1} + \beta_0 f_n). \tag{3.18}$$

Starting from the identity

$$y(x_{n+2}) - y(x_n) \equiv \int_{x_n}^{x_{n+2}} y'(x)\,\mathrm{d}x, \tag{3.19}$$

we replace $y'(x)$ by $f(x, y(x))$ and, having an eye to the data we wish to involve on the right-hand side of (3.18), approximate f by the unique vector interpolant of degree 2 in x passing through the three points (x_{n+2}, f_{n+2}), (x_{n+1}, f_{n+1}), (x_n, f_n) in $\mathbb{R}^{m+1}$. Referring to §1.10, the appropriate interpolant is given by the Newton–Gregory backward interpolation formula

$$I_2(x) = I_2(x_{n+2} + rh) =: P_2(r) = [1 + r\nabla + \tfrac{1}{2}r(r+1)\nabla^2]f_{n+2}, \tag{3.20}$$

enabling us to approximate the right-hand side of (3.19) by

$$\int_{x_n}^{x_{n+2}} I_2(x)\,\mathrm{d}x = \int_{-2}^{0} [f_{n+2} + r\nabla f_{n+2} + \tfrac{1}{2}r(r+1)\nabla^2 f_{n+2}]h\,\mathrm{d}r.$$

On evaluating the definite integral and expanding ∇f_{n+2} and $\nabla^2 f_{n+2}$ in terms of

f_{n+2}, f_{n+1}, f_n, (3.19) yields

$$y(x_{n+2}) - y(x_n) \approx \frac{h}{3}(f_{n+2} + 4f_{n+1} + f_n). \tag{3.21}$$

We can now define y_{n+2} and y_n to be approximations to $y(x_{n+2})$ and $y(x_n)$ respectively, such that the approximate equality in (3.21) becomes an exact equality, thus obtaining Simpson's Rule.

The reader will have noted that the above derivation of Simpson's Rule is virtually identical with that of Simpson's Rule for quadrature, that is, for the numerical approximation of the definite integral $\int_{x_n}^{x_{n+2}} f(x)\,\mathrm{d}x$; indeed all Newton–Cotes quadrature formulae can be interpreted as linear multistep methods. However, there is an important distinction in the way in which quadrature rules and linear multistep methods are applied. If Simpson's Rule is used to evaluate the definite integral $\int_{x_0}^{x_n} f(x)\,\mathrm{d}x$, then it is successively applied to the sub-intervals $[x_0, x_2]$, $[x_2, x_4]$, $[x_4, x_6]$, etc., *which sub-intervals do not overlap*; the error in integration over the whole interval is simply the sum of the errors over each sub-interval. In contrast, if Simpson's Rule is used to integrate an initial value problem, then it is successively applied to the sub-intervals $[x_0, x_2]$, $[x_1, x_3]$, $[x_2, x_4]$, etc., *which sub-intervals do overlap*. The accumulation of error is now much more complicated, and it should not be totally unexpected that Simpson's Rule, an excellent method for quadrature, turns out (as we shall see later) to be a bad method for integrating initial value problems.

The above procedure can be used to derive only linear multistep methods for which $\rho(\zeta) = \zeta^k - \zeta^{k-q}$ for some integer q, $0 \leqslant q < k$. Note that this class contains more than the linear multistep equivalents of the standard Newton–Cotes quadrature formulae. For example, the 2-step Adams–Moulton method could be derived by replacing (3.19) by

$$y(x_{n+2}) - y(x_{n+1}) \equiv \int_{x_{n+1}}^{x_{n+2}} y'(x)\,\mathrm{d}x,$$

but retaining the approximation (3.20) for f. Likewise, the 2-step Adams–Bashforth method could be derived by reducing the degree of the interpolant P to one and avoiding the involvement of f_{n+2}.

The second approach is somewhat more direct, in that it interpolates the data (x_n, y_n) rather than (x_n, f_n), but in a *Hermite* or *osculatory* sense; this means that the interpolant is required not only to take prescribed values at the interpolation points, but to have prescribed slopes at such points as well. Over the span $[x_n, x_{n+2}]$ of the method, let $I(x)$ be such a Hermite interpolant (with vector coefficients). That is, we require $I(x)$ to satisfy

$$I(x_{n+j}) = y_{n+j}, \quad I'(x_{n+j}) = f_{n+j}, \qquad j = 0, 1, 2.$$

There are six (vector) conditions in all; if we allowed $I(x)$ to have six free (vector) parameters, then all that would happen would be that the six conditions would specify $I(x)$ uniquely. Instead, we choose $I(x)$ to have five free parameters, use any five of the six conditions to specify $I(x)$, and substitute the result in the sixth condition; in other words, we find the *eliminant* of the five free parameters between the six conditions. Choosing $I(x)$ to be a polynomial, then in order to achieve five free vector parameters, we must choose the degree to be four, that is choose

$$I(x) = x^4 a + x^3 b + x^2 c + xd + e, \tag{3.22}$$

where a, b, c, d, e are m-dimensional vector parameters. A straightforward calculation shows that the eliminant is indeed Simpson's Rule. Note that if the solution of the initial value problem happens to be a polynomial of degree $\leqslant 4$, then $I(x)$ and $y(x)$ become identical, and Simpson's Rule would be exact. This is consistent with the readily established fact that Simpson's Rule has order 4, and error constant $-\frac{1}{90}$, so that the difference operator associated with Simpson's Rule expands to give

$$\mathscr{L}[z(x); h] = -\tfrac{1}{90} h^5 z^{(5)} + \cdots.$$

If $z(x)$ is replaced by $y(x)$, a polynomial of degree $\leqslant 4$, then clearly $y^{(q)}(x) = 0$ for all $q \geqslant 5$, so that $\mathscr{L}[y(x); h] \equiv 0$, implying that the method is exact.

The link between linear multistep methods and polynomial interpolation is a revealing one. We can anticipate that linear multistep methods will perform badly in situations where polynomial interpolation would perform badly—a point we shall return to in Chapter 6 where we discuss stiffness. However, useful though the correspondence between linear multistep methods and polynomial interpolation is, that correspondence is not one-to-one. If, in the above derivation, the reader cares to replace the quadratic polynomial $I(x)$ given by (3.22) by the *cubic spline* $S(x)$ defined by

$$S(x) = \begin{cases} x^3 a + x^2 b + xc + d & \text{if } x \in [x_n, x_{n+1}] \\ x^3 a + x^2 b + xc + d + (x - x_{n+1})^3 e & \text{if } x \in [x_{n+1}, x_{n+2}], \end{cases}$$

which also has the requisite number (five) of free vector parameters a, b, c, d, e, then again Simpson's Rule emerges.

Exercises

3.3.1. Starting from the identity

$$y(x_{n+2}) - y(x_{n+1}) \equiv \int_{x_{n+1}}^{x_{n+2}} y'(x)\,\mathrm{d}x$$

derive the 2-step Adams–Bashforth and Adams–Moulton methods.

3.3.2. At the end of the above section, we indicate that a single application of Simpson's Rule is equivalent to local Hermite interpolation by a cubic spline. Let $S_n(x)$ indicate such a cubic spline applied in the interval $[x_n, x_{n+2}]$. By considering the relationships between $S_n(x)$ and $S_{n+1}(x)$ at the overlapping points, show that integrating an initial value problem from $x = a$ to $x = b$ by repeated applications of Simpson's Rule is equivalent to *global* Hermite interpolation by a (global) cubic spline (that is, a function which is cubic in each of the sub-intervals $[x_n, x_{n+1}]$ in $[a, b]$ and has continuous first and second derivatives in $[a, b]$).

3.3.3. There is an alternative way of establishing the relationship between Simpson's Rule and Hermite interpolation by a cubic spline. Consider the interpolant $I_n(x) = \gamma_n x^3 + a_{n2} x^2 + a_{n1} x + a_{n0}$. Impose the conditions $I_n(x_{n+j}) = y_{n+j}$, $I'_n(x_{n+j}) = f_{n+j}$, $j = 0, 1$, and eliminate a_{n2}, a_{n1}, a_{n0} from the resulting four conditions to get a formula which will involve γ_n. Write down this formula again, with n replaced by $n+1$ and add, to get a two-step method involving the parameter $\gamma_n + \gamma_{n+1}$. Choosing the value of this parameter in order that $I''_n(x_{n+1}) = I''_{n+1}(x_{n+1})$ should again produce Simpson's Rule.

3.3.4. (i) Let the exact solution of the initial value problem be locally represented in the interval $[x_n, x_{n+1}]$ by the cubic interpolant $I(x) = a_3x^3 + a_2x^2 + a_1x + a_0$. Find the eliminant of the four coefficients a_i, $i = 0, 1, 2, 3$, between the five conditions

$$I(x_{n+j}) = y_{n+j}, \quad j = 0, 1, \quad I^{(q)}(x_n) = f_n^{(q-1)}, \quad q = 1, 2, 3 \tag{1}$$

to obtain an explicit method (the Taylor algorithm of order three).

(ii) Repeat (i) but with $I(x)$ replaced by the rational function $R(x) = (a_2x^2 + a_1x + a_0)/(x + a_3)$, applying the same conditions (1). The result will be a new explicit method. Why is the method derived in (i) applicable to an m-dimensional problem, while that derived in (ii) is applicable only to a scalar problem?

(iii) Suggest circumstances in which the method found in (ii) might be expected to perform better than that found in (i). Illustrate by applying both methods to the scalar problem, $y' = 1 + y^2$, $y(0) = 1$, $0 \leqslant x \leqslant 0.75$. (The exact solution is $y(x) = \tan(x + \pi/4)$.)

3.4 THE FIRST DAHLQUIST BARRIER

A natural question to ask is what is the highest order that can be achieved by a convergent linear k-step method. In seeking high order, the consistency condition is automatically satisfied, but we meet a very real barrier in attempting to satisfy the root condition. This barrier has become known as the *first Dahlquist barrier*, since it was originally investigated in the seminal paper of Germund Dahlquist (1956); this paper was the first to bring strict mathematical analysis to the problem of the convergence of numerical solutions to initial value problems, and ushered in a new era in the subject.

The linear k-step method (3.1) has $2k + 2$ free coefficients $\alpha_j, \beta_j, j = 0, 1, \ldots, k$, of which one, α_k is specified by (3.2) to be 1. There are thus $2k + 1$ free parameters ($2k$, if the method is constrained to be explicit). From (3.14) and (3.15), it follows that if the method is to have order p, then $p + 1$ linear equations in $\alpha_j, \beta_j, j = 0, 1, \ldots, k$, must be satisfied. Thus the highest order we can expect from a linear k-step method is $2k$ if the method is implicit, and $2k - 1$ if it is explicit. Linear k-step methods achieving such orders are called *maximal*. However, maximal methods, in general, fail to satisfy the root condition and are thus zero-unstable. The first Dahlquist barrier is encapsulated in the following theorem (Dahlquist (1956); see also Henrici (1962)):

Theorem 3.1 No zero-stable linear k-step method can have order exceeding $k + 1$ *when k is odd and* $k + 2$ *when k is even.*

A zero-stable linear k-step method of order $k + 2$ is called an *optimal* method; naturally k must be even and the method implicit. It can be shown that all of the spurious roots of the first characteristic polynomial of an optimal method lie on the unit circle, a situation that gives rise to some stability difficulties which we shall investigate in §3.8. The result is that optimal methods do not perform well, and so it would be incorrect to deduce from Theorem 3.1 that zero-stable k-step methods of order $k + 1$, where k is odd, are overshadowed by the optimal methods that can be achieved when k is even.

Simpson's Rule occupies a unique position in this hierarchy. It has stepnumber 2 and order 4, and is thus both maximal and optimal.

3.5 LOCAL TRUNCATION ERROR AND GLOBAL TRUNCATION ERROR

In §3.2 we used the power of h in the first non-vanishing term in the Taylor expansion of the residual R_{n+k}, defined by (2.7), to define the order of a linear multistep method. It is natural to use the residual itself as a finer measure of accuracy, giving rise to the following definition:

Definition *The* **local truncation error** *or* **LTE** *of the method (3.1) at* x_{n+k}, *denoted by* T_{n+k} *is defined by*

$$T_{n+k} = \mathscr{L}[y(x_n); h], \tag{3.23}$$

where $\mathscr{L}$ *is the associated difference operator defined by (3.13) and* $y(x)$ *is the exact solution of the initial value problem (2.2).*

T_{n+k} is thus seen to be identical with R_{n+k}. The local nature of T_{n+k} can be seen if we make the following somewhat artificial assumption, known as the *localizing assumption*. We assume that $y_{n+j} = y(x_{n+j}), j = 0, 1, \ldots, k-1$, that is, that all of the back values are exact; let us denote by $\tilde{y}_{n+k}$ the value at x_{n+k} generated by the method when the localizing assumption is in force. It follows from (3.13) that

$$\begin{aligned}\sum_{j=0}^{k} \alpha_j y(x_n + jh) &= h \sum_{j=0}^{k} \beta_j y'(x_n + jh) + \mathscr{L}[y(x_n); h] \\ &= h \sum_{j=0}^{k} \beta_j f(x_n + jh, y(x_n + jh)) + T_{n+k},\end{aligned}$$

since $y(x)$ satisfies the differential system $y' = f(x, y)$. The value $\tilde{y}_{n+k}$ given by the method satisfies

$$\tilde{y}_{n+k} + \sum_{j=0}^{k-1} \alpha_j y_{n+j} = h\beta_k f(x_{n+k}, \tilde{y}_{n+k}) + h \sum_{j=0}^{k-1} \beta_j f(x_{n+j}, y_{n+j}),$$

and on subtracting and using the localizing assumption we obtain

$$y(x_{n+k}) - \tilde{y}_{n+k} = h\beta_k[f(x_{n+k}, y(x_{n+k})) - f(x_{n+k}, \tilde{y}_{n+k})] + T_{n+k}. \tag{3.24}$$

We now apply the mean value theorem to the right side of (3.24). Using the notation of §1.3, Case 4, we have that

$$f(x_{n+k}, y(x_{n+k})) - f(x_{n+k}, \tilde{y}_{n+k}) = \bar{J}(x_{n+k}, \eta_{n+k})[y(x_{n+k}) - \tilde{y}_{n+k}],$$

where $\bar{J}$ is the Jacobian matrix of f with respect to y, and the notation implies that each row of $\bar{J}$ is evaluated at different mean values η_{n+k}, each lying in the internal part of the line segment joining $y(x_{n+k})$ to $\tilde{y}_{n+k}$ in $\mathbb{R}^m$. Equation (3.24) now yields

$$[I - h\beta_k \bar{J}(x_{n+k}, \eta_{n+k})][y(x_{n+k}) - \tilde{y}_{n+k}] = T_{n+k}. \tag{3.25}$$

Thus if the method is explicit ($\beta_k = 0$) then the LTE at x_{n+k} is simply the difference between the exact and the numerical solutions at x_{n+k} (subject, of course, to the localizing assumption). If the method is implicit, then to a first approximation (that is, ignoring the $0(h)$ term on the left side) the same is true.

We note that the definition (3.23) of LTE demands only that $y(x) \in C^1[a,b]$, and this is guaranteed by Theorem 1.1 of §1.4. If, however, we are prepared to assume that $y(x) \in C^{p+1}[a,b]$, where p is the order of the method, then by (3.14) we have

$$T_{n+k} = C_{p+1} h^{p+1} y^{(p+1)}(x_n) + 0(h^{p+2}) \tag{3.26}$$

and it follows from (3.25) that for both explicit and implicit methods

$$y(x_{n+k}) - \tilde{y}_{n+k} = C_{p+1} h^{p+1} y^{(p+1)}(x_n) + 0(h^{p+2}).$$

The term $C_{p+1} h^{p+1} y^{(p+1)}(x_n)$ is referred to as the *principal local truncation error* or PLTE.

If no localizing assumption is made, then the difference between the exact and the numerical solution is the *accumulated* or *global error*.

Definition The **global truncation error** *or* **GTE** *of the method (3.1) at* x_{n+k}, *denoted by* E_{n+k} *is defined by*

$$E_{n+k} = y(x_{n+k}) - y_{n+k}.$$

The LTE and the starting errors accumulate to produce the GTE, but this accumulation process is very complicated, and we cannot hope to obtain any usable general expression for the GTE. However, some insight into the accumulation process can be gleaned by looking at an example.

Consider the Mid-point Rule

$$y_{n+2} - y_n = 2hf_{n+1} \tag{3.27}$$

applied to the scalar problem

$$y' = y, \quad y(0) = 1 \tag{3.28}$$

whose exact solution is $y(x) = \exp(x)$. Since we wish to see the effect of starting errors, we choose as starting values $0(h^q)$ perturbations of the exact starting values, and take

$$y_0 = 1 + \omega_0 h^q, \quad y_1 = \exp(h) + \omega_1 h^q, \quad q \geqslant 1. \tag{3.29}$$

The method (3.27) has order 2 and error constant $\frac{1}{3}$, and it follows from (3.26) that the LTE at x_n is

$$T_n = \tfrac{1}{3}h^3 \exp(x_{n-2}) + 0(h^4) = \tfrac{1}{3}h^3 \exp(x_n) + 0(h^4). \tag{3.30}$$

To get an expression for the GTE, we must first attempt to solve the difference equation obtained when (3.27) is applied to (3.28), namely

$$y_{n+2} - 2hy_{n+1} - y_n = 0. \tag{3.31}$$

It is easily checked that the solution of (3.31) which satisfies the initial conditions (3.29) is given by

$$\left.\begin{aligned} y_n &= \frac{\Omega(r_2)r_1^n - \Omega(r_1)r_2^n}{r_1 - r_2} \\ \text{where}\quad \Omega(r) &= \exp(h) - r + (\omega_1 - r\omega_0)h^q \end{aligned}\right\} \tag{3.32}$$

and r_1, r_2 are the roots of $r^2 - 2hr - 1 = 0$. Now,

$$\left.\begin{aligned} r_1 &= h + \surd(1+h^2) = 1 + h + h^2/2 + 0(h^4) = \exp(h)[1 - h^3/6 + 0(h^4)] \\ r_2 &= h - \surd(1+h^2) = -1 + h - h^2/2 + 0(h^4) = -\exp(-h)[1 + h^3/6 + 0(h^4)]. \end{aligned}\right\} \tag{3.33}$$

Using the fact that $nh = x_n$, we obtain

$$\left.\begin{aligned} r_1^n &= \exp(x_n)[1 - h^2x_n/6 + 0(h^3)] \\ r_2^n &= (-1)^n \exp(-x_n)[1 + h^2x_n/6 + 0(h^3)]. \end{aligned}\right\} \tag{3.34}$$

Further, since $r_1 - r_2 = 2 + 0(h^2)$, we find from (3.32) and (3.33) that

$$\begin{aligned} \Omega(r_1) &= (\omega_1 - \omega_0)h^q + 0(h^3) + 0(h^{q+1}) \\ &= (r_1 - r_2)[(\omega_1 - \omega_0)h^q/2 + 0(h^3) + 0(h^{q+1})] \end{aligned}$$

and

$$\begin{aligned} \Omega(r_2) &= r_1 - r_2 + (\omega_1 + \omega_0)h^q + 0(h^3) + 0(h^{q+1}) \\ &= (r_1 - r_2)[1 + (\omega_1 + \omega_0)h^q/2 + 0(h^3) + 0(h^{q+1})]. \end{aligned}$$

Thus, (3.32) gives

$$\begin{aligned} y_n = {} & \exp(x_n)[1 - \tfrac{1}{6}h^2x_n + \tfrac{1}{2}(\omega_1 + \omega_0)h^q] \\ & - \tfrac{1}{2}(-1)^n \exp(-x_n)(\omega_1 - \omega_0)h^q + 0(h^3) + 0(h^{q+1}). \end{aligned}$$

Since $y(x_n) = \exp(x_n)$, we have that the global truncation error at x_n is

$$\begin{aligned} E_n &= y(x_n) - y_n \\ &= \tfrac{1}{6}h^2x_n \exp(x_n) - \tfrac{1}{2}h^q[(\omega_1 + \omega_0)\exp(x_n) - (-1)^n(\omega_1 - \omega_0)\exp(-x_n)] \\ &\quad + 0(h^3) + 0(h^{q+1}). \end{aligned} \tag{3.35}$$

Two points of importance emerge from this example. On comparing (3.30) with (3.35) (and ignoring for the moment $0(h^q)$ terms in the latter), we see that while the local truncation error is $0(h^3)$, the global truncation error is $0(h^2)$; that is, one power of h has been lost owing to the process of accumulation. We can see exactly where this loss occurred, namely, on going from (3.33) to (3.34). Secondly, on comparing (3.29) with

(3.35) we see that there is no loss of order in the starting errors due to accumulation. Further, if we were to mimic what we would do in practice and choose $q \geqslant 3$, so that the starting errors were at least of the same order as the LTE, then the starting errors would not influence the leading term in E_n; indeed we could afford to take $q = 2$ and still not alter the order of E_n.

Exercises

3.5.1. Using (i) Euler's Rule and (ii) the Trapezoidal Rule, verify the validity of (3.25) for the scalar initial value problem $y' = \lambda y$, $y(0) = 1$, for a general steplength h. (Use the exact solution of the problem to impose the localizing assumption.)

3.5.2. Consider the application of the method (3.17) of §3.2 to the scalar initial value problem $y' = y$, $y(0) = 1$. Using the results of Exercise 2.5.5* of §2.5 show that the relationships between the GTE, the LTE and the starting errors, established in this section for the Mid-point Rule, also hold for (3.17).

3.6 ERROR BOUNDS

It is possible—at the cost of some quite heavy analysis—to establish bounds for both the local and the global truncation errors. However, as we shall show presently, these bounds are of no practical value, and so we shall only summarize the results here, giving references where full derivations may be found.

Referring to equation (3.26), it is tempting to conjecture that, by analogy with the Lagrange form of remainder for a Taylor series, the local truncation error can be expressed (with the notation of §1.3) in the form

$$T_{n+k} = C_{p+1} h^{p+1} \bar{y}^{(p+1)}(\xi_n), \qquad \xi_n \in (x_n, x_{n+k}), \tag{3.36}$$

whence we would have the bound

$$\| T_{n+k} \| \leqslant |C_{p+1}| h^{p+1} Y \tag{3.37}$$

where

$$Y = \max_{x \in [a,b]} \| y^{(p+1)}(x) \|, \tag{3.38}$$

the bound holding over the range of integration $[a, b]$. However, it turns out that (3.36) holds for some linear multistep methods but not for others. The *influence function* or *Peano kernel*, $G(s)$, of the method is defined by

$$G(s) = \sum_{j=0}^{k} [\alpha_j (j-s)_+^p - p\beta_j (j-s)_+^{p-1}]$$

where p is the order of the method and the function z_+ is defined by

$$z_+ = \begin{cases} z & \text{if } z \geqslant 0 \\ 0 & \text{if } z < 0. \end{cases}$$

If and only if $G(s)$ is of constant sign in the interval $[0,k]$ of s does (3.36) hold; this is the case for all Adams methods. However, whether or not $G(s)$ changes sign in $[0,k]$, a bound akin to (3.37) holds, namely

$$\|T_{n+k}\| \leqslant Gh^{p+1}Y \tag{3.39}$$

where Y is given by (3.38) and

$$G = \begin{cases} |C_{p+1}| & \text{if } G(s) \text{ does not change sign in } [0,k] \\ [\int_0^k |G(s)|\,ds]/p! & \text{if } G(s) \text{ changes sign in } [0,k]. \end{cases}$$

(Full details can be found in, for example, Lambert (1973).)

The global error bound we are about to quote can take account of the effect of local round-off error. Let us denote by $\{\tilde{y}\}$ the sequence generated by the method when a local round-off error is committed at each step, that round-off error being bounded by Kh^{q+1}; that is, $\{\tilde{y}_n\}$ is given by

$$\sum_{j=0}^{k} \alpha_j \tilde{y}_{n+j} = h \sum_{j=0}^{k} \beta_j f(x_{n+j}, \tilde{y}_{n+j}) + \theta_n K h^{q+1}, \qquad |\theta_n| \leqslant 1. \tag{3.40}$$

We similarly adapt the definition of global error by defining $\hat{E}_{n+k} := y(x_{n+k}) - \tilde{y}_{n+k}$. We introduce the notation

$$A = \sum_{j=0}^{k} |\alpha_j| \quad B = \sum_{j=0}^{k} |\beta_j| \quad \delta = \max_{\mu=0,1,\ldots,k-1} \|\hat{E}_\mu\|$$

$$\Gamma = \left[\sup_{s=0,1,\ldots} |\gamma_s|\right] \Big/ (1 - h|\beta_k|L)$$

where

$$1/(1 + \alpha_{k-1}\zeta + \cdots + \alpha_0\zeta^k) = \gamma_0 + \gamma_1\zeta + \gamma_2\zeta^2 + \cdots$$

and L is the Lipschitz constant of the differential system in the problem being solved. We note that δ is a bound on the starting errors, and that for all Adams methods $\Gamma = 1/(1 - h|\beta_k|L)$.

Then, provided that

$$h|\beta_k|L < 1 \tag{3.41}$$

the global error (including round-off) is bounded as follows:

$$\|\hat{E}_n\| \leqslant \Gamma\{Ak\delta + (x_n - a)(h^p GY + h^q K)\} \exp\{\Gamma LB(x_n - a)\} \tag{3.42}$$

for all $x_n \in [a,b]$. For a derivation of this bound, see Henrici (1962, 1963).

Several comments can be made about the bound given by (3.42).

(a) The condition (3.41) is clearly satisfied for all explicit methods; for implicit methods, it coincides with (3.8) of §3.1, which was the condition for the implicit difference equation to have a unique solution.
(b) The effect of the three sources of error, LTE, starting error and round-off error are

clearly seen. We note from (3.39) that, if the LTE is bounded by $Gh^{p+1}Y$, then the corresponding term in the global bound is bounded by Gh^pY. This mirrors exactly the behaviour we observed in the example of §3.5. That example showed that the actual global truncation error, for a particular example, was $O(h^p)$, and (3.42) shows that for all linear multistep methods applied to a general problem, it cannot be worse than $O(h^p)$; it follows that, in general, the global truncation error of a pth-order linear multistep method is $O(h^p)$.

(c) If the bound δ is $O(h^{p+1})$ (the natural choice) then the starting errors have a second-order effect on the error bound; one can afford to take $\delta = O(h^p)$ without altering the order of the bound. These conclusions again reflect those drawn from the example of §3.5.

(d) Just as for the LTE, if round-off errors are locally bounded by Kh^{q+1}, then they are globally bounded by Kh^q. However, this is not a realistic model for what happens in practice. We know of no computers where the user can ask for arbitrarily small levels of round-off; we choose to work in single-, double- or triple-length arithmetic, and have no further control over the level of local round-off. A more realistic assumption would thus be to replace Kh^{q+1} in (3.40) by ε, a *fixed* bound on local round-off error. This clearly has the effect of replacing the term Kh^q in (3.42) by ε/h, which leads to an interesting conclusion: as $h \to 0$, the bound initially decreases due to the term in h^p, but eventually increases (to infinity) due to the term ε/h, thus corroborating what common sense tells us, that, in general, convergence to the exact solution can never be achieved in practice with a computer that works in finite arithmetic.

(e) Let us ignore round-off by setting $K = 0$ (or $\varepsilon = 0$). Then if, in accordance with the definition of convergence, we assume $\delta \to 0$ as $h \to 0$, we have that $\|E_n\| \to 0$ as $h \to 0$ if $p \geqslant 1$; that is, the method is convergent if it is consistent. But what has happened to the condition of zero-stability, which we know to be necessary for convergence? The answer is that if the method is zero-unstable, then it can be shown that the $\{\gamma_s\}$ used to define Γ become unbounded, so that $\Gamma = \infty$ and convergence is lost.

(f) Can the bound be used in practice to give the user a helpful guarantee on how accurate the computed solution is? The answer is no! All of the terms in (3.42) except L and Y are functions of the coefficients of the method only, and are readily computed. L can be taken to be the maximum value that $\|\partial f/\partial y\|$ takes in $[a, b]$, and this could be estimated *a posteriori* by evaluating $\partial f/\partial y$ on the numerical solution $\{y_n\}$ rather then on the (unknown) exact solution $y(x)$. Y could similarly be estimated *a posteriori* by the (error-prone) process of numerically differentiating the solution $\{y_n\}$ $p+1$ times. However, the real reason why the bound is of no practical value lies in the fact that it is nearly always excessively conservative. For example, we note from (3.42) that the bound grows exponentially with x_n and this applies even if the solution decays, in which case we would expect the actual global error also to decay (see the examples of §2.7).

We conclude by illustrating comment (f) above by applying the bound (3.42) to Example 3 of §2.7 where the method

$$y_{n+3} + \tfrac{1}{4}y_{n+2} - \tfrac{1}{2}y_{n+1} - \tfrac{3}{4}y_n = \frac{h}{8}(19f_{n+2} + 5f_n)$$

was applied in the interval [0, 1] to the problem

$$\begin{aligned} {}^1y' &= {}^2y & {}^1y(0) &= \tfrac{1}{2} \\ {}^2y' &= {}^2y({}^2y - 1)/{}^1y & {}^2y(0) &= -3 \end{aligned}$$

whose exact solution is ${}^1y(x) = [1 + 3\exp(-8x)]/8$, ${}^2y(x) = -3\exp(-8x)$. We ignore round-off error, and since exact starting values were used in §2.7 we set $\delta = 0$. The bound given by (3.42) now reads

$$\| E_n \| \leqslant \Gamma x_n h^p G Y \exp(\Gamma L B x_n), \tag{3.43}$$

where, to be consistent with the results of §2.7, we take $\|\cdot\| = \|\cdot\|_2$ throughout. The method has order 3 and error constant 17/48; a tedious calculation establishes that $G(s)$ does not change sign for $s \in [0, 3]$, so that $G = 17/48$. By constructing and solving a difference equation for $\{\gamma_s\}$, we find that $\Gamma = 1$. Using the exact solution, we have that

$$y^{(4)}(x) = \exp(-8x)[1536, 12\,228]^{\mathrm{T}},$$

and $\| y^{(4)}(x) \|_2$ takes its maximum value of 12 384 at $x = 0$. The maximum value of $\| \partial f/\partial y \|_2$ also occurs at $x = 0$ and is 50.001. B takes the value 3, so that (3.43) gives

$$\| E_n \| \leqslant 4386 x_n h^3 \exp(150.003 x_n) \tag{3.44}$$

and we see at once that this bound is hopelessly pessimistic. From §2.7, when $h = 0.0125$, the actual global errors at $x = 0.2$ and $x = 1.0$ were 1.1×10^{-4} and 9.6×10^{-7} respectively; the bounds on the global error given by (3.44) at these points are 1.8×10^{10} and 1.2×10^{65} respectively. A customer is unlikely to be impressed by being told that there is a cast-iron guarantee that the errors in the numerical solution are everywhere less than 1.2×10^{65}.

Despite its inability to give useful practical results, the bound (3.42) is none the less helpful in our understanding of how local errors propagate.

Exercise

3.6.1. Construct the influence function $G(s)$ for the method

$$y_{n+2} - (1 + \alpha) y_{n+1} + \alpha y_n = \tfrac{1}{2} h[(3 - \alpha) f_{n+1} - (1 + \alpha) f_n], \qquad \alpha \neq -5.$$

Find the range of α for which $G(s)$ does not change sign for $s \in [0, 2]$ and demonstrate that for α in that range,

$$\frac{1}{p!}\int_0^2 |G(s)| \mathrm{d}s = |C_{p+1}|.$$

3.7 LOCAL ERROR

The LTE, as defined in §3.5, is useful in analysing local errors, but the localizing assumption, necessary for an interpretation in terms of the difference between exact and numerical solutions, is highly artificial, and there is no way that it can be implemented in practice. Some authors propose, as an alternative measure of local accuracy, the local error or LE defined as follows:

Definition *Let* $u(x)$ *be the solution of the initial value problem*

$$u' = f(x, u), \quad u(x_{n+k-1}) = y_{n+k-1}.$$

Then the **local error**, LE_{n+k} *at* x_{n+k} *is defined by*

$$\mathrm{LE}_{n+k} := u(x_{n+k}) - y_{n+k}.$$

The nomenclature in the literature is a little confused in this area and some authors use 'local error' to have other meanings (e.g. Hairer, Nørsett and Wanner (1980)). If LE_{n+k} is expanded in powers of h, then the leading term in this expansion is called the *principal local error* or PLE. Note that no localizing assumption arises in this definition; y_{n+k-1} and y_{n+k} are the actual computed values, in which truncation error will have accumulated. The situation is perhaps clarified by Figure 3.1, which illustrates the situation in the case $k = 3$; for a typical component ${}^t y$, the points marked $\times$ denote the numerical solution $\{{}^t y_n\}$, those marked $\square$ the back values ${}^t y_{n+j}$, $j = 0, 1, 2$ under the localizing assumption, and the point marked $+$ the value ${}^t \tilde{y}_{n+k}$ (in the notation of §3.5). It is sometimes claimed that, because of the absence of any localizing assumptions, the LE is a more natural measure of local accuracy than is the LTE. We shall challenge this view later in this section.

By (3.26), for a method of order p the LTE is $0(h^{p+1})$. Intuition suggests that the LE is also $0(h^{p+1})$, since it is free of accumulation of error. An interesting question arises: is the PLTE the same as the PLE? We approach this question by first considering what happens for an example. The convergent method

$$y_{n+2} - (1 + \alpha)y_{n+1} + \alpha y_n = \frac{h}{2}[(3 - \alpha)f_{n+1} - (1 + \alpha)f_n], \tag{3.45}$$

where α, $-1 \leqslant \alpha < 1$, is a parameter, has order 2 and error constant $C_3 = (5 + \alpha)/12$. If

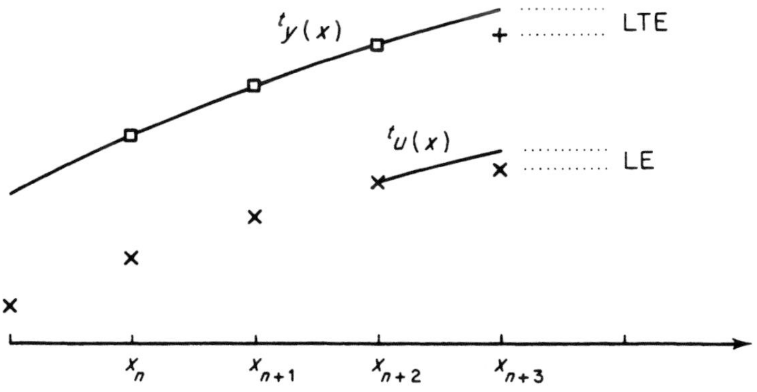

Figure 3.1

we apply it to the scalar problem

$$y' = y, \quad y(0) = 1, \tag{3.46}$$

whose exact solution is $y(x) = \exp(x)$, then the local truncation error at x_{n+2} is

$$\mathrm{LTE}_{n+2} = \tfrac{1}{12}(5 + \alpha)h^3 \exp(x_n) + 0(h^4). \tag{3.47}$$

Applying (3.45) to (3.46) generates the difference equation

$$y_{n+2} - \left[1 + \alpha + \frac{h}{2}(3 - \alpha)\right] y_{n+1} + \left[\alpha + \frac{h}{2}(1 + \alpha)\right] y_n = 0. \tag{3.48}$$

We are interested in seeing whether local error is affected by starting errors, so we take the initial conditions for (3.48) to be $0(h^q)$ perturbations of the exact conditions:

$$y_0 = 1 + \omega_0 h^q, \quad y_1 = \exp(h) + \omega_1 h^q, \qquad q \geqslant 1. \tag{3.49}$$

As for the example of §3.5, we find that the solution of (3.48) satisfying (3.49) may be written

$$\left.\begin{aligned} y_n &= \frac{\Omega(r_2)r_1^n - \Omega(r_1)r_2^n}{r_1 - r_2} \\ \text{where} \qquad & \\ \Omega(r) &= \exp(h) - r + (\omega_1 - r\omega_0)h^q \end{aligned}\right\} \tag{3.50}$$

and r_1, r_2 are now the roots of

$$r^2 - \left[1 + \alpha + \frac{h}{2}(3 - \alpha)\right] r + \alpha + \frac{h}{2}(1 + \alpha) = 0.$$

By expanding the discriminant of this quadratic in powers of h, or by using sums and products of roots (or preferably by both!) we find the following expansions for r_1 and r_2:

$$\left.\begin{aligned} r_1 &= 1 + h + \frac{1}{2}h^2 - \frac{1}{4}\frac{1 + \alpha}{1 - \alpha}h^3 + 0(h^4), \\ r_2 &= \alpha + \frac{1}{2}(1 - \alpha)h - \frac{1}{2}h^2 + \frac{1}{4}\frac{1 + \alpha}{1 - \alpha}h^3 + 0(h^4). \end{aligned}\right\} \tag{3.51}$$

The function u in the definition of local error satisfies $u' = u$, $u(x_{n+1}) = y_{n+1}$, and is therefore $u(x) = y_{n+1}\exp(x - x_{n+1})$. It follows that

$$\mathrm{LE}_{n+2} = u(x_{n+2}) - y_{n+2} = y_{n+1}\exp(h) - y_{n+2}.$$

On substituting for y_{n+1}, y_{n+2} from (3.50) and using the expansions (3.51), we find after some manipulation that

$$\mathrm{LE}_{n+2} = \frac{1}{12}\frac{5 + \alpha}{1 - \alpha}h^3 \exp(x_n) - \left[\frac{1}{12}\frac{5 + \alpha}{1 - \alpha}h^3 + (\omega_1 - \omega_0)h^q\right] r_2^{n+1} + 0(h^4) + 0(h^{q+1}), \tag{3.52}$$

where $r_2 = \alpha + 0(h)$.

Comparing (3.52) with (3.47), we see that for general $\alpha\in[-1,1)$, the LE and the LTE are very different. In particular, note that when $\alpha=-1$ the factor r_2^{n+1} in the second term of (3.52) does not decay as $n\to\infty$. We conclude that for general linear multistep methods PLE $\neq$ PLTE.

Does the above example persuade us that LE is to be preferred to LTE as a measure of local accuracy? Whilst the definition of LE does appear to indicate that it is the more natural measure of local accuracy, the results of the above example can be interpreted as being a little disturbing. Global error arises through three factors, (i) the starting errors, (ii) the 'local' errors committed at each step, and (iii) the manner in which these two errors are propagated; here we are searching for the most acceptable definition of (ii). From (3.47) and (3.52) we first note that the LE, unlike the LTE, is influenced by the starting errors (and the PLE is so influenced if we make the natural choice $q=3$). But any measure of *local* accuracy should not be concerned with what happened in the starting phase; LE is straying from factor (ii) to factor (i). Secondly, in the example we demanded that $-1\leqslant\alpha<1$, in order to ensure that the method was zero-stable; the analysis would certainly need modification in the case $\alpha=1$, but it holds good for $\alpha>1$. If we take $\alpha>1$ then the method is zero-unstable and the global error will be unbounded. From (3.52) we see from the term in r_2^{n+1} that the LE will also become unbounded as n increases. One could argue that the LE is rightly warning us that the numerical solution is bad. However, a measure of *local* accuracy need not, indeed should not reflect zero-instability, which as we have seen in §2.5 is *precisely* to do with the adverse accumulation of local errors and not with the size of those local errors themselves; so LE also strays from factor (ii) to factor (iii). One could say that LE pokes its nose into what is not its business!

Having made some possibly controversial remarks, let us now make some conciliatory ones by noting that LE is usually defined in the context of codes which almost invariably use Adams methods. We observe from (3.47) and (3.52) that if $\alpha=0$, when the method becomes the 2-step Adams–Bashforth method, a remarkable simplification takes place. Firstly,

$$5+\alpha=5=(5+\alpha)/(1-\alpha),$$

and secondly, from (3.51),

$$r_2=\tfrac{1}{2}h-\tfrac{1}{2}h^2+\tfrac{1}{4}h^3+0(h^4)=\tfrac{1}{2}h[\exp(-h)+0(h^3)]$$

whence $r_2^{n+1}=(h/2)^{n+1}+0(h^{n+2})$, and provided $q\geqslant 3$, we have

$$\mathrm{LE}_{n+2}=\tfrac{5}{12}h^3\exp(x_{n+1})+0(h^4)=\tfrac{5}{12}h^3\exp(x_n)+0(h^4)$$

and $\mathrm{PLE}_{n+2}=\mathrm{PLTE}_{n+2}$.

We are tempted to conjecture that *for all methods of Adams type*, PLTE = PLE; we can prove as follows that this conjecture is indeed true, under natural conditions on the starting errors.

Consider the general linear k-step method of Adams type of order $p\geqslant 1$,

$$y_{n+k}-y_{n+k-1}=h\sum_{j=0}^{k}\beta_j f_{n+j} \tag{3.53}$$

applied to the problem $y' = f(x, y)$, $y(a) = \eta$, with starting values which are in error by $0(h^{p+1})$. The method is convergent and from comment (b) in §3.6, we know that the global error satisfies $E_n = 0(h^p)$, $n = k, k+1, \dots$; it follows from our assumption on the starting errors that

$$E_n = 0(h^p), \quad n = 0, 1, \dots. \tag{3.54}$$

Assuming that the exact solution $y(x)$ is sufficiently differentiable, then, if C_{p+1} is the error constant, we have that

$$y(x_{n+k}) - y(x_{n+k-1}) = h \sum_{j=0}^{k} \beta_j f(x_{n+j}, y(x_{n+j})) + C_{p+1} h^{p+1} y^{(p+1)}(x_n) + 0(h^{p+2})$$

and on subtracting (3.53) and using the mean value theorem with the notation of §1.3, we obtain

$$E_{n+k} - E_{n+k-1} = h \sum_{j=0}^{k} \beta_j \overline{\frac{\partial f}{\partial y}}(x_{n+j}, \eta_{n+j}) E_{n+j} + 0(h^{p+1}), \qquad n = 0, 1, \dots, \tag{3.55}$$

where the bar indicates that each row of the Jacobian $\partial f/\partial y$ is evaluated at possibly different values of η_{n+j}, all of which lie in the line segment in $\mathbb{R}^m$ from $y(x_{n+j})$ to y_{n+j}. It follows from (3.54) and (3.55) that

$$E_{n+k} - E_{n+k-1} = 0(h^{p+1}), \qquad n = 0, 1, \dots,$$

and in view of our hypothesis that the starting errors are $0(h^{p+1})$ we may write

$$E_{n+1} - E_n = 0(h^{p+1}), \qquad n = 0, 1, \dots. \tag{3.56}$$

Let $u(x)$ be the solution of the initial value problem

$$u' = f(x, u), \quad u(x_{n+k-1}) = y_{n+k-1}. \tag{3.57}$$

Then, by definition of the LTE of the method

$$u(x_{n+k}) - u(x_{n+k-1}) = h \sum_{j=0}^{k} \beta_j f(x_{n+j}, u(x_{n+j})) + C_{p+1} h^{p+1} u^{(p+1)}(x_n) + 0(h^{p+2})$$

and on subtracting (3.53) and again using the mean value theorem we have

$$\begin{aligned} u(x_{n+k}) - y_{n+k} = h \sum_{j=0}^{k} \beta_j \overline{\frac{\partial f}{\partial y}}(x_{n+j}, \zeta_{n+j})[u(x_{n+j}) - y_{n+j}] \\ + C_{p+1} h^{p+1} u^{(p+1)}(x_n) + 0(h^{p+2}), \qquad n = 0, 1, \dots \end{aligned}$$

From the definition of LE_{n+k}, $u(x_{n+k-1}) = y_{n+k-1}$ and it follows that

$$\begin{aligned} \left[I - h\beta_k \overline{\frac{\partial f}{\partial y}}(x_{n+k}, \zeta_{n+k})\right] \mathrm{LE}_{n+k} = h \sum_{j=0}^{k-2} \beta_j \overline{\frac{\partial f}{\partial y}}(x_{n+j}, \zeta_{n+j})[u(x_{n+j}) - y_{n+j}] \\ + C_{p+1} h^{p+1} u^{(p+1)}(x_n) + 0(h^{p+2}), \quad n = 0, 1, \dots. \end{aligned} \tag{3.58}$$

Now, for $j = 0, 1, \ldots, k-2$,

$$u(x_{n+j}) = u(x_{n+k-1}) + \sum_{s=1}^{p} \frac{(j-k+1)^s}{s!} h^s u^{(s)}(x_{n+k-1}) + 0(h^{p+1})$$

$$= y_{n+k-1} + \sum_{s=1}^{p} \frac{(j-k+1)^s}{s!} h^s f^{(s-1)}(x_{n+k-1}, y_{n+k-1}) + 0(h^{p+1}),$$

where $f^{(s-1)}(x, y) = [d^{s-1}/dx^{s-1}]f(x, y)$, $s = 1, 2, \ldots, p$. Also,

$$y(x_{n+j}) = y(x_{n+k-1}) + \sum_{s=1}^{p} \frac{(j-k+1)^s}{s!} h^s f^{(s-1)}(x_{n+k-1}, y(x_{n+k-1})) + 0(h^{p+1}).$$

On subtracting and applying the mean value theorem we obtain, using (3.54)

$$u(x_{n+j}) - y(x_{n+j}) = -[I + 0(h)]E_{n+k-1} + 0(h^{p+1}) = -E_{n+k-1} + 0(h^{p+1}).$$

Hence, for $j = 0, 1, \ldots, k-2$, $n = 0, 1, \ldots,$

$$\begin{aligned} u(x_{n+j}) - y_{n+j} &= u(x_{n+j}) - y(x_{n+j}) + y(x_{n+j}) - y_{n+j} \\ &= -E_{n+k-1} + E_{n+j} + 0(h^{p+1}) \\ &= -E_{n+k-1} + E_{n+k-2} - E_{n+k-2} + \cdots - E_{n+j+1} + E_{n+j} + 0(h^{p+1}) \\ &= 0(h^{p+1}), \end{aligned}$$

by (3.56). Hence from (3.58) we obtain that

$$\mathrm{LE}_{n+k} = C_{p+1} h^{p+1} u^{(p+1)}(x_n) + 0(h^{p+2}),$$

whereas

$$\mathrm{LTE}_{n+k} = C_{p+1} h^{p+1} y^{(p+1)}(x_n) + 0(h^{p+2}).$$

Since $y(x) - u(x) = 0(h^p)$, it follows that $\mathrm{PLE}_{n+k} = \mathrm{PLTE}_{n+k}$, $n = 0, 1, \ldots,$ thus proving the conjecture.

The above result can be extended a little. Assume that a linear multistep method satisfies the root condition and, in addition, $\zeta = 1$ is the *only* root of $\rho(\zeta)$ on the unit circle; such methods are sometimes said to satisfy the *strong root condition*. Then it can be shown (Lambert, 1990) that if the solution of the problem satisfies $y(x) \in C^{p+1}[a, b]$, where $p \geqslant 1$ is the order of the method, and the starting values are in error by $0(h^p)$, then PLTE = PLE if and only if $\sum_{j=0}^{k} \beta_j = 1$. Clearly, Adams methods satisfy this last condition and the strong root condition.

We have said nothing so far on how we would, in practice, estimate the LTE (or the LE). From (3.26) we have that the PLTE is given by

$$\mathrm{PLTE}_{n+k} = C_{p+1} h^{p+1} y^{(p+1)}(x_n).$$

It would be possible to replace the exact solution $y(x)$ by the numerical solution $\{y_n\}$ and hence estimate $y^{(p+1)}(x_n)$ by the process of numerical differentiation—an inaccurate process, especially when high derivatives are sought. However, linear multistep methods are normally implemented in the form of predictor–corrector pairs, and it turns out that in this form a much more satisfactory estimate of the PLTE is available at virtually no computational cost. We therefore defer the question of estimating the PLTE to Chapter 4.

Exercise

3.7.1*. Repeat the analysis performed in the above section for the method (3.45) applied to the problem (3.46), but with the explicit method (3.45) replaced by the implicit method

$$y_{n+2} - (1+\alpha)y_{n+1} + \alpha y_n = \frac{h}{12}[(5+\alpha)f_{n+2} + 8(1-\alpha)f_{n+1} - (1-5\alpha)f_n].$$

Show that the PLTE and PLE coincide if and only if $\alpha = 0$, when the method becomes the 2-step Adams–Moulton method.

3.8 LINEAR STABILITY THEORY

Let us refer back to Example 3 of §2.7, where a convergent linear multistep method was applied to the test problem (2.24). The numerical results showed that there appeared to exist some value h^* of the steplength such that for fixed $h > h^*$ the error increased as x increased, whereas for fixed $h < h^*$ it decreased. Further, it can happen (as in Example 4, which is not, however, a linear multistep method) that for *all* fixed positive values of h, the errors produced by a *convergent* linear multistep method increase as x increases. In such situations, clearly the local errors are accumulating in an adverse fashion; in other words, we are dealing with a stability phenomenon. The only form of stability we have considered so far is zero-stability, which controls the manner in which errors accumulate, *but only in the limit as* $h \rightarrow 0$. What is needed is a stability theory which applies when h takes a fixed non-zero value.

In attempting to set up such a theory, we follow the spirit of §2.6, where we discussed the syntax of a stability definition, and seek some simple test system, all of whose solutions tend to zero as x tends to infinity. We then attempt to find conditions for the numerical solutions to behave similarly. The simplest such test system is the linear constant coefficient homogeneous system

$$y' = Ay \tag{3.59}$$

where the eigenvalues λ_t, $t = 1, 2, \ldots, m$ of the constant $m \times m$ matrix A (assumed distinct) satisfy

$$\operatorname{Re}\lambda_t < 0, \qquad t = 1, 2, \ldots, m. \tag{3.60}$$

The general solution of (3.59) takes the form

$$y(x) = \sum_{t=0}^{m} \varkappa_t \exp(\lambda_t x) c_t \tag{3.61}$$

(see §1.6) and it follows from (3.60) that all solutions $y(x)$ of (3.59) satisfy

$$\| y(x) \| \to 0 \qquad \text{as } x \to \infty.$$

We now ask what conditions must be imposed in order that the numerical solutions $\{y_n\}$ generated when a linear multistep method is applied to (3.59) satisfy

$$\| y_n \| \to 0 \qquad \text{as } n \to \infty. \tag{3.62}$$

Let the linear multistep method

$$\sum_{j=0}^{k} \alpha_j y_{n+j} = h \sum_{j=0}^{k} \beta_j f_{n+j} \tag{3.63}$$

be applied to (3.59). The resulting difference system for $\{y_n\}$ is

$$\sum_{j=0}^{k} (\alpha_j I - h\beta_j A) y_{n+j} = 0, \tag{3.64}$$

where I is the $m \times m$ unit matrix. Since the eigenvalues of A are assumed distinct, there exists a non-singular matrix Q such that

$$Q^{-1}AQ = \Lambda := \operatorname{diag}[\lambda_1, \lambda_2, \ldots, \lambda_m].$$

We now define z_n by

$$y_n = Qz_n \tag{3.65}$$

and on pre-multiplying (3.64) by Q^{-1}, we obtain

$$\sum_{j=0}^{k} (\alpha_j I - h\beta_j \Lambda) z_{n+j} = 0.$$

Since I and Λ are both diagonal matrices, this system is *uncoupled*, that is we may write it as

$$\sum_{j=0}^{k} (\alpha_j - h\beta_j \lambda_t)\,{}^t z_{n+j} = 0, \qquad t = 1, 2, \ldots, m, \tag{3.66}$$

where $z_n = [{}^1 z_n, {}^2 z_n, \ldots, {}^m z_n]^T$. Since the eigenvalues of A are in general complex, we note that each equation in (3.66) is a *complex* linear constant coefficient homogeneous difference equation. By (3.65), $\| y_n \| \to 0$ as $n \to \infty$, if and only if $\| z_n \| \to 0$ as $n \to \infty$, and hence (3.62) is satisfied if and only if all solutions $\{{}^t z_n\}$ of (3.66) satisfy

$$|{}^t z_n| \to 0 \qquad \text{as } n \to \infty, \quad t = 1, 2, \ldots, m. \tag{3.67}$$

Now, from §1.7 we know that the general solution of each of the difference equations in (3.66) takes the form

$${}^t z_n = \sum_{s=1}^{m} \varkappa_{ts} r_s^n, \quad t = 1, 2, \ldots, m, \tag{3.68}$$

where the $\varkappa_{ts}$ are arbitrary complex constants and r_s, $s = 1, 2, \ldots, m$ are the roots, assumed distinct, of the characteristic polynomial

$$\sum_{j=0}^{k} (\alpha_j - h\beta_j \lambda_t) r^j.$$

This polynomial can conveniently be written in terms of the first and second characteristic polynomials ρ, σ of the method (see equation (3.4)) as

$$\left.\begin{aligned} \pi(r, \hat{h}) &:= \rho(r) - \hat{h}\sigma(r) \\ \text{where} \qquad \hat{h} &:= h\lambda \end{aligned}\right\} \tag{3.69}$$

and λ, a complex parameter, represents *any* of the eigenvalues λ_t, $t = 1, 2, \ldots, m$ of A.

The polynomial $\pi(r, \hat{h})$ is called the *stability polynomial* of the method. Clearly, (3.67) and consequently (3.62) are satisfied if all the roots r_s $(= r_s(\hat{h}))$, $s = 1, 2, \ldots, k$ of $\pi(r, \hat{h})$ satisfy $|r_s| < 1$, and we are motivated to make the following definition.

Definition *The linear multistep method (3.63) is said to be* **absolutely stable** *for given* $\hat{h}$ *if for that* $\hat{h}$ *all the roots of the stability polynomial (3.69) satisfy* $|r_s| < 1$, $s = 1, 2, \ldots, k$, *and to be* **absolutely unstable** *for that* $\hat{h}$ *otherwise.*

Clearly we are interested in knowing for what products of h and λ the method is absolutely stable, whence the following definition:

Definition *The linear multistep method (3.63) is said to have* **region of absolute stability** $\mathscr{R}_A$, *where* $\mathscr{R}_A$ *is a region of the complex* $\hat{h}$*-plane, if it is absolutely stable for all* $\hat{h} \in \mathscr{R}_A$. *The intersection of* $\mathscr{R}_A$ *with the real axis is called the* **interval of absolute stability**.

Note that the interval of absolute stability is relevant to the case of the *scalar* test equation $y' = \lambda y$, λ real.

We construct below the syntax diagram for absolute stability.

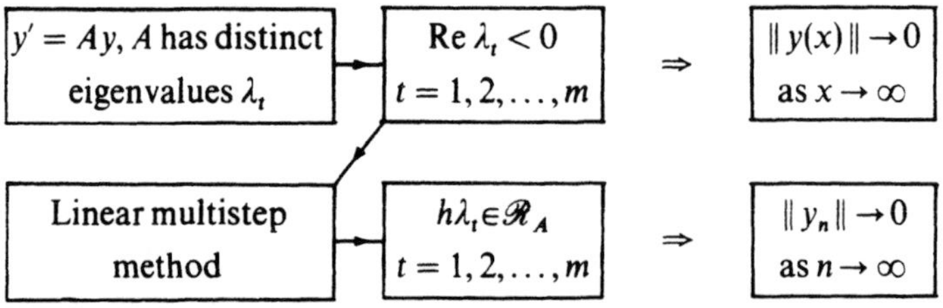

The region of absolute stability, $\mathscr{R}_A$, is a function of the method and the complex parameter $\hat{h}$ only, so that for each linear multistep method we are able to plot the region $\mathscr{R}_A$ in the complex $\hat{h}$-plane. If the eigenvalues of A are known it is then possible to choose h sufficiently small for $h\lambda_t \in \mathscr{R}_A$ to hold for $t = 1, 2, \ldots, m$.

We note from (3.69) that when $h = 0$ the stability polynomial $\pi(r, \hat{h})$ reduces to the first characteristic polynomial ρ. Recall from §2.5 that for a consistent linear multistep method $\rho(\zeta)$ always has a root $\zeta_1 = 1$, and this we called the principal root; if the method is zero-stable, this root must be simple. Now the roots of a polynomial are continuous functions of the coefficients of the polynomial, and it follows that there must exist a root r_1 of π which has the property that $r_1 \to \zeta_1 = 1$ as $h \to 0$. The following argument tells us *how* r_1 approaches 1 as $h \to 0$.

Let the linear multistep method satisfy the root condition, have associated difference operator $\mathscr{L}$ and have order $p > 1$. Then $\mathscr{L}[z(x); h] = 0(h^{p+1})$ for any sufficiently differentiable function $z(x)$; $\exp(\lambda x)$, where $\lambda \in \mathbb{C}$, is such a function, and we may write

$$\mathscr{L}[\exp(\lambda x); h] = \sum_{j=0}^{k} \{\alpha_j \exp[\lambda(x + jh)] - h\beta_j \lambda \exp[\lambda(x + jh)]\} = 0(h^{p+1}).$$

On dividing by $\exp(\lambda x)$ we obtain

$$\sum_{j=0}^{k} \{\alpha_j [\exp(\hat{h})]^j - \hat{h}\beta_j [\exp(\hat{h})]^j\} = 0(\hat{h}^{p+1})$$

which may be written

$$\pi(\exp(\hat{h}), \hat{h}) = 0(\hat{h}^{p+1}). \tag{3.70}$$

The polynomial $\pi(r, \hat{h})$ can be reconstructed in terms of its roots r_s, $s = 1, 2, \dots, k$ as

$$\pi(r, \hat{h}) = (1 - \hat{h}\beta_k)(r - r_1)(r - r_2) \cdots (r - r_k). \tag{3.71}$$

(Note that, by the following argument, the factor $1 - \hat{h}\beta_k$ can never be zero: let $\tau(A)$ be the spectral radius of A. The Lipschitz constant L of the function $f = Ay$ is $L = \| A \| \geqslant \tau(A) \geqslant |\lambda|$ (where λ is any eigenvalue of A). Hence the condition (3.8), $h < 1/|\beta_k| L$, implies that $\hat{h} < 1/|\beta_k|$.)

On setting $r = \exp(\hat{h})$ in (3.71) and using (3.70), we obtain

$$[\exp(\hat{h}) - r_1][\exp(\hat{h}) - r_2] \cdots [\exp(\hat{h}) - r_k] = 0(\hat{h}^{p+1}).$$

Since, as $\hat{h} \to 0$, $\exp(\hat{h}) \to 1$, and $r_s \to \zeta_s$, $s = 1, 2, \dots, k$, the first factor on the left side tends to zero as $h \to 0$, and no other factor can do so since, by the roots condition, ζ_1 is the only roots of $\rho(\zeta)$ located at $+1$. It follows that

$$r_1 = \exp(\hat{h}) + 0(\hat{h}^{p+1}). \tag{3.72}$$

It immediately follows that for small $\hat{h}$ with $\operatorname{Re} \hat{h} > 0$, $|r_1| > 1$, and the method is absolutely unstable. In other words, *the region of absolute stability of any convergent linear multistep method cannot contain the positive real axis in the neighbourhood of the origin.* Note that since the above argument is asymptotic (as $\hat{h} \to 0$), we cannot conclude that the region of absolute stability does not contain part of the positive real axis for large $|\hat{h}|$ or that the boundary of the region does not intrude into the positive half-plane away from the origin.

The most convenient method for finding regions of absolute stability is the *boundary locus technique*. The region $\mathscr{R}_A$ of the complex $\hat{h}$-plane is defined by the requirement that for all $\hat{h} \in \mathscr{R}_A$ all of the roots of $\pi(r, \hat{h})$ have modulus less than 1. Let the contour $\partial \mathscr{R}_A$ in the complex $\hat{h}$-plane be defined by the requirement that for all $\hat{h} \in \partial \mathscr{R}_A$ *one* of the roots of $\pi(r, \hat{h})$ has modulus 1, that is, is of the form $r = \exp(i\theta)$. Since the roots of a polynomial are continuous functions of its coefficients it follows that the boundary of $\mathscr{R}_A$ must consist of $\partial\mathscr{R}_A$ (or of part of $\partial\mathscr{R}_A$; some parts of $\partial\mathscr{R}_A$ could, for example, correspond to $\pi(r, \hat{h})$ having one root of modulus 1, some of the remaining roots having modulus less than 1 and some having modulus greater than 1). Thus, for all $\hat{h}\in\partial\mathscr{R}_A$, the identity

$$\pi(\exp(i\theta), \hat{h}) = \rho(\exp(i\theta)) - \hat{h}\sigma(\exp(i\theta)) = 0 \tag{3.73}$$

must hold. This equation is readily solved for $\hat{h}$, and we have that the locus of $\partial\mathscr{R}_A$ is given by

$$\hat{h} = \hat{h}(\theta) = \rho(\exp(i\theta))/\sigma(\exp(i\theta)). \tag{3.74}$$

For the Adams–Moulton method with $k = 1$ (the Trapezoidal Rule), the contour $\partial\mathscr{R}_A$ is particular geometrical shape, but in most cases we simply use (3.74) to plot $\hat{h}(\theta)$ for a range of $\theta\in[0, 2\pi]$, and link consecutive plotted points by straight lines to get a representation of $\partial\mathscr{R}_A$. The contours $\partial\mathscr{R}_A$ so obtained for the Adams–Moulton methods of stepnumbers 1 to 4 are shown in Figure 3.2 and those for the Adams–Bashforth

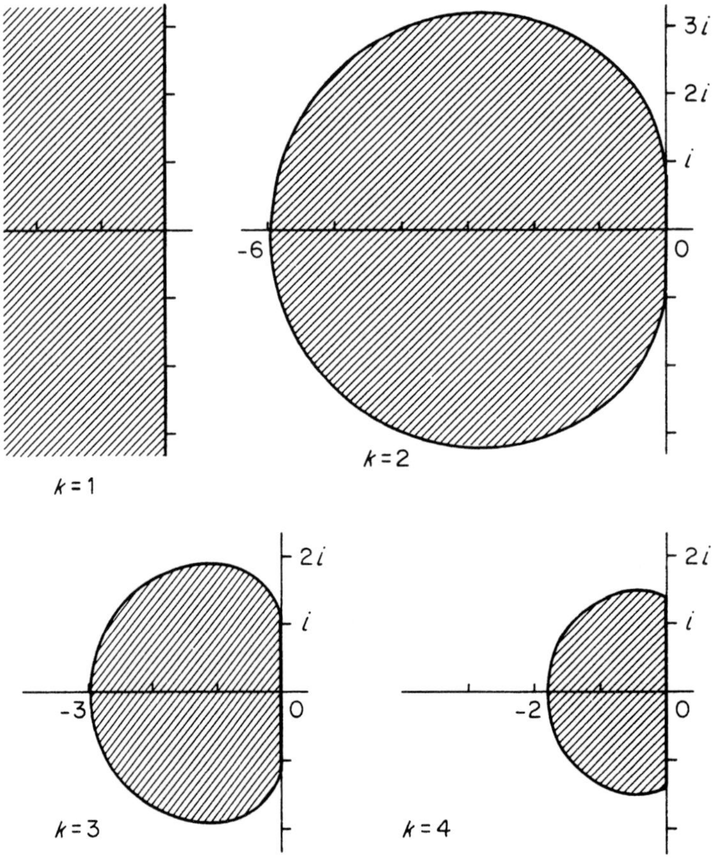

Figure 3.2 Regions of absolute stability for k-step Adams–Moulton methods.

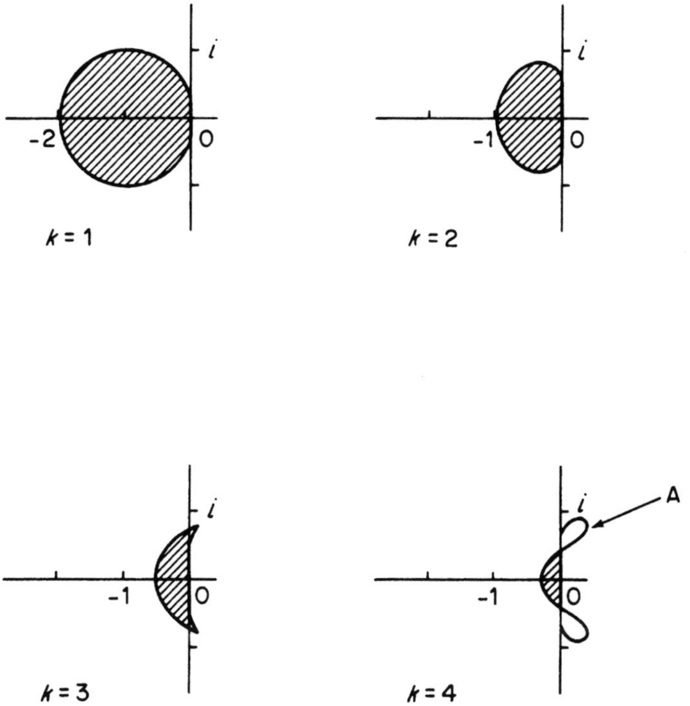

Figure 3.3 Regions of absolute stability for k-step Adams–Bashforth methods.

methods of stepnumbers 1 to 4 in Figure 3.3. Having found $\partial\mathscr{R}_A$, we have to deduce what $\mathscr{R}_A$ is. For the Adams–Moulton methods with $k = 2, 3, 4$ and the Adams–Bashforth methods with $k = 1, 2, 3$, $\partial\mathscr{R}_A$ is a simple closed contour; to see that the shaded regions in Figures 3.2 and 3.3, that is the *interiors* of the regions bounded by $\partial\mathscr{R}_A$, are indeed the regions of absolute stability, all we need do is observe that, from (3.72), all linear multistep methods are necessarily absolutely unstable for small positive values of $\mathrm{Re}(\hat{h})$. For the Adams–Moulton method with $k = 1$ (the Trapezoidal Rule), the contour $\partial\mathscr{R}_A$ is simple but no longer closed; it is indeed the whole of the imaginary axis. The same argument shows that the region of absolute stability is the whole of the negative half-plane $\mathrm{Re}(\hat{h}) < 0$.

Things are not quite so simple for the Adams–Bashforth method with $k = 4$, where the contour $\partial\mathscr{R}_A$ is closed but no longer simple, since it crosses itself at two points (which are just to the right of the imaginary axis). The argument we have used above establishes that the region of absolute stability contains the shaded region, but it is not clear whether it also contains the two loops lying in the positive half-plane. The point marked A in Figure 3.3 lying on the loop of $\partial\mathscr{R}_A$ in the first quadrant corresponds to taking $\theta = \pi/2$, which, by (3.74), implies that $\hat{h} = 0.272 + 0.578\mathrm{i}$ at A. With this value for $\hat{h}$, $\pi(r, \hat{h})$ turns out to have roots

$$\mathrm{i}, \quad 1.076 + 0.744\mathrm{i}, \quad 0.357 + 0.051\mathrm{i}, \quad 0.190 - 0.470\mathrm{i}.$$

(Note that since π is a polynomial with complex coefficients, its complex roots no longer appear as complex conjugate pairs.) The first root has modulus 1 (justifying A as a point on $\partial\mathscr{R}_A$) but the second root has modulus greater than 1. By continuity, the interior of the loop cannot be a region of absolute stability. By symmetry, the same holds for the loop in the fourth quadrant, and we conclude that the shaded region in Figure 3.3 constitutes the entire region of absolute stability.

We note in passing that Figures 3.2 and 3.3 prompt two conjectures: that implicit methods have larger regions of absolute stability than explicit methods and that, as order increases, the regions of absolute stability shrink. The first is, in a general sense, true of all numerical methods; the second is not (cf. explicit Runge–Kutta methods, to be discussed in Chapter 5).

It is of some interest to apply the boundary locus method to one further example, namely Simpson's Rule, given by

$$\rho(r) = r^2 - 1, \quad \sigma(r) = \tfrac{1}{3}(r^2 + 4r + 1).$$

Applying (3.74), we find after a little manipulation that the locus of $\partial\mathscr{R}_A$ is given by

$$\hat{h}(\theta) = \frac{3\mathrm{i}\sin\theta}{2 + \cos\theta}$$

and therefore lies wholly on the imaginary axis. For all θ, the function $3\sin\theta/(2+\cos\theta)$ lies between $-\sqrt{3}$ and $\sqrt{3}$, and it follows that $\partial\mathscr{R}_A$ is that part of the imaginary axis running from $-\sqrt{3}\mathrm{i}$ to $\sqrt{3}\mathrm{i}$. We know that for $\hat{h}\in\partial\mathscr{R}_A$, $\pi(r,\hat{h})$ has a root of modulus 1 (in this instance, both roots have modulus 1) and we conclude that Simpson's Rule has an empty region of absolute stability. All optimal methods have regions of absolute stability which are either empty or essentially useless in that they do not contain the negative real axis in the neighbourhood of the origin (see Stetter (1973), page 268 for fuller details). In essence, by squeezing out the maximum possible order (subject to zero-stability) from a linear multistep method of given stepnumber, the absolute stability region gets squeezed flat. This is why optimal methods are not favoured.

The interval of absolute stability can, of course, be deduced directly from the region, but sometimes we want to find only the interval, in which case quicker methods are available. It is then appropriate to take $\hat{h}\in\mathbb{R}$, in which case $\pi(r,\hat{h})$ becomes a real polynomial. The criterion for absolute stability is then that $\pi(r,\hat{h})$ be a Schur polynomial, and the Routh–Hurwitz criterion, discussed in §1.9, can be applied. Consider, for example, the 3-step Adams–Moulton method given by

$$\rho(r) = r^3 - r^2, \quad \sigma(r) = (9r^3 + 19r^2 - 5r + 1)/24$$

whence, setting $H := \hat{h}/24$ for convenience, we obtain

$$\pi(r,\hat{h}) = (1 - 9H)r^3 - (1 + 19H)r^2 + 5Hr - H.$$

On applying the transformation $r = (1+z)/(1-z)$ we obtain

$$(1-z)^3\pi((1+z)/(1-z),\hat{h}) = a_0z^3 + a_1z^2 + a_2z + a_3,$$

where

$$a_3 = -24H > 0 \text{ if } H < 0,$$
$$a_2 = 2 - 48H > 0 \text{ if } H < \tfrac{1}{24},$$
$$a_1 = 4 - 16H > 0 \text{ if } H < \tfrac{1}{4},$$
$$a_0 = 2 + 16H > 0 \text{ if } H > -\tfrac{1}{8}.$$

The conditions $a_j > 0$, $j = 0, 1, 2, 3$, are thus satisfied iff $H \in (-\frac{1}{8}, 0)$. The remaining condition $a_1 a_2 > a_0 a_3$ is satisfied iff

$$144H^2 - 22H + 1 > 0,$$

which condition (the discriminant of the left side being negative) is easily seen to be satisfied for all H. It follows that $\pi(r, \hat{h})$ is Schur iff $H \in (-\frac{1}{8}, 0)$, or equivalently $\hat{h} \in (-3, 0)$. The interval of absolute stability is thus $(-3, 0)$, a result corroborated by Figure 3.2.

Linear stability theory supports a further crop of definitions concerning *relative stability*, a topic nowadays considered less significant than hitherto. The rationale for this development is as follows. Recall the result (3.72) that

$$r_1 = \exp(\hat{h}) + 0(\hat{h}^{p+1}). \tag{3.72}$$

A consequence of this is that when we use a linear multistep method to solve numerically the test system (3.59) in the case when A has some eigenvalues with positive real parts then, for sufficiently small h, the numerical solution $\{y_n\}$ will have the property that $\|y_n\| \to \infty$ as $n \to \infty$. This is as it should be, since from (3.61) we know that $\|y(x)\| \to \infty$ as $x \to \infty$ when any of the λ_t have positive real part; the situation is acceptable, provided that $\|y(x)\|$ and $\|y_n\|$ tend to ∞ at approximately the same rate. Let λ^* be the eigenvalue of A with greatest (positive) real part. It follows from (3.61) that the growth of $\|y(x)\|$ will be dominated by the term $\exp[(\text{Re}\,\lambda^*)x]$. Now, from (3.72),

$$r_1(h\lambda^*)^n = [\exp(h\lambda^*)]^n + 0(h^p) = \exp[\lambda^*(x_n - a)] + 0(h^p). \tag{3.75}$$

Relative stability is concerned with whether $r_1(h\lambda^*)^n$ or $r_s(h\lambda^*)^n$ (for some $s \in [2, 3, \ldots, k]$) dominates the numerical solution. If the former holds then, in view of (3.75), that is acceptable. The concept also has some relevance to the case of decaying solutions, since one might not be happy with a numerical solution which decayed, but not as fast as did the theoretical solution. Among many criteria that have been proposed to encapsulate this notion are the following:

$$\text{criterion A:} \quad |r_s| < |r_1|, \qquad s = 2, 3, \ldots, k$$
$$\text{criterion B:} \quad |r_s| < |\exp(\hat{h})|, \qquad s = 2, 3, \ldots, k,$$
$$\text{criterion C:} \quad |r_s| < |\exp(\hat{h})|, \qquad s = 1, 2, \ldots, k.$$

We can clearly define regions and intervals of relative stability based on each of these criteria. A is probably the most sensible criterion, but is hard to apply. In view of (3.72), B is not very different from A, at least for small $|\hat{h}|$, and is a little easier to apply. C is much easier to apply, since a change of argument in the stability polynomial from r to

R, where $r = R|\exp(\hat{h})|$ implies that C will be satisfied if $\pi(R|\exp(\hat{h})|, \hat{h})$ is a Schur polynomial in R. Unfortunately, C can give bizarre results, reflecting the fact that, for substantial regions of $\hat{h}$, $|r_1|$ may be greater than $|\exp(\hat{h})|$, although close to it. For example, the intervals of relative stability for the 2-step Adams–Moulton method, using each of these criteria, are:

$$\begin{aligned} &\text{A:} \quad (-1.50, \infty) \\ &\text{B:} \quad (-1.39, \infty) \\ &\text{C:} \quad (+2.82, \infty). \end{aligned}$$

Confusion over definitions may have been a minor factor in the decline of interest in relative stability, but the major reason is probably the fact that it is absolute stability, not relative stability, that is relevant to the problem of stiffness, to be discussed later in this book.

It will not have escaped the reader's notice that the theory propounded above is highly restrictive, in that it applies only to the test system $y' = Ay$, whereas in practice we usually find ourselves dealing with the general system $y' = f(x, y)$. Attempts can be made to extend the applicability of linear stability theory to general systems by constructing an approximate system of difference equations for the global error $E_n := y(x_n) - y_n$. As we shall see presently, such attempts can give rise to very misleading results. One starts by noting that the definition of the local truncation error T_{n+k} implies that

$$\sum_{j=0}^{k} \alpha_j y(x_n + jh) = h \sum_{j=0}^{k} \beta_j f(x_{n+j}, y(x_{n+j})) + T_{n+k}.$$

The sequence $\{y_n\}$ generated by the method satisfies

$$\sum_{j=0}^{k} \alpha_j y_{n+j} = h \sum_{j=0}^{k} \beta_j f(x_{n+j}, y_{n+j}),$$

and on subtracting and using the mean value theorem (with the notation of §1.3) we obtain

$$\sum_{j=0}^{k} \alpha_j E_{n+j} = h \sum_{j=0}^{k} \beta_j \overline{\frac{\partial f}{\partial y}}(x_{n+j}, \zeta_{n+j}) E_{n+j} + T_{n+k}. \tag{3.76}$$

This difference system for $\{E_n\}$ is deceptive; it looks linear but is indeed nonlinear since the values ζ_{n+j} all lie on the line segment from y_{n+j} to $y(x_{n+j})$, and hence E_{n+j} is an unknown nonlinear function of ζ_{n+j}. We cannot handle this system, so we force it to be linear constant coefficient by making the assumption

$$\frac{\partial f}{\partial y} = J, \text{ a constant matrix.} \tag{3.77}$$

We make the further assumption that $T_{n+k} = T$, a constant vector, so that (3.76) now reads

$$\sum_{j=0}^{k} [\alpha_j I - h\beta_j J] E_{n+j} = T \tag{3.78}$$

sometimes known as the *linearized error equation*. Since the constant term T plays no part in determining whether the norms of the solutions of (3.78) grow or decay as $n \to \infty$, it can be ignored, and (3.78) is essentially the same system as (3.64) with A replaced by J and y_{n+j} by E_{n+j}. The subsequent analysis holds, and we conclude that $\| E_n \| \to 0$ as $n \to \infty$ if $h\lambda_t \in \mathscr{R}_A$, where λ_t, $t = 1, 2, \ldots, m$ are now the eigenvalues of J, and $\mathscr{R}_A$ is the region of absolute stability of the method. A similar extension in the case of relative stability is clearly possible.

The flaw in this argument lies in the assumption (3.77). It is simply not true in general that the eigenvalues of J, even if J is taken to be piecewise constant (that is, recalculated from time to time as the computation proceeds), always correctly represent, even in a qualitative sense, the behaviour of the solutions of the nonlinear system (3.76). (It is, of course, true when $f(x, y) = Ay$, A a constant matrix, in which case the difference systems for the solution and for the errors are essentially the same.) We shall have more to say about 'frozen Jacobian' assumptions in Chapter 7, but meanwhile let us look at two examples.

First, consider the problem used in the numerical examples of §2.7,

$$\left.\begin{aligned} & y' = f(x, y), \quad y(0) = \eta, \qquad x \in [0, 1] \\ \text{where} \qquad & y = [u, v]^{\mathrm{T}}, \\ \text{and} \qquad & f(x, y) = [v, v(v-1)/u]^{\mathrm{T}}, \qquad \eta = [1/2, -3]^{\mathrm{T}} \end{aligned}\right\} \tag{3.79}$$

with exact solution

$$u(x) = [1 + 3\exp(-8x)]/8, \quad v(x) = -3\exp(-8x). \tag{3.80}$$

The Jacobian of the system is

$$\partial f/\partial y = \begin{bmatrix} 0 & 1 \\ -v(v-1)/u^2 & (2v-1)/u \end{bmatrix}.$$

Its eigenvalues are real, and can be expressed in closed form as

$$\lambda_1 = (v-1)/u, \qquad \lambda_2 = v/u. \tag{3.81}$$

Substituting for u and v from the exact solution (3.80), we obtain

$$\lambda_1 = -8, \qquad \lambda_2 = -24/[3 + \exp(8x)]. \tag{3.82}$$

In the interval $[0, 1]$ of integration, λ_2 increases from -6 to nearly zero, and we note that for all $x > 0$ both eigenvalues lie in the interval $[-8, 0]$.

The linear multistep method of Example 3 in §2.7 was

$$y_{n+3} + \tfrac{1}{4}y_{n+2} - \tfrac{1}{2}y_{n+1} - \tfrac{3}{4}y_n = \frac{h}{8}(19f_{n+2} + 5f_n). \tag{3.83}$$

Since the eigenvalues of the Jacobian of the system to be solved are real, we need compute only the interval rather than the region of absolute stability for this method. This interval turns out to be $(-\tfrac{1}{3}, 0)$, and hence we can satisfy the condition

$$h\lambda_t \in \mathscr{R}_A, \qquad t = 1, 2$$

by choosing h such that $-8h$ lies in $(-\frac{1}{3},0)$; that is, by choosing $h<h^*=\frac{1}{24}=0.0417$. From Table 2.4 of §2.7, we see that the global errors do indeed decrease for $h<h^*$ and increase for $h>h^*$. For this example the linearized error equation approach gives sensible results.

We do not need to look far for a counter-example, however. Consider the same system (3.79), but with new initial values given by

$$\eta=[-\tfrac{1}{4},3]^{\mathrm{T}}.$$

The exact solution is now

$$u(x)=[1-3\exp(-8x)]/8, \qquad v(x)=3\exp(-8x). \tag{3.84}$$

Since the system is unchanged, the eigenvalues of the Jacobian are still given by (3.81), but when we substitute the exact solutions (3.84) for u and v we obtain

$$\lambda_1=-8, \qquad \lambda_2=-24/[3-\exp(8x)].$$

Comparing with (3.82), we see that while λ_1 is unchanged, the behaviour of λ_2 for positive x is radically changed. At $x=\tilde{x}=(\ln 3)/8\approx 0.137$, λ_2 is infinite. For $x\in[0,\tilde{x})$, λ_2 is negative, but $|\lambda_2|$ becomes extremely large as $x\to\tilde{x}$; for $x>\tilde{x}$, λ_2 is positive. Were we to compute a solution of this new initial value problem using method (3.83), whose interval of absolute stability is $(-\frac{1}{3},0)$, then the theory based on the linearized error equation would predict that in order to avoid error growth, we would have to take sharply decreasing values of h as x approached $\tilde{x}$, and that error growth would be unavoidable for $x>\tilde{x}$. In practice, nothing of the sort happens, and the table of errors for numerical solutions in the interval $[0,1]$ with a range of steplengths is virtually identical with those given in Table 2.4 of §2.7 for the original initial value problem. What is happening is that the linearized error equation, faced with the impossible task of attempting to mimic the behaviour of the true nonlinear error equation, throws up one eigenvalue, λ_1, which does the best it can in predicting in a general sort of way the exponential decay $\exp(-8x)$, but the other eigenvalue, λ_2, is meaningless; in the case of the original initial value problem, it happened not to get in the way.

The literature of the 1960s and 1970s contains many results (some of which are reported in Lambert (1973)) on absolute and relative stability, but these results are of less significance nowadays. The reasons for this are partly that there has been a steadily growing appreciation of the limitations of linear stability theory and the emergence of a much more satisfactory theory of nonlinear stability (which we shall discuss in Chapter 7), but mainly because modern codes for the numerical solution of initial value problems do not actually test for absolute or relative stability. Quite apart from the possibility of bizarre results, as illustrated above, such a procedure would be hopelessly uneconomic; it would require frequent updatings of the Jacobian *and* of its entire spectrum of eigenvalues, a heavy computational task for a large system. Instead, these codes rely on their monitoring of the local truncation error to alert them to any instability; if the estimate of the LTE becomes too large, the step is aborted and the steplength reduced.

This is not to say that linear stability theory is of no value. A method which cannot handle satisfactorily the linear test system $y'=Ay$ is not a suitable candidate for incorporation into an automatic code. More precisely, linear stability theory provides a useful yardstick (if one can have a yardstick in the complex plane!) by which different linear multistep methods (or classes of such methods) can be compared as candidates for inclusion in an automatic code. There is ample computational evidence that methods with large regions of absolute/relative stability out-perform those with small regions.

One specific result of linear stability theory, already referred to, is worthy of note, namely that all optimal methods have regions of absolute stability which either are empty or do not contain the negative real axis in the neighbourhood of the origin. This means that when such methods are applied to the test system $y' = Ay$, where the eigenvalues of A have negative real parts, the numerical solution will satisfy $\| y_n \| \to \infty$ as $n \to \infty$, for all sufficiently small positive h, whereas the exact solution satisfies $\| y(x) \| \to 0$ as $x \to \infty$. Example 4 of §2.7 (although not a linear multistep method) has an empty region of absolute stability, and such behaviour is exemplified by the numerical results quoted. (Once again, it is fortuitous that linear stability works well for the particular nonlinear problem in question; it would not be difficult to produce one for which the theory was much less satisfactory.) The mechanism by which methods with empty regions of absolute stability are none the less convergent has already been explained in the comments on Example 4 in §2.7.

Exercises

3.8.1. Use the boundary locus method to show that the region of absolute stability of the method

$$y_{n+2} - y_n = \tfrac{1}{2}h(f_{n+1} + 3f_n)$$

is a circle centre $(-\tfrac{2}{3}, 0)$ and radius $\tfrac{2}{3}$. Check this result by using the Routh–Hurwitz criterion to show that the interval of absolute stability of the method is the diameter of this circle.

3.8.2. Using Figure 3.2, find the approximate maximum steplength that will achieve absolute stability when the 3-step Adams–Moulton method is applied to the scalar problem $y'' + 20y' + 200y = 0$, $y(0) = 1$, $y'(0) = -10$, recast as a first-order system. Test your answer numerically.

3.8.3*. Find an expression for the locus of $\partial\mathscr{R}_A$ for the method

$$y_{n+2} - (1+\alpha)y_{n+1} + \alpha y_n = \frac{h}{12}[(5+\alpha)f_{n+2} + 8(1-\alpha)f_{n+1} - (1+5\alpha)f_n]$$

Deduce the interval of absolute stability when $\alpha \neq -1$; check your result by using the Routh–Hurwitz criterion to find the interval of absolute stability. Find $\mathscr{R}_A$ in the cases (i) $\alpha = 1$, (ii) $\alpha = -1$.

3.8.4*. A linear multistep method is defined by its first and second characteristic polynomials ρ, σ. Show that if

$$\operatorname{Re}[\rho(\exp(i\theta))\sigma(\exp(-i\theta))] = 0 \qquad (1)$$

then the method is absolutely stable either for no $\hat{h}$ or for all $\hat{h}$ with $\operatorname{Re}(\hat{h}) < 0$. Show that the most general zero-stable linear 2-step method of order at least 2 which satisfies (1) is

$$y_{n+2} - y_n = h[\beta f_{n+2} + 2(1-\beta)f_{n+1} + \beta f_n],$$

and that it is absolutely stable for all $\hat{h}$ with $\operatorname{Re}(\hat{h}) < 0$ if and only if $\beta > \tfrac{1}{2}$.

3.8.5*. Consider the method

$$y_{n+2} - (1+\alpha)y_{n+1} + \alpha y_n = \tfrac{1}{2}h(1-\alpha)(f_{n+1} + f_n), \qquad -1 < \alpha < 1.$$

(i) Show that the order is independent of α.
(ii) Find the locus of $\partial\mathscr{R}_A$ in the form $\eta^2 = F(\xi)$, where $\hat{h} = \xi + i\eta$.
(iii) Hence sketch $\partial\mathscr{R}_A$ and show that it divides the complex plane into three regions; ascertain which of these regions are regions of absolute stability.
(iv) Deduce that the interval of absolute stability is independent of α.
(v) Construct a suitable two-dimensional linear constant coefficient problem and use it to devise and carry out a numerical experiment to corroborate your findings in (iii).

3.8.6*. A consistent linear 2-step method for the numerical solution of the *second-order* problem

$$y'' = f(x, y), \quad y(a) = \eta, \quad y'(a) = \tilde{\eta} \tag{1}$$

is defined by

$$y_{n+2} - 2y_{n+1} + y_n = h^2[\beta f_{n+2} + (1-2\beta)f_{n+1} + \beta f_n]. \tag{2}$$

We are interested in the question of whether the numerical solution $\{y_n\}$ given by (2) is periodic when the exact solution $y(x)$ of (1) is periodic. Accordingly, we choose as test equation the scalar equation

$$y'' = -\mu^2 y, \tag{3}$$

whose solutions are periodic, of period $2\pi/\mu$. In a development analogous to linear stability theory, we say that (2) has an *interval of periodicity* $(0, H_0^2)$, where $H = h\mu$, if the numerical solution of the difference equation resulting from applying (2) to (3) is periodic for all $H^2 \in (0, H_0^2)$.

(i) Find the interval of periodicity of (2) in each of the cases $\beta = 0$, $\beta = 5/6$ and $\beta = 1/(2 - 2\cos\varphi)$, $0 < \varphi < 2\pi$
(ii) If (2) is applied to (3) with a steplength h for which $H^2 \in (0, H_0^2)$, show that the numerical solution has approximately the correct period in the following sense. Let h be such that there exists an integer m such that $mh = 2\pi/\mu$; then show that

$$y_{n+m} = y_n + O(h^{p+1}) \qquad \text{for all } n,$$

where

$$\mathscr{L}[z(x); h] := z(x+2h) - 2z(x+h) + z(x) - h^2[\beta z''(x+2h) + (1-2\beta)z''(x+h) + \beta z''(x)] = O(h^{p+2}).$$

(*Hint*: Put $z(x) = \exp(i\mu x)$ in $\mathscr{L}[z(x); h]$.)

3.9 ADAMS METHODS IN BACKWARD DIFFERENCE FORM

Adams methods constitute a sub-family of linear multistep methods defined by

$$y_{n+k} - y_{n+k-1} = h\sum_{j=0}^{k} \beta_j f_{n+j}. \tag{3.85}$$

These methods have a long history, the explicit Adams–Bashforth methods having been first introduced in a numerical investigation of capillary attraction (Bashforth and Adams, 1883); the implicit Adams–Moulton methods first appeared in connection with problems of ballistics (Moulton, 1926). Today they still remain easily the most popular family of linear multistep methods, and form the basis of almost all predictor–corrector codes for non-stiff initial value problems.

There are good reasons for this popularity. Firstly, in comparison with many other families of linear multistep methods, Adams methods have good regions of absolute stability; this is to be expected since as $h \to 0$ the roots r_s of the stability polynomial satisfy $r_s \to \zeta_s = 0, s = 2, 3, \dots, k$. Secondly, the Adams methods have a definite advantage in the situation where the steplength is changed during the computation. We shall discuss the implementation of step changes in the next chapter, but it is clear that when the steplength is changed there is a problem, in that the back values are no longer at the appropriate values of x. One solution is to use interpolation to establish the necessary back values, and for general linear multistep methods this would mean interpolation of the existing back values of y followed by function evaluations to obtain the back values of f. For Adams methods, there is clearly never a need to interpolate the back values of y, and direct interpolation of the back values of f is enough. Lastly, Adams methods are capable of being expressed in terms of backward differences in a form that greatly eases the problems of implementing them in an automatic code; we now derive these alternative forms. In this context, it is convenient to rewrite (3.85) in the equivalent form

$$y_{n+1} - y_n = h \sum_{j=0}^{k} \beta_j f_{n+j-k+1}. \tag{3.86}$$

We start by considering the explicit Adams–Bashforth methods. In the next chapter, we need to distinguish between implicit and explicit linear multistep methods which appear in the same context; we do this by attaching the superscript * to all symbols relating to explicit methods. It is thus appropriate to do this for all symbols relating to the Adams–Bashforth methods. Recall from §3.3 that certain classes of linear multistep methods (including the class of Adams methods) could be derived by a process of polynomial interpolation. Analogously to (3.19), we consider the identity

$$y(x_{n+1}) - y(x_n) = \int_{x_n}^{x_{n+1}} y'(x)\,\mathrm{d}x. \tag{3.87}$$

We replace $y'(x)$ by $f(x, y(x))$ and seek a polynomial interpolant of the data

$$(x_n, f_n), \quad (x_{n-1}, f_{n-1}), \dots, \quad (x_{n-k+1}, f_{n-k+1}).$$

By §1.10, such an interpolant, in terms of backward differences, is given by

$$I^*_{k-1}(x) = I^*_{k-1}(x_n + rh) =: P^*_{k-1}(r) = \sum_{i=0}^{k-1} (-1)^i \binom{-r}{i} \nabla^i f_n. \tag{3.88}$$

Approximating the integrand on the right side of (3.87) by $I^*_{k-1}(x)$ and proceeding as in §3.3, we obtain

$$y_{n+1} - y_n = \int_0^1 P_{k-1}^*(r)h\,dr = h\sum_{i=0}^{k-1} \gamma_i^* \nabla^i f_n, \tag{3.89}$$

where

$$\gamma_i^* = (-1)^i \int_0^1 \binom{-r}{i} dr. \tag{3.90}$$

It is important to note that the γ_i^* are independent of k. We could use (3.90) to evaluate the $\gamma_i^*, i = 0, 1, 2, \ldots$, but there is a more constructive way to proceed. We seek a *generating function* for the γ_i^*, that is, a function of a dummy variable t which, when expanded in powers of t will have the γ_i^* as coefficients. That is, we seek a function $G^*(t)$ such that

$$G^*(t) = \sum_{i=0}^{\infty} \gamma_i^* t^i = \sum_{i=0}^{\infty} (-t)^i \int_0^1 \binom{-r}{i} dr = \int_0^1 \left[\sum_{i=0}^{\infty} (-t)^i \binom{-r}{i}\right] dr.$$

The integrand on the furthest right side can be recognized as the expansion of the function $(1-t)^{-r}$, whose integral with respect to r is $-(1-t)^{-r}/\ln(1-t)$. Hence,

$$G^*(t) = \frac{1}{\ln(1-t)}\left[\frac{-1}{1-t} + 1\right] = \frac{-t}{(1-t)\ln(1-t)} \tag{3.91}$$

which we rewrite in the form

$$G^*(t)\left[\frac{-\ln(1-t)}{t}\right] = \frac{1}{1-t}. \tag{3.92}$$

Now

$$\frac{-\ln(1-t)}{t} = 1 + \frac{t}{2} + \frac{t^2}{3} + \frac{t^3}{4} + \cdots$$

and

$$\frac{1}{1-t} = 1 + t + t^2 + t^3 + \cdots$$

and it follows form (3.92) that the γ_i^* are given by

$$(\gamma_0^* + \gamma_1^* t + \gamma_2^* t^2 + \gamma_3^* t^3 + \cdots)\left(1 + \frac{t}{2} + \frac{t^2}{3} + \frac{t^3}{4} + \cdots\right) = 1 + t + t^2 + t^3 + \cdots.$$

Equating coefficients of t^i gives the following rèlation, from which the γ_i^* can be readily calculated:

$$\gamma_i^* + \frac{\gamma_{i-1}^*}{2} + \frac{\gamma_{i-2}^*}{3} + \cdots + \frac{\gamma_0^*}{i+1} = 1, \qquad i = 0, 1, 2, \ldots.$$

The first few γ_i^* are easily seen to be

$$\gamma_0^* = 1, \qquad \gamma_1^* = \tfrac{1}{2}, \qquad \gamma_2^* = \tfrac{5}{12}, \qquad \gamma_3^* = \tfrac{3}{8}.$$

Thus the family of Adams–Bashforth methods can be written as

$$y_{n+1} - y_n = h(f_n + \tfrac{1}{2}\nabla f_n + \tfrac{5}{12}\nabla^2 f_n + \tfrac{3}{8}\nabla^3 f_n + \cdots). \tag{3.93}$$

Truncating the series on the right side after k terms and expanding the backward differences in terms of function values, gives the following:

$$k = 1: \quad y_{n+1} - y_n = hf_n$$

$$k = 2: \quad y_{n+1} - y_n = \frac{h}{2}(3f_n - f_{n-1})$$

$$k = 3: \quad y_{n+1} - y_n = \frac{h}{12}(23f_n - 16f_{n-1} + 5f_{n-2})$$

$$k = 4: \quad y_{n+1} - y_n = \frac{h}{24}(55f_n - 59f_{n-1} + 37f_{n-2} - 9f_{n-3}).$$

These are the standard k-step Adams–Bashforth methods, $k = 1, 2, 3, 4$, but in the form (3.86) rather than the form (3.85).

The importance of the fact that the γ_i^* are independent of k is now clear. By storing only the four numbers γ_i^*, $i = 0, 1, 2, 3$, we effectively store all four k-step Adams–Bashforth methods with $k = 1, 2, 3, 4$. Further, if we wish to replace a k-step Adams–Bashforth method by a $(k - 1)$-step, we merely drop the last term in the series on the right side of (3.93); if we wish to replace it by a $(k + 1)$-step method, we add an extra term, constructing the additional backward difference from the previously calculated values. This ability easily to change the stepnumber (and therefore the order) is an essential property of the algorithms we shall discuss in the next chapter.

We have established that the k-step Adams–Bashforth method is given by

$$y_{n+1} - y_n = h \sum_{i=0}^{k-1} \gamma_i^* \nabla^i f_n.$$

The difference between the values for y_{n+1} given by the $(k + 1)$-step and the k-step Adams–Bashforth methods is therefore

$$h\gamma_k^* \nabla^k f_n = h^{k+1}\gamma_k^* y^{(k+1)}(x_n) + 0(h^{k+2}),$$

(by §1.2, using the fact that $y' = f$), a result which strongly suggests that the k-step Adams–Bashforth method has order k and error constant γ_k^*. We can establish this more formally as follows.

Let $\mathscr{L}^*$ be the linear difference operator associated with the k-step Adams–Bashforth method, and let $z(x)$ be a sufficiently differentiable function. Then, by (3.13),

$$\mathscr{L}^*[z(x); h] = z(x_{n+1}) - z(x_n) - h \sum_{i=0}^{k-1} \gamma_i^* \nabla^i z'(x_n),$$

where $\nabla z'(x_n) = z'(x_n) - z'(x_{n-1})$, etc. Hence we may write

$$\mathscr{L}^*[z(x); h] = \int_{x_n}^{x_{n+1}} z'(x)\,\mathrm{d}x - h \sum_{i=0}^{k-1} \gamma_i^* \nabla^i z'(x_n).$$

Let $\tilde{I}_{k-1}(x) = \tilde{I}_{k-1}(x_n + rh) =: \tilde{P}_{k-1}(r)$ interpolate the data

$$(x_n, z'(x_n)), (x_{n-1}, z'(x_{n-1})), \ldots, (x_{n-k+1}, z'(x_{n-k+1})).$$

Then

$$\tilde{P}_{k-1}(r) = \sum_{i=0}^{k-1} (-1)^i \binom{-r}{i} \nabla^i z'(x_n)$$

and, by (1.31), the interpolation error is given by

$$z'(x) - \tilde{I}_{k-1}(x) = (-1)^k \binom{-r}{k} h^k \bar{z}^{(k+1)}(\xi(r)),$$

where $\xi(r)$ is in internal point of the smallest interval containing $x_n, x_{n-1}, \ldots, x_{n-k+1}$ and $x_n + rh$, and the bar over $z^{(k+1)}$ indicates that each component of $z^{(k+1)}$ is evaluated at a different value of $\xi(r)$. Hence

$$\mathscr{L}^*[z(x_n); h] = \int_0^1 \left[\sum_{i=0}^{k-1} (-1)^i \binom{-r}{i} \nabla^i z'(x_n) + (-1)^k \binom{-r}{k} h^k \bar{z}^{(k+1)}(\xi(r)) \right] h\,dr - h \sum_{i=0}^{k-1} \gamma_i^* \nabla^i z'(x_n).$$

It follows from (3.90) that

$$\mathscr{L}^*[z(x_n); h] = h^{k+1} \int_0^1 (-1)^k \binom{-r}{k} \bar{z}^{(k+1)}(\xi(r))\,dr.$$

Noting that $(-1)^k \binom{-r}{k}$ does not change sign for $r \in [0, 1]$, we can apply the generalized mean value theorem for integrals (see §1.3) to obtain

$$\begin{aligned}\mathscr{L}^*[z(x_n); h] &= h^{k+1} \bar{z}^{(k+1)}(\hat{\xi}) \int_0^1 (-1)^k \binom{-r}{k} dr \\ &= h^{k+1} \gamma_k^* \bar{z}^{(k+1)}(\hat{\xi}),\end{aligned}$$

by (3.90).

It follows that the k-step Adams–Bashforth method has order k and error constant γ_k^*.

By an analogous approach, the implicit Adams–Moulton methods can also be expressed in terms of backwards differences of f. We start from the same identity (3.87), but this time, after replacing y' by f, we seek a (vector) polynomial interpolant of the data

$$(x_{n+1}, f_{n+1}), \quad (x_n, f_n), \ldots, \quad (x_{n+k-1}, f_{n+k-1}).$$

Note that there are now $k+1$ data points rather than k, so that the appropriate interpolant, replacing (3.88), is

$$I_k(x) = I_k(x_{n+1} + rh) =: P_k(r) = \sum_{i=0}^{k} (-1)^i \binom{-r}{i} \nabla^i f_{n+1}.$$

We obtain in place of (3.89) and (3.90)

$$y_{n+1} - y_n = \int_{-1}^{0} P_k(r)h\,\mathrm{d}r = h\sum_{i=0}^{k} \gamma_i \nabla^i f_{n+1}$$

where

$$\gamma_i = (-1)^i \int_{-1}^{0} \binom{-r}{i} \mathrm{d}r.$$

The generating function $G(t)$ for the γ_i is given by

$$G(t) = \sum_{i=0}^{\infty} \gamma_i t^i = \sum_{i=0}^{\infty} (-t)^i \int_{-1}^{0} \binom{-r}{i} \mathrm{d}r = \int_{-1}^{0} \left[\sum_{i=0}^{\infty} (-t)^i \binom{-r}{i}\right] \mathrm{d}r.$$

The integrand is the same as in the case of the Adams–Bashforth methods, but the limits of integration are different. We easily find that

$$G(t) = \frac{-t}{\ln(1-t)} \tag{3.94}$$

and it follows that

$$(\gamma_0 + \gamma_1 t + \gamma_2 t^2 + \gamma_3 t^3 + \cdots)\left(1 + \frac{t}{2} + \frac{t^2}{3} + \frac{t^3}{4} + \cdots\right) = 1$$

whence

$$\gamma_i + \frac{\gamma_{i-1}}{2} + \frac{\gamma_{i-2}}{3} + \cdots + \frac{\gamma_0}{i+1} = \begin{cases} 1 & \text{if } i = 0 \\ 0 & \text{if } i = 1, 2, \ldots. \end{cases}$$

The first few γ_i are seen to be

$$\gamma_0 = 1, \quad \gamma_1 = -\tfrac{1}{2}, \quad \gamma_2 = -\tfrac{1}{12}, \quad \gamma_3 = -\tfrac{1}{24}, \quad \gamma_4 = -\tfrac{19}{720}.$$

Thus the family of Adams–Moulton methods can be written as

$$y_{n+1} - y_n = h(f_{n+1} - \tfrac{1}{2}\nabla f_{n+1} - \tfrac{1}{12}\nabla^2 f_{n+1} - \tfrac{1}{24}\nabla^3 f_{n+1} - \tfrac{19}{720}\nabla^4 f_{n+1} + \cdots). \tag{3.95}$$

Truncating the series on the right side after $k+1$ terms (contrast with truncating after k terms in (3.93)) and expanding the backward differences in terms of function values gives the following:

$$k = 1: \quad y_{n+1} - y_n = \frac{h}{2}(f_{n+1} + f_n)$$

$$k = 2: \quad y_{n+1} - y_n = \frac{h}{12}(5f_{n+1} + 8f_n - f_{n-1})$$

$$k = 3: \quad y_{n+1} - y_n = \frac{h}{24}(9f_{n+1} + 19f_n - 5f_{n-1} + f_{n-2})$$

$$k = 4: \quad y_{n+1} - y_n = \frac{h}{720}(251f_{n+1} + 646f_n - 264f_{n-1} + 106f_{n-2} - 19f_{n-3}).$$

These are the standard k-step Adams–Moulton methods, $k = 1, 2, 3, 4$ in the form (3.86).

Note that, formally, we do not include in the class of Adams–Moulton methods the method obtained by truncating the right side of (3.95) after just one term, that is, the method

$$y_{n+1} - y_n = hf_{n+1},$$

known as the *Backward Euler method*. It could be argued that to do so would be confusing, since we would then have two 1-step Adams–Moulton methods, the Backward Euler method and the Trapezoidal Rule. Nevertheless, there is advantage in regarding the Backward Euler method as the unique Adams–Moulton method of order 1.

By an argument exactly analogous to that used for the Adams–Bashforth methods, we can establish that the k-step Adams–Moulton method has order $k + 1$ and error constant γ_{k+1}. (Note that order $k + 1$, rather than k, is consistent with the fact that we truncated the series in (3.95) after $k + 1$ terms, whereas that in (3.93) was truncated after k terms.)

We can summarize the results we have obtained as follows, where p is the order and C_{p+1} the error constant:

$$\left.\begin{array}{llll} k\text{-step Adams–Bashforth:} & y_{n+1} - y_n = h\sum_{i=0}^{k-1} \gamma_i^* \nabla^i f_n, & p^* = k, & C_{k+1}^* = \gamma_k^* \\ k\text{-step Adams–Moulton:} & y_{n+1} - y_n = h\sum_{i=0}^{k} \gamma_i \nabla^i f_{n+1}, & p = k + 1, & C_{k+2} = \gamma_{k+1}. \end{array}\right\} \tag{3.96}$$

We conclude this section by observing that the history of the application of Adams methods has a somewhat ironic flavour. In the pre-computer days computations had to be done by hand, with only a (non-programmable) mechanical calculator to help with the arithmetic. It was standard practice in all step-by-step computations to keep up-dating a table of the differences (including higher differences) of the numerical solution, since this was a good way of spotting the inevitable arithmetic errors that crept in. (Such a difference table amplifies errors in an identifiable pattern.) Thus, difference tables were an accepted adjunct to *all* step-by-step computations. In the case when an initial value problem was being solved by an Adams method, it was natural to use the backward difference form, since the differences were all to hand. If one computed with, say, a kth-order Adams–Bashforth, then the differences $\nabla^i f_n$, $i = 0, 1, \ldots, k - 1$, were utilized in the method, and the difference $\nabla^k f_n$ gave an indication of the local accuracy. If, as the computation proceeded, the differences $\nabla^k f_n$ became too large, one would simply start adding the term $h\gamma_k^* \nabla^k f_n$ to the right side of the method; if $\nabla^{k-1} f_n$ became too small, one would drop the term $h\gamma_{k-1}^* \nabla^{k-1} f_n$ from the right side. In other words, the kth-order method would be replaced by a $(k + 1)$th- or a $(k - 1)$th-order method as the occasion demanded. When programmable computers first became available such arbitrary changes of method were somewhat frowned upon, and it was accepted practice to compute with a fixed method and rely on changes of steplength (exceedingly unpleasant to implement in a hand computation) to control accuracy. The irony is that it is now accepted that the key to high efficiency in modern codes for initial value problems is the ability to vary both the steplength *and* the order of the method. Such codes almost always use Adams methods (in predictor–corrector form), and the implementation of order changes is precisely that employed in the old days of hand computation. *Plus ça change, plus c'est la même chose*!

3.10 PROPERTIES OF THE ADAMS COEFFICIENTS

In the next chapter we shall be much concerned with the important role that Adams methods play in predictor–corrector theory. In that context, we shall need a number of results concerning the coefficients which define the Adams methods, and it is convenient to gather these together in this section. Some of these properties are needed to enable us to move from backward difference form to standard form, and call for a little additional notation.

Let the kth-order Adams–Bashforth method in standard form be defined by the characteristic polynomials $\rho_k^*(r), \sigma_k^*(r)$ and the kth-order Adams–Moulton method by $\rho_k(r)$, $\sigma_k(r)$. It is important to note that the subscript k denotes the *order*, *not the stepnumber*, of the method. For $k \geqslant 2$, the kth-order Adams–Bashforth method has stepnumber k and is explicit whereas the kth-order Adams Moulton has stepnumber $k-1$ and is implicit. Hence for $k \geqslant 2$,

$$\begin{gathered} \rho_k^*(r) = r^k - r^{k-1}, \qquad \rho_k(r) = r^{k-1} - r^{k-2}, \\ \sigma_k^*(r) \text{ and } \sigma_k(r) \text{ have degree } k-1. \end{gathered} \tag{3.97}$$

Thus, for example,

$$\begin{aligned} \rho_2^*(r) &= r^2 - r, & \sigma_2^*(r) &= \tfrac{1}{2}(3r-1), \\ \rho_2(r) &= r - 1, & \sigma_2(r) &= \tfrac{1}{2}(r+1), \end{aligned}$$

which define the second-order Adams–Bashforth and Adams–Moulton methods. Note the anomalous situation when $k = 1$; we have already agreed to regard the Backward Euler method as the Adams–Moulton method of order 1, but it does not satisfy (3.97). For this reason, some of the properties we are about to list hold only for $k \geqslant 2$.

Property 1 $\gamma_j^* = \sum_{i=0}^{j} \gamma_i, \qquad j = 0, 1, 2, \ldots$

Property 2 $\gamma_j^* - \gamma_{j-1}^* = \gamma_j, \qquad j = 1, 2, 3, \ldots$

Property 3 $\sum_{j=0}^{k-1} (\gamma_j \nabla^j f_{n+1} - \gamma_j^* \nabla^j f_n) = \gamma_{k-1}^* \nabla^k f_{n+1}, \qquad k \geqslant 1$

Property 4 The leading coefficient in $\sigma_k(r)$ is γ_{k-1}^*, $\quad k \geqslant 1$

Property 5 $\begin{cases} \sigma_{k+1}^*(r) = r\sigma_k^*(r) + \gamma_k^*(r-1)^k, & k \geqslant 1 \\ \sigma_{k+1}(r) = r\sigma_k(r) + \gamma_k(r-1)^k, & k \geqslant 2 \end{cases}$

Property 6 $\gamma_{k-1}^* \sigma_{k+1}(r) = \gamma_k^* r\sigma_k(r) - \gamma_k \sigma_k^*(r), \qquad k \geqslant 2$

Property 7 $\sigma_{k+1}^*(r) = \sigma_{k+1}(r) + (r-1)\sigma_k^*(r), \qquad k \geqslant 1$

Property 8 $r\sigma_k(r) - \sigma_k^*(r) = \gamma_{k-1}^*(r-1)^k, \qquad k \geqslant 2.$

Proof of Properties 1 and 2 Recall the generating functions $G^*(t)$ and $G(t)$ for the coefficients $\{\gamma_i^*\}$ and $\{\gamma_i\}$, defined by (3.91) and (3.94). It follows that

$$G(t)/(1-t) = G^*(t),$$

whence

$$(\gamma_0 + \gamma_1 t + \gamma_2 t^2 + \cdots)(1 + t + t^2 + \cdots) = \gamma_0^* + \gamma_1^* t + \gamma_2^* t^2 + \cdots.$$

On equating coefficients of t^j, Property 1 results. Property 2 follows immediately. □

Proof of Property 3

$$f_n = f_{n+1} - (f_{n+1} - f_n) = (1 - \nabla) f_{n+1}.$$

The left side of Property 3 can thus be written as

$$\sum_{j=0}^{k-1} [\gamma_j \nabla^j f_{n+1} - \gamma_j^* \nabla^j (1-\nabla) f_{n+1}] = S(\nabla) f_{n+1}$$

where

$$S(\nabla) := \sum_{j=0}^{k-1} [\gamma_j^* \nabla^{j+1} + (\gamma_j - \gamma_j^*)\nabla^j]$$

$$= \gamma_{k-1}^* \nabla^k + \sum_{j=0}^{k-2} \gamma_j^* \nabla^{j+1} + \sum_{j=1}^{k-1} (\gamma_j - \gamma_j^*)\nabla^j + \gamma_0 - \gamma_0^*.$$

Put $j = i - 1$ in the first summation and $j = i$ in the second to get

$$S(\nabla) = \gamma_{k-1}^* \nabla^k + \sum_{i=1}^{k-1} [\gamma_{i-1}^* \nabla^i + (\gamma_i - \gamma_i^*)\nabla^i] + \gamma_0 - \gamma_0^*.$$

By Property 2, the second term on the right side is zero, and the third term vanishes since $\gamma_0 = \gamma_0^*$ $(=1)$. Hence

$$S(\nabla) = \gamma_{k-1}^* \nabla^k,$$

and Property 3 is proved. □

Proof of Property 4 By writing the kth-order Adams–Bashforth and Adams–Moulton methods in standard and in backward difference form and equating the results, we obtain

$$\left.\begin{aligned} \sigma_k^*(E) f_{n-k+1} &= \sum_{j=0}^{k-1} \gamma_j^* \nabla^j f_n, \qquad k \geqslant 1 \\ \sigma_k(E) f_{n-k+1} &= \sum_{j=0}^{k-1} \gamma_j \nabla^j f_n, \qquad k \geqslant 2. \end{aligned}\right\} \tag{3.98}$$

The leading coefficient in $\sigma_k(r)$ is the coefficient of f_n on the left side of the second of (3.98) which, from the right side, is $\sum_{j=0}^{k-1} \gamma_j = \gamma_{k-1}^*$, by Property 1. Hence Property 4 is established for $k \geqslant 2$; that it also holds for $k = 1$ is readily checked. □

Proof of Property 5 Since $\nabla = 1 - E^{-1}$, the first of (3.98) may be written in the form

$$\sigma_k^*(E) f_{n-k+1} = \sum_{j=0}^{k-1} \gamma_j^* (1 - E^{-1})^j E^{k-1} f_{n-k+1}, \qquad k \geqslant 1. \tag{3.99}$$

Replacing k by $k+1$ in (3.99) gives

$$\sigma_{k+1}^*(E) f_{n-k} = \sum_{j=0}^{k} \gamma_j^* (1 - E^{-1})^j E^k f_{n-k}$$

or

$$E^{-1} \sigma_{k+1}^*(E) f_{n-k+1} = \sum_{j=0}^{k} \gamma_j^* (1 - E^{-1})^j E^{k-1} f_{n-k+1}.$$

On subtracting (3.99) we obtain

$$[E^{-1} \sigma_{k+1}^*(E) - \sigma_k^*(E)] f_{n-k+1} = \gamma_k^* (1 - E^{-1})^k E^{k-1} f_{n-k+1}$$

whence

$$\sigma_{k+1}^*(E) - E\sigma_k^*(E) = \gamma_k^* (E - 1)^k,$$

which establishes the first part of Property 5. The proof of the second part is identical except that, in view of (3.98), the result holds only for $k \geqslant 2$. □

Proof of Property 6 By (3.98) and the fact that $E^{-1} = 1 - \nabla$, we have, for $k \geqslant 2$

$$\begin{aligned}
&[\gamma_k^* \sigma_k(E) - E^{-1} \gamma_k \sigma_k^*(E)] f_{n-k+1} \\
&= \left[\gamma_k^* \sum_{j=0}^{k-1} \gamma_j \nabla^j - \gamma_k \sum_{j=0}^{k-1} \gamma_j^* (1 - \nabla) \nabla^j \right] f_n \\
&= \left[\gamma_k^* \sum_{j=0}^{k-1} \gamma_j \nabla^j - \gamma_k \left\{ \gamma_0^* + \sum_{j=1}^{k-1} (\gamma_j^* - \gamma_{j-1}^*) \nabla^j - \gamma_{k-1}^* \nabla^k \right\} \right] f_n \\
&= \left[\gamma_k^* \sum_{j=0}^{k-1} \gamma_j \nabla^j - \gamma_k \sum_{j=0}^{k-1} \gamma_j \nabla^j + \gamma_k \gamma_{k-1}^* \nabla^k \right] f_n \qquad \text{(by Property 2 and } \gamma_0^* = \gamma_0) \\
&= \left[\gamma_{k-1}^* \sum_{j=0}^{k-1} \gamma_j \nabla^j + \gamma_k \gamma_{k-1}^* \nabla^k \right] f_n \qquad \text{(by Property 2)} \\
&= \gamma_{k-1}^* \sum_{j=0}^{k} \gamma_j \nabla^j f_n = \gamma_{k-1}^* \sigma_{k+1}(E) f_{n-k} \qquad \text{(by (3.98))}.
\end{aligned}$$

Hence

$$[E\gamma_k^* \sigma_k(E) - \gamma_k \sigma_k^*(E)] f_{n-k+1} = \gamma_{k-1}^* \sigma_{k+1}(E) f_{n-k+1}$$

which establishes Property 6. □

Proof of Property 7 Eliminating $(r-1)^k$ from the two identities in Property 5 gives

$$\gamma_k^* \sigma_{k+1}(r) - \gamma_k \sigma_{k+1}^*(r) = r[\gamma_k^* \sigma_k(r) - \gamma_k \sigma_k^*(r)], \qquad k \geqslant 2.$$

Subtracting this result from Property 6 and using Property 2 gives

$$-\gamma_k\sigma_{k+1}(r) + \gamma_k\sigma^*_{k+1}(r) = (r-1)\gamma_k\sigma^*_k(r), \qquad k \geqslant 2$$

which establishes Property 7 in the case $k \geqslant 2$; that the property also holds for $k = 1$ is readily checked. □

Proof of Property 8 Properties 5 and 6 imply that

$$\gamma^*_{k-1}[r\sigma_k(r) + \gamma_k(r-1)^k] = \gamma^*_k r\sigma_k(r) - \gamma_k\sigma^*_k(r), \qquad k \geqslant 2$$

whence, by Property 2

$$\gamma^*_{k-1}\gamma_k(r-1)^k = \gamma_k r\sigma_k(r) - \gamma_k\sigma^*_k(r).$$

Dividing through by γ_k gives Property 8. □

It is of interest to note that Property 5, a recurrence relationship for the polynomials $\sigma^*_k(r)$, $k = 1, 2, \ldots$ and $\sigma_k(r)$, $k = 2, 3, \ldots$, provides a very efficient means of generating the Adams methods in standard form from the coefficients $\{\gamma^*_j\}$, $\{\gamma_j\}$ which define the same methods in backward difference form. Recall the first few γ^*_k, γ_k:

k	0	1	2	3	4
γ^*_k	1	$\frac{1}{2}$	$\frac{5}{12}$	$\frac{3}{8}$	
γ_k	1	$-\frac{1}{2}$	$-\frac{1}{12}$	$-\frac{1}{24}$	$-\frac{19}{720}$

The Adams–Bashforth methods in standard form are generated as shown below.

$\sigma^*_1(r)$ (Euler's Rule)				1
$r\sigma^*_1(r)$			r	
$\gamma^*_1(r-1)$			$\frac{1}{2}r$	$-\frac{1}{2}$
$\sigma^*_2(r)$			$\frac{3}{2}r$	$-\frac{1}{2}$
$r\sigma^*_2(r)$		$\frac{3}{2}r^2$	$-\frac{1}{2}r$	
$\gamma^*_2(r-1)^2$		$\frac{5}{12}r^2$	$-\frac{5}{6}r$	$+\frac{5}{12}$
$\sigma^*_3(r)$		$\frac{23}{12}r^2$	$-\frac{4}{3}r$	$+\frac{5}{12}$
$r\sigma^*_3(r)$	$\frac{23}{12}r^3$	$-\frac{4}{3}r^2$	$+\frac{5}{12}r$	
$\gamma^*_3(r-1)^3$	$\frac{3}{8}r^3$	$-\frac{9}{8}r^2$	$+\frac{9}{8}r$	$-\frac{3}{8}$
$\sigma^*_4(r)$	$\frac{55}{24}r^3$	$-\frac{59}{24}r^2$	$+\frac{37}{24}r$	$-\frac{3}{8}$
⋮				

Similarly, the Adams–Moulton methods are generated as shown below.

$\sigma_1^*(r)$ (Trapezoidal Rule)				$\frac{1}{2}r$	$+\frac{1}{2}$
$r\sigma_2(r)$			$\frac{1}{2}r^2$	$+\frac{1}{2}r$	
$\gamma_2(r-1)^2$			$-\frac{1}{12}r^2$	$+\frac{1}{6}r$	$-\frac{1}{12}$
$\sigma_3(r)$			$\frac{5}{12}r^2$	$+\frac{2}{3}r$	$-\frac{1}{12}$
$r\sigma_3(r)$		$\frac{5}{12}r^3$	$+\frac{2}{3}r^2$	$-\frac{1}{12}r$	
$\gamma_3(r-1)^3$		$-\frac{1}{24}r^3$	$+\frac{1}{8}r^2$	$-\frac{1}{8}r$	$+\frac{1}{24}$
$\sigma_4(r)$		$\frac{3}{8}r^3$	$+\frac{19}{24}r^2$	$-\frac{5}{24}r$	$+\frac{1}{24}$
$r\sigma_4(r)$	$\frac{3}{8}r^4$	$+\frac{19}{24}r^3$	$-\frac{5}{24}r^2$	$+\frac{1}{24}r$	
$\gamma_4(r-1)^4$	$-\frac{19}{720}r^4$	$+\frac{19}{180}r^3$	$-\frac{19}{120}r^2$	$+\frac{19}{180}r$	$-\frac{19}{720}$
$\sigma_5(r)$	$\frac{251}{720}r^4$	$+\frac{323}{360}r^3$	$-\frac{11}{30}r^2$	$+\frac{53}{360}r$	$-\frac{19}{720}$
$\vdots$					

3.11 GENERAL LINEAR MULTISTEP METHODS IN BACKWARD DIFFERENCE FORM

We have seen that *all* Adams–Bashforth methods of order up to k can be generated in backward difference form if the $k+1$ numbers γ_i^*, $i=0,1,\dots,k$ are known; likewise, *all* Adams–Moulton methods of order up to $k+1$ can be generated if the $k+1$ numbers γ_i, $i=0,1,2,\dots,k$ are known. It is natural to ask whether general linear multistep methods can similarly be compactly expressed in terms of backward differences. This is indeed possible, and turns out to be an efficient way of computing the coefficients of classes of linear multistep methods.

The key is to extend the class of Adams methods to the more general form

$$y_{n+1}-y_n=h(\gamma_0^s f_{n+s}+\gamma_1^s\nabla f_{n+s}+\gamma_2^s\nabla^2 f_{n+s}+\cdots),\quad s=0,1,2,\dots. \tag{3.100}$$

Clearly, putting $s=0$ in (3.100) gives the class of Adams–Bashforth methods (where $\gamma_i^*=\gamma_i^0$), and putting $s=1$ gives the class of Adams–Moulton methods (where $\gamma_i \doteq \gamma_i^1$). For $s>1$, the methods retain the Adams left side, but are 'over-implicit'. The technique for finding the coefficients $\{\gamma_i^s\}$ is a straightforward modification of that used to find $\{\gamma_i^*\}$ and $\{\gamma_i\}$ for the conventional Adams methods. Starting from the identity (3.87), we now seek an interpolant for f on the set of $k+s$ data points

$$(x_{n+s},f_{n+s}),(x_{n+s-1},f_{n+s-1}),\dots,(x_n,f_n),\dots,(x_{n-k+1},f_{n-k+1}).$$

The required interpolant, of degree $k+s-1$, is

$$I_{k+s-1}(x) = I_{k+s-1}(x_{n+s} + rh) =: P^s_{k+s-1}(r) = \sum_{i=0}^{k+s-1} (-1)^i \binom{-r}{i} \nabla^i f_{n+s}$$

and in place of (3.89) and (3.90) we obtain

$$y_{n+1} - y_n = \int_{-s}^{1-s} P^s_{k+s-1}(r) h \, \mathrm{d}r = h \sum_{i=0}^{k+s-1} \gamma_i^s \nabla^i f_{n+s}$$

where

$$\gamma_i^s = (-i)^i \int_{-s}^{1-s} \binom{-r}{i} \mathrm{d}r. \tag{3.101}$$

We note that the integrands in (3.90) and (3.101) are the same; only the limits of integration differ. The argument following (3.90) holds, the only change being that the generating function $G^s(t)$, defined by

$$G^s(t) = \sum_{i=0}^{\infty} \gamma_i^s t^i$$

now simplifies to

$$\begin{aligned} G^s(t) &= \left. \frac{-(1-t)^{-r}}{\ln(1-t)} \right|_{-s}^{1-s} \\ &= \frac{-t(1-t)^{s-1}}{\ln(1-t)}. \end{aligned} \tag{3.102}$$

Rewriting this in the form

$$\frac{-\ln(1-t)}{t} G^s(t) = (1-t)^{s-1}$$

or

$$(\gamma_0^s + \gamma_1^s t + \gamma_2^s t^2 + \cdots)(1 + t/2 + t/3 + \cdots) = (1-t)^{s-1} \tag{3.103}$$

and equating coefficients of powers of t enables us to compute the coefficients γ_i^s. However, it is easier to use an obvious generalization of Property 1 of §3.10. It follows from (3.102) that

$$G^s(t) = \frac{1}{1-t} G^{s+1}(t), \qquad s = 0, 1, \ldots$$

or

$$\gamma_0^s + \gamma_1^s t + \gamma_2^s t^2 + \cdots = (1 + t + t^2 + \cdots)(\gamma_0^{s+1} + \gamma_1^{s+1} t + \gamma_2^{s+1} t^2 + \cdots).$$

Equating the coefficients of t^j gives

$$\gamma_j^s = \sum_{i=0}^{j} \gamma_i^{s+1}, \qquad j = 0, 1, 2, \ldots, s = 0, 1, 2, \ldots$$

whence

$$\gamma_j^s - \gamma_{j-1}^s = \gamma_j^{s+1}, \qquad j = 1, 2, 3, \ldots, s = 0, 1, 2, \ldots \tag{3.104}$$

Table 3.1 The Adams array: γ_j^s, $s = 0, 1, \dots, 7$

	$j=0$	$j=1$	$j=2$	$j=3$	$j=4$	$j=5$	$j=6$	$j=7$
$s=0$	1	$\frac{1}{2}$	$\frac{5}{12}$	$\frac{3}{8}$	$\frac{251}{720}$	$\frac{95}{288}$	$\frac{19087}{60480}$	$\frac{5257}{17280}$
$s=1$	1	$\frac{-1}{2}$	$\frac{-1}{12}$	$\frac{-1}{24}$	$\frac{-19}{720}$	$\frac{-3}{160}$	$\frac{-863}{60480}$	$\frac{-275}{24192}$
$s=2$	1	$\frac{-3}{2}$	$\frac{5}{12}$	$\frac{1}{24}$	$\frac{11}{720}$	$\frac{11}{1440}$	$\frac{271}{60480}$	$\frac{13}{4480}$
$s=3$	1	$\frac{-5}{2}$	$\frac{23}{12}$	$\frac{-3}{8}$	$\frac{-19}{720}$	$\frac{-11}{1440}$	$\frac{-191}{60480}$	$\frac{-191}{120960}$
$s=4$	1	$\frac{-7}{2}$	$\frac{53}{12}$	$\frac{-55}{24}$	$\frac{251}{720}$	$\frac{3}{160}$	$\frac{271}{60480}$	$\frac{191}{120960}$
$s=5$	1	$\frac{-9}{2}$	$\frac{95}{12}$	$\frac{-161}{24}$	$\frac{1901}{720}$	$\frac{-95}{288}$	$\frac{-863}{60480}$	$\frac{-13}{4480}$
$s=6$	1	$\frac{-11}{2}$	$\frac{149}{12}$	$\frac{-351}{24}$	$\frac{6731}{720}$	$\frac{-4277}{1440}$	$\frac{19087}{60480}$	$\frac{275}{24192}$
$s=7$	1	$\frac{-13}{2}$	$\frac{215}{12}$	$\frac{-649}{24}$	$\frac{17261}{720}$	$\frac{-1971}{160}$	$\frac{198721}{60480}$	$\frac{-5257}{17280}$

a direct generalization of Properties 1 and 2 of the Adams–Bashforth/Adams–Moulton methods. From (3.103), it is clear that

$$\gamma_0^s = 1, \quad s = 0, 1, 2, \dots,$$

so that (3.104), together with a knowledge of the Adams–Bashforth coefficients $\{\gamma_j^0\}$, enables us readily to write down a two-dimensional array of the coefficients γ_j^s, $s = 0, 1, 2, \dots, j = 0, 1, 2, \dots$; we shall christen this array the *Adams array*. The array is shown for $s, j = 0, 1, 2, \dots, 7$ in Table 3.1.

We note in passing that the columns of the Adams array, as opposed to the rows, possess a fair amount of structure arising from (3.104). Thus we note that the first $j+1$ entries of the $(j+1)$th column are symmetric if j is even, and antisymmetric if j is odd. Further, *all* of the entries in the $(j+1)$th column satisfy the following identity:

$$\sum_{i=0}^{j} (-1)^i \binom{j}{i} \gamma_j^{s+i} = 1, \qquad s = 0, 1, 2, \dots, j = 0, 1, 2, \dots.$$

Finally, we note the relationships between the main diagonal and the first row, between the diagonal above the main and the second row, etc.

The order p and error constant C_{p+1} of members of the class (3.100) can be established by a direct extension of the analysis given in §3.9 for Adams–Bashforth and Adams–Moulton methods. We are now in a position to define formally the *k-step s-Adams method* as follows:

$$\left.\begin{aligned} &\textbf{Case } s = 0: \quad y_{n+1} - y_n = h \sum_{i=0}^{k-1} \gamma_i^0 \nabla^i f_n, \qquad k \geqslant 1, \quad p = k, \quad C_{p+1} = \gamma_k^0 \\ &\textbf{Case } s \geqslant 1: \quad y_{n+1} - y_n = h \sum_{i=0}^{k} \gamma_i^s \nabla^i f_{n+s}, \qquad k \geqslant s, \quad p = k+1, \quad C_{p+1} = \gamma_{k+1}^s. \end{aligned}\right\} \quad (3.105)$$

(Note that the restriction $k \geqslant s$ when $s \geqslant 1$ is consistent with the fact that in §3.9 the Backward Euler method did not fit the Adams–Moulton pattern.)

We can use (3.105) to express general classes of linear multistep methods in backward difference form. In attempting to list the coefficients for the general class of linear multistep methods, one faces the practical problem of deciding how many parameters to include. If one attempts to include all possible methods, then the number of parameters becomes unmanageable; for example the *general* class of 6-step implicit zero-stable methods contains 12 parameters. Low-order methods of such classes are of little interest, and a reasonable compromise is to include just enough parameters to allow complete control of the location of the spurious roots of the first characteristic polynomial $\rho(\zeta)$. That is, we retain $k-1$ parameters in a k-step method. This results in explicit methods having order k (the maximum possible subject to zero-stability) and in implicit methods having order $k+1$ (the maximum possible subject to zero-stability, if k is odd; if k is even the maximum is $k+2$, see §3.4). It is convenient to choose these $k-1$ parameters in the following way: recalling that consistency demands that $\rho(\zeta)$ has a root at $+1$, we write $\rho(\zeta)$ for a k-step method in the form

$$\rho(\zeta) = (\zeta - 1)(\zeta^{k-1} + A_1\zeta^{k-2} + \cdots + A_{k-2}\zeta + A_{k-1}) \tag{3.106}$$

where the $A_j, j = 1, 2, \ldots, k-1$ are the parameters to be retained. Clearly, if $\rho(\zeta)$ is to satisfy the root condition, these parameters have to be chosen so that the polynomial

$$\hat{\rho}(\zeta) := \zeta^{k-1} + A_1\zeta^{k-2} + \cdots + A_{k-2}\zeta + A_{k-1}$$

has all its roots in or on the unit circle, has no multiple roots on the unit circle *and* does not have a root at $+1$. The left side of the linear k-step method becomes

$$\begin{aligned}\sum_{j=0}^{k} \alpha_j y_{n+j} &= y_{n+k} - y_{n+k-1} + A_1(y_{n+k-1} - y_{n+k-2}) + \cdots + A_{k-2}(y_{n+2} - y_{n+1}) \\ &\quad + A_{k-1}(y_{n+1} - y_n) &(3.107)\\ &= \sum_{s=0}^{k-1} A_s \nabla y_{n+k-s}, \qquad A_0 := 1, \qquad k \geqslant 2. &(3.108)\end{aligned}$$

(Note that the condition $k \geqslant 2$ is not restrictive; if $k = 1$, there are no free parameters, and the methods become the one-step Adams methods, explicit or implicit.)

Let $hR_n(k, s)$ denote the right side of the methods given by (3.105); that is, let

$$R_n(k, s) := \sum_{i=0}^{k-\omega} \gamma_i^s \nabla^i f_{n+s}, \qquad \omega = \begin{cases} 1 & \text{if } s = 0 \\ 0 & \text{if } s \geqslant 1. \end{cases} \tag{3.109}$$

The subscript n denotes that the method is being applied at x_n to give a value for y at x_{n+1}. Thus, for example, the k-step s-Adams method (with $s \geqslant 1$) shifted one steplength to the right is given by

$$\begin{aligned} y_{n+2} - y_{n+1} &= hR_{n+1}(k, s) \\ &= h\sum_{i=0}^{k} \gamma_i^s \nabla^i f_{n+1+s}. \end{aligned}$$

From (3.105), (3.107) and (3.109), the family of implicit linear k-step methods with $k-1$ free parameters A_i, $i=1,2,\ldots,k-1$, can be written in backward difference form as

$$\begin{aligned}&y_{n+k}-y_{n+k-1}+A_1(y_{n+k-1}-y_{n+k-2})+\cdots+A_{k-1}(y_{n+1}-y_n)\\&=h[R_{n+k-1}(k,1)+A_1R_{n+k-2}(k,2)+\cdots+A_{k-1}R_n(k,k)],\qquad k\geqslant 2.\end{aligned}$$

By (3.108) and (3.109), this class can be written more formally as

$$\sum_{s=0}^{k-1}A_s\nabla y_{n+1-s}=h\sum_{s=0}^{k-1}A_s\sum_{i=0}^{k}\gamma_i^{s+1}\nabla^i f_{n+1},\qquad A_0:=1,\quad k\geqslant 2,\tag{3.110}$$

where we have left-shifted the methods by $k-1$ steplengths (that is, replaced n by $n-k+1$) so that the class of methods is presented in a form analogous to that of the Adams–Moulton methods.

The order of (3.110) is $k+1$, and the error constant is

$$C_{k+2}=\sum_{s=0}^{k-1}A_s\gamma_{k+1}^{s+1}.$$

In the case when k is even, it is possible to stretch the order to $k+2$, and still achieve zero-stability, by suitably choosing the A_s such that $C_{k+2}=0$. In this case the error constant is given by

$$C_{k+3}=\sum_{s=0}^{k}A_s\gamma_{k+2}^{s+1}.$$

The family of explicit k-step linear methods with $k-1$ free parameters can similarly be written in backward difference form as

$$\begin{aligned}&y_{n+k}-y_{n+k-1}+A_1(y_{n+k-1}-y_{n+k-2})+\cdots+A_{k-1}(y_{n+1}-y_n)\\&=h[R_{n+k-1}(k,0)+A_1R_{n+k-2}(k-1,1)+\cdots+A_{k-1}R_n(k-1,k-1)],\qquad k\geqslant 2\end{aligned}$$

or, in a form analogous to (3.110),

$$\sum_{s=0}^{k-1}A_s\nabla y_{n+1-s}=h\sum_{s=0}^{k-1}A_s\sum_{i=0}^{k-1}\gamma_i^s\nabla^i f_n,\qquad A_0:=1,\quad k\geqslant 2.\tag{3.111}$$

The order of (3.111) is k and the error constant is

$$C_{k+1}=\sum_{s=0}^{k-1}A_s\gamma_k^s.$$

It is not possible to increase the order past k and still achieve zero-stability.

We have thus been able to express in terms of backward differences the class of implicit k-step linear methods of order $k+1$ and the class of explicit k-step linear methods of order k. If we wish to obtain such classes in standard form, rather than in backward difference form, then, obviously, we could express the backward differences in (3.110) and (3.111) in terms of function values. It is, however, somewhat easier first to express the s-Adams methods themselves in terms of function values. This is done in Table 3.2 for $s = 0, 1, \ldots, 5$ and $s \leqslant k - t \leqslant 5$, where $t = 1$ if $s = 0$ and $t = 0$ if $s \geqslant 1$. For convenience of presentation, we have left-shifted the methods so that the k-step s-Adams method is expressed in the form

$$y_{n-s+1} - y_{n-s} = h \sum_{s=0}^{k-t} B_{n-s} f_{n-s}, \qquad t = \begin{cases} 1 & \text{if } s = 0 \\ 0 & \text{if } s \geqslant 1. \end{cases}$$

Further, in order to make the table more readable, we have expressed the coefficients B_{n-s} for each method in the form $B_{n-s} = b_{n-s}/d$, where b_{n-s} and d are integers. In Table 3.2 the coefficients b_{n-s} are listed (under columns headed f_{n-s}) together with the denominator d, stepnumber k, order p and error constant C_{p+1}.

We illustrate the application of Table 3.2 by using it first to construct a family of

Table 3.2 Coefficients of s-Adams methods

	f_n	f_{n-1}	f_{n-2}	f_{n-3}	f_{n-4}	f_{n-5}	d	k	p	C_{p+1}
0-Adams	1						1	1	1	$\frac{1}{2}$
	3	−1					2	2	2	$\frac{5}{12}$
$(y_{n+1} - y_n)/h =$	23	−16	5				12	3	3	$\frac{3}{8}$
	55	−59	37	−9			24	4	4	$\frac{251}{720}$
	1901	−2774	2616	−1274	251		720	5	5	$\frac{95}{288}$
	4277	−7923	9982	−7298	2877	−475	1440	6	6	$\frac{19\,087}{60\,480}$
1-Adams	1	1					2	1	2	$\frac{-1}{12}$
	5	8	−1				12	2	3	$\frac{-1}{24}$
$(y_n - y_{n-1})/h =$	9	19	−5	1			24	3	4	$\frac{-19}{720}$
	251	646	−264	106	−19		720	4	5	$\frac{-3}{160}$
	475	1427	−798	482	−173	27	1440	5	6	$\frac{-863}{60\,480}$
2-Adams	−1	8	5				12	2	3	$\frac{1}{24}$
	−1	13	13	−1			24	3	4	$\frac{11}{720}$
$(y_{n-1} - y_{n-2})/h =$	−19	346	456	−74	11		720	4	5	$\frac{11}{1440}$
	−27	637	1022	−258	77	−11	1440	5	6	$\frac{271}{60\,480}$
3-Adams	1	−5	19	9			24	3	4	$\frac{-19}{720}$
$(y_{n-2} - y_{n-3})/h =$	11	−74	456	346	−19		720	4	5	$\frac{-11}{1440}$
	11	−93	802	802	−93	11	1440	5	6	$\frac{-191}{60\,480}$
4-Adams										
$(y_{n-3} - y_{n-4})/h =$	−19	106	−264	646	251		720	4	5	$\frac{3}{160}$
	−11	77	−258	1022	637	−27	1440	5	6	$\frac{271}{60\,480}$
5-Adams										
$(y_{n-4} - y_{n-5})/h =$	27	−173	482	−798	1427	475	1440	5	6	$\frac{-863}{60\,480}$

implicit linear 4-step methods of order 6 (the maximum possible, subject to zero-stability). Following the discussion preceding (3.110), we see that the family of implicit linear 4-step methods of order 5 can be constructed by forming a linear combination of the 4-step s-Adams methods for $s = 1, 2, 3, 4$. From Table 3.2, this procedure yields

$$\begin{aligned} &y_n - y_{n-1} + A_1(y_{n-1} - y_{n-2}) + A_2(y_{n-2} - y_{n-3}) + A_3(y_{n-3} - y_{n-4}) \\ &\quad = \frac{h}{720}[(251 - 19A_1 + 11A_2 - 19A_3)f_n + (646 + 346A_1 - 74A_2 + 106A_3)f_{n-1} \\ &\quad + (-264 + 456A_1 + 456A_2 - 264A_3)f_{n-2} + (106 - 74A_1 + 346A_2 + 646A_3)f_{n-3} \\ &\quad + (-19 + 11A_1 - 19A_2 + 251A_3)f_{n-4}. \end{aligned} \tag{3.112}$$

(The reader may wish to ascertain that the same formula results from setting $k = 4$ in (3.110).) Again from Table 3.2, the order of (3.112) is 5 and the error constant is

$$C_6 = \tfrac{3}{160}(A_3 - 1) + \tfrac{11}{1440}(A_1 - A_2).$$

The order rises to 6 if we choose A_1, A_2 and A_3 such that $C_6 = 0$, that is if

$$A_3 = 1 + \tfrac{11}{27}(A_2 - A_1). \tag{3.113}$$

Substituting for A_3 from (3.113) in (3.112) gives the required 2-parameter family of implicit 4-step methods of order 6. The error constant is, from Table 3.2,

$$\begin{aligned} C_7 &= (-863 + 271A_1 - 191A_2 + 271A_3)/60\,480 \\ &= (-999 + 271A_1 - 136A_2)/102\,060, \end{aligned} \tag{3.114}$$

on substituting for A_3 from (3.113). One must, of course, choose the parameters A_1 and A_2 so that the method is zero-stable. For example, we can construct a 1-parameter family of symmetric methods by choosing $A_2 = A_1$, whence, by (3.113), $A_3 = 1$. The first characteristic polynomial now factorizes thus

$$\rho(\zeta) = (\zeta - 1)(\zeta + 1)[\zeta^2 + (A_1 - 1)\zeta + 1],$$

and it is easily ascertained from Figure 1.1 of §1.9 that zero-stability is achieved if $-1 < A_1 < 3$. We have thus identified a 1-parameter family of zero-stable symmetric 4-step methods which have order 6, given by

$$\begin{aligned} &y_n + (A_1 - 1)(y_{n-1} - y_{n-3}) - y_{n-4} \\ &\quad = \frac{h}{90}[(29 - A_1)(f_n + f_{n-4}) + (-66 + 144A_1)f_{n-2} + (94 + 34A_1)(f_{n-1} + f_{n-3})] \end{aligned}$$

with error constant $C_7 = (-37 + 5A_1)/3780$. We note in passing that setting $A_1 = 11/19$ (an acceptable value for zero-stability) produces the special case of *Quade's method*,

$$y_n - \frac{8}{19}(y_{n-1} - y_{n-3}) - y_{n-4} = \frac{6h}{19}(f_n + 4f_{n-1} + 4f_{n-3} + f_{n-4})$$

a method from the 1950s. (See Exercises 3.2.2 and 2.5.2.)

Families of explicit linear k-step methods of order k can likewise be constructed by taking a linear combination of the k-step 0-Adams method and the $(k-1)$-step s-Adams methods, $s = 1, 2, \ldots, k-1$. Thus, for example, the 2-parameter family of explicit 3-step methods of order 3 is seen from Table 3.2 to be

$$y_{n+1} + (A_1 - 1)y_n + (A_2 - A_1)y_{n-1} - A_2 y_{n-2}$$
$$= \frac{h}{12}[(23 + 5A_1 - A_2)f_n + (-16 + 8A_1 + 8A_2)f_{n-1} + (5 - A_1 + 5A_2)f_{n-2}],$$

with error constant $C_4 = (9 - A_1 + A_2)/24$.

Exercise

3.11.1. Use Table 3.2 to construct a $(k-1)$-parameter family of explicit linear k-step methods for $k = 2, 3, 4$. Show that for each family, the order of the methods is k and that orders greater than k cannot be obtained if the methods are to be zero-stable.

3.12 THE BACKWARD DIFFERENTIATION FORMULAE

As we shall see later in this book, the regions of absolute stability of the Adams–Moulton methods, though reasonably sized, turn out to be inadequate to cope with the problem of stiffness, where stability rather than accuracy is paramount. A class of implicit linear k-step methods with regions of absolute stability large enough to make them relevant to the problem of stiffness is the class of *Backward Differentiation Formulae* or *BDF*, defined by

$$\sum_{j=0}^{k} \alpha_j y_{n+j} = h\beta_k f_{n+k}. \tag{3.115}$$

This class can be seen (in a hand-waving sort of way) to be a dual of the class of Adams–Moulton methods. The latter is characterized by having the simplest possible (subject to consistency) first characteristic polynomials $\rho(\zeta) = \zeta^k - \zeta^{k-1}$, whereas the BDF have the simplest possible second characteristic polynomials $\sigma(\zeta) = \beta_k \zeta^k$. Moreover, there is a certain duality between the techniques for deriving the Adams–Moulton methods and the BDF in backward difference form. Instead of starting from the identity (3.87)

$$y(x_{n+1}) - y(x_n) = \int_{x_n}^{x_{n+1}} y'(x)\,dx$$

replacing y' by f and integrating the polynomial interpolant of the back values of f, we start from the differential system itself

$$y' = f(x, y) \tag{3.116}$$

and differentiate the polynomial interpolant of the back values of y. By §1.10, the data

$$(x_{n+1}, y_{n+1}), (x_n, y_n), \ldots, (x_{n-k+1}, y_{n-k+1})$$

is interpolated by the polynomial $I_k(x)$ of degree k given by

$$I_k(x) = I_k(x_{n+1} + rh) =: P_k(r) = \sum_{i=0}^{k} (-1)^i \binom{-r}{i} \nabla^i y_{n+1}.$$

The left side of (3.116) is replaced by the derivative of this interpolant at $x = x_{n+1}$, given by

$$I'_k(x_{n+1}) = \frac{1}{h} P'_k(r)|_{r=0} = \frac{1}{h} \sum_{i=0}^{k} (-1)^i \frac{d}{dr} \binom{-r}{i} \bigg|_{r=0} \nabla^i y_{n+1}$$

and the right side is replaced by f_{n+1}, giving the Backward Differentiation Formulae in the following backward difference form:

$$\left.\begin{aligned} &\sum_{i=0}^{k} \delta_i \nabla^i y_{n+1} = hf_{n+1} \\ \text{where} \qquad & \\ &\delta_i = (-1)^i \frac{d}{dr} \binom{-r}{i} \bigg|_{r=0} \end{aligned}\right\} \tag{3.117}$$

The δ_i are easily found by direct evaluation to be

$$\delta_0 = 0, \quad \delta_i = 1/i, \qquad i = 1, 2, \ldots. \tag{3.118}$$

(A generating function is no longer necessary, but it is easily seen to be $G^{\mathrm{BDF}}(t) = -\ln(1-t)$.) In order to put the methods given by (3.117) and (3.118) in the standard form (3.115), we first divide through by $\sum_{i=0}^{k} \delta_i$ (in order that $\alpha_k = 1$), and appropriately right-shift each method. Recalling that $\delta_0 = 0$, we get

$$\tau_k \sum_{i=1}^{k} \delta_i \nabla^i y_{n+k} = h\tau_k f_{n+k}, \tag{3.119}$$

where

$$\tau_k = 1 \bigg/ \sum_{i=1}^{k} \delta_i. \tag{3.120}$$

On expanding the differences on the left side of (3.119), we get the class of BDF methods in the standard form (3.115).

By an argument analogous to that used for the Adams–Moulton methods in §3.10, we find that the order of the k-step BDF is k and the error constant is

$$C_{k+1} = -\tau_k \delta_{k+1}. \tag{3.121}$$

Clearly, the BDF do not have zero-stability built in, in the way that the Adams methods do, and it is necessary to examine, for each k, the roots of the first characteristic polynomial. It turns out that for $k = 1, 2, \ldots, 6$, the methods are zero-stable, but that for

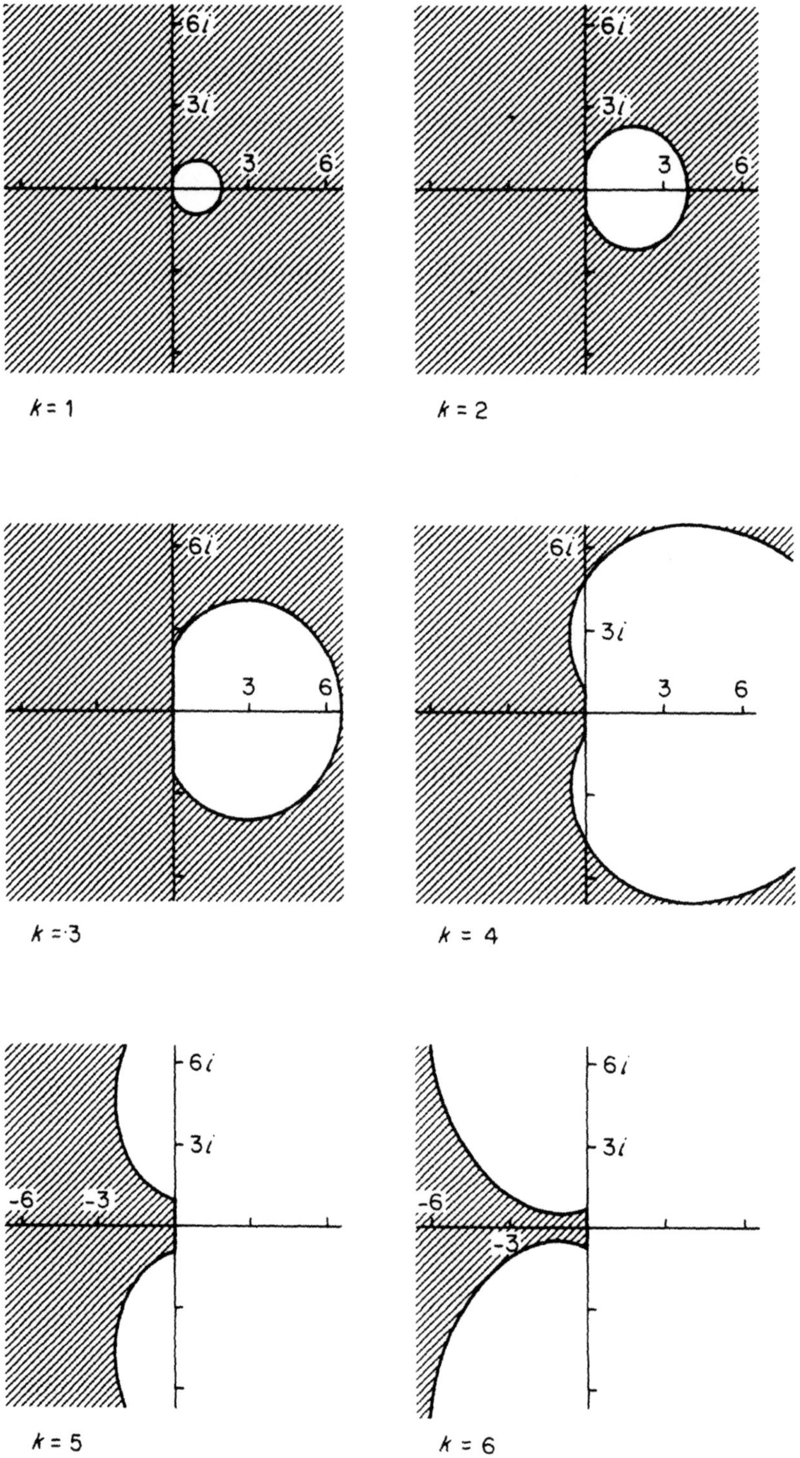

Figure 3.4 Regions of absolute stability for the k-step BDF.

Table 3.3 Coefficients of the BDF

k	α_6	α_5	α_4	α_3	α_2	α_1	α_0	β_k	p	C_{p+1}
1						1	-1	1	1	$-\frac{1}{2}$
2					1	$-\frac{4}{3}$	$\frac{1}{3}$	$\frac{2}{3}$	2	$-\frac{2}{9}$
3				1	$-\frac{18}{11}$	$\frac{9}{11}$	$-\frac{2}{11}$	$\frac{6}{11}$	3	$-\frac{3}{22}$
4			1	$-\frac{48}{25}$	$\frac{36}{25}$	$-\frac{16}{25}$	$\frac{3}{25}$	$\frac{12}{25}$	4	$-\frac{12}{125}$
5		1	$-\frac{300}{137}$	$\frac{300}{137}$	$-\frac{200}{137}$	$\frac{75}{137}$	$-\frac{12}{137}$	$\frac{60}{137}$	5	$-\frac{10}{137}$
6	1	$-\frac{360}{147}$	$\frac{450}{147}$	$-\frac{400}{147}$	$\frac{225}{147}$	$-\frac{72}{147}$	$\frac{10}{147}$	$\frac{60}{147}$	6	$-\frac{20}{343}$

$k \geqslant 7$ the methods are all zero-unstable (see Cryer, 1972). The coefficients of the BDF in the form (3.115), together with the error constants are given in Table 3.3 for $k = 1, 2, \ldots, 6$.

The important feature of the BDF is the size of their regions of absolute stability. These are shown in Figure 3.4. We note that for $1 \leqslant k \leqslant 6$ these regions contain the whole of the negative real axis, and that for $k = 1, 2$, they contain the whole of the negative half-plane. (For $k = 3$, the boundary of the region marginally invades the negative half-plane near the points $\pm i$.) These properties are significant in the context of stiffness.

4 Predictor–Corrector Methods

4.1 PREDICTOR–CORRECTOR MODES

Suppose that we wish to solve the standard initial value problem by an implicit linear multistep method. Then at each step we have to solve for y_{n+k} the implicit system

$$y_{n+k} + \sum_{j=0}^{k-1} \alpha_j y_{n+j} = h\beta_k f(x_{n+k}, y_{n+k}) + h \sum_{j=0}^{k-1} \beta_j f_{n+j}. \tag{4.1}$$

We normally do this by the fixed point iteration

$$y_{n+k}^{[\nu+1]} + \sum_{j=0}^{k-1} \alpha_j y_{n+j} = h\beta_k f(x_{n+k}, y_{n+k}^{[\nu]}) + h \sum_{j=0}^{k-1} \beta_j f_{n+j}, \qquad y_{n+k}^{[0]} \text{ arbitrary}, \ \nu = 0, 1, \ldots. \tag{4.2}$$

which, by (3.8), will converge to the unique solution of (4.1) provided that

$$h < 1/(|\beta_k| L),$$

where L is the Lipschitz constant of f with respect to y. For non-stiff problems, this restriction on h is not significant; in practice, considerations of accuracy put a much more restrictive constraint on h. Although (4.2) will converge for arbitrary $y_{n+k}^{[0]}$, each iteration calls for one evaluation of the function f, and computation can obviously be saved if we can provide as good a guess as possible for $y_{n+k}^{[0]}$. This is conveniently done by using a separate *explicit* linear multistep method to provide the initial guess, $y_{n+k}^{[0]}$. We call this explicit method the *predictor* and the implicit method (4.1) the *corrector*; the two together comprise a *predictor–corrector pair*. There will turn out to be advantage in having the predictor and the corrector of the same order, which usually means that the stepnumber of the predictor has to be greater than that of the corrector. Rather than deal with the complication of having two different stepnumbers, we take the stepnumber of the predictor, which we shall call k, to be the stepnumber of the pair, and no longer demand of the corrector that the second of the conditions (3.2), namely $|\alpha_0| + |\beta_0| \neq 0$, holds. Thus, for example, we regard

$$y_{n+2} - y_{n+1} = \frac{h}{2}(3f_{n+1} - f_n), \qquad y_{n+2} - y_{n+1} = \frac{h}{2}(f_{n+2} + f_{n+1})$$

as a predictor–corrector pair with stepnumber 2, even though the corrector is essentially a 1-step method. We shall always distinguish between the predictor and the corrector by attaching asterisks to the coefficients (and to any other parameters, such as order and error constant) of the predictor. Thus the general k-step predictor–corrector or PC pair is

$$\left.\begin{aligned}\sum_{j=0}^{k} \alpha_j^* y_{n+j} &= h \sum_{j=0}^{k-1} \beta_j^* f_{n+j}\\ \sum_{j=0}^{k} \alpha_j y_{n+j} &= h \sum_{j=0}^{k} \beta_j f_{n+j}.\end{aligned}\right\} \tag{4.3}$$

There are various ways, or *modes*, in which the pair (4.3) can be implemented. Firstly, we could use the predictor to give the first guess $y_{n+k}^{[0]}$, then allow the iteration (4.2) to proceed until we achieve convergence (in practice, until some criterion like $\| y_{n+k}^{[\nu+1]} - y_{n+k}^{[\nu]} \| < \varepsilon$, where ε is of the order of round-off error, is satisfied). This is called the mode of *correcting to convergence*. In this mode, the predictor plays a very ancillary role, and the local truncation error and linear stability characteristics of the predictor–corrector pair are those of the corrector alone. What makes this mode unattractive in practice is that we cannot tell in advance how many iterations of the corrector—and therefore how many function evaluations—will be involved at each step. In writing an algorithm based on the mode of iterating to convergence, we are in effect writing a blank cheque. In general this is to be avoided; in the special case of *real-time* problems, it can be downright dangerous. An example of a real-time problem would be an automatic landing system for an aircraft; such a system can be modelled by a system of ordinary differential equations, the solution of which determines the appropriate settings of the control surfaces and throttles of the aircraft; it is not much use if the numerical procedure for solving the differential system takes so long to compute on an on-board computer that it ends up by telling the control system what these settings should have been a few seconds ago! In such situations, it is paramount that the computing time should be predictable, and that is never possible in the mode of correcting to convergence.

A much more acceptable procedure is to state in advance just how many iterations of the corrector are to be permitted at each step. Normally this number is small, usually 1 or 2. The local truncation error and linear stability characteristics of the predictor–corrector method in such a finite mode depend on both the predictor and the corrector; we investigate this dependence in later sections. A useful mnemonic for describing modes of this sort can be constructed by using P and C to indicate one application of the predictor or the corrector respectively, and E to indicate one evaluation of the function f, given x and y. Note that if the system of differential equations is of large dimension then a function evaluation can represent a significant amount of computing; thus it is usual to regard the number of evaluations per step as a rough indication of the computing effort demanded by the method. Suppose we apply the predictor to evaluate $y_{n+k}^{[0]}$, evaluate $f_{n+k}^{[0]} = f(x_{n+k}, y_{n+k}^{[0]})$, and then apply (4.2) just once to obtain $y_{n+k}^{[1]}$. The mode is then described as PEC. If we call the iteration a second time to obtain $y_{n+k}^{[2]}$, which obviously involves the further evaluation $f_{n+k}^{[1]} = f(x_{n+k}, y_{n+k}^{[1]})$, then the mode is described as PECEC or P(EC)^2. There is one further decision we have to make. At the end of the P(EC)^2 step we have a value $y_{n+k}^{[2]}$ for y_{n+k} and a value $f_{n+k}^{[1]}$ for $f(x_{n+k}, y_{n+k})$. We may choose to update the value of f by making a further evaluation $f_{n+k}^{[2]} = f(x_{n+k}, y_{n+k}^{[2]})$; the mode would then be described as P(EC)^{2}E. The two classes of modes P(EC)$^\mu$E and P(EC)$^\mu$ can be written as a single mode P(EC)$^\mu$E^{1-t}, where μ is positive integer and

$t = 0$ or 1, and are defined by

$P(EC)^{\mu}E^{1-t}$:

$$
\left.
\begin{array}{lll}
\text{P:} & y_{n+k}^{[0]} + \sum_{j=0}^{k-1} \alpha_j^* y_{n+j}^{[\mu]} = h \sum_{j=0}^{k-1} \beta_j^* f_{n+j}^{[\mu-t]} & \\
(\text{EC})^{\mu}: & \left.\begin{array}{l} f_{n+k}^{[\nu]} = f(x_{n+k}, y_{n+k}^{[\nu]}) \\ y_{n+k}^{[\nu+1]} + \sum_{j=0}^{k-1} \alpha_j y_{n+j}^{[\mu]} = h\beta_k f_{n+k}^{[\nu]} + h \sum_{j=0}^{k-1} \beta_j f_{n+j}^{[\mu-t]} \end{array}\right\} & \nu = 0, 1, \ldots, \mu - 1 \\
\text{E}^{(1-t)}: & f_{n+k}^{[\mu]} = f(x_{n+k}, y_{n+k}^{[\mu]}), \quad \text{if } t = 0. &
\end{array}
\right\} \qquad (4.4)
$$

Alternatively, the predictor and corrector may be written in the form (3.5) as

$$\rho^*(E)y_n = h\sigma^*(E)f_n, \quad \rho(E)y_n = h\sigma(E)f_n,$$

respectively, where ρ^*, ρ and σ have degree k and σ^* has degree $k-1$ at most. With this notation, the mode $P(EC)^{\mu}E^{1-t}$ may be defined by

$P(EC)^{\mu}E^{1-t}$:

$$
\left.
\begin{array}{lll}
\text{P:} & E^k y_n^{[0]} + [\rho^*(E) - E^k] y_n^{[\mu]} = h\sigma^*(E) f_n^{[\mu-t]} & \\
(\text{EC})^{\mu}: & \left.\begin{array}{l} E^k f_n^{[\nu]} = f(x_{n+k}, E^k y_n^{[\nu]}) \\ E^k y_n^{[\nu+1]} + [\rho(E) - E^k] y_n^{[\mu]} = h\beta_k E^k f_n^{[\nu]} + h[\sigma(E) - \beta_k E^k] f_n^{[\mu-t]} \end{array}\right\} & \nu = 0, 1, \ldots, \mu - 1 \\
\text{E}^{(1-t)}: & E^k f_n^{[\mu]} = f(x_{n+k}, E^k y_n^{[\mu]}), \quad \text{if } t = 0. &
\end{array}
\right\} \qquad (4.5)
$$

4.2 THE LOCAL TRUNCATION ERROR OF PREDICTOR–CORRECTOR METHODS

If the predictor–corrector pair is applied in the mode of correcting to convergence then the local truncation error is clearly that of the corrector alone. If, however, the pair is applied in $P(EC)^{\mu}E^{1-t}$ mode, $t = 0, 1$, then the local truncation error of the corrector will be polluted by that of the predictor. In this section we investigate the level of this pollution.

Let the predictor and corrector defined by (4.3) have associated linear difference operators $\mathscr{L}^*$ and $\mathscr{L}$, orders p^* and p and error constants $C_{p^*+1}^*$ and C_{p+1} respectively. As always when dealing with local truncation error, we make the localizing assumption $y_{n+j}^{[\nu]} = y(x_{n+j}), j = 0, 1, \ldots, k-1$, and indicate by $\tilde{y}_{n+k}^{[\nu]}$ approximations to y at x_{n+k} generated when the localizing assumption is in force. We also assume that $y(x) \in C^{\hat{p}+1}$, where $\hat{p} = \max(p^*, p)$. It then follows from §3.5 that

$$
\left.
\begin{array}{l}
\mathscr{L}^*[y(x); h] = C_{p^*+1}^* h^{p^*+1} y^{(p^*+1)}(x) + 0(h^{p^*+2}) \\
\mathscr{L}[y(x); h] = C_{p+1} h^{p+1} y^{(p+1)}(x) + 0(h^{p+2}).
\end{array}
\right\} \qquad (4.6)
$$

Following the analysis of §3.5, we have that for the predictor

$$\sum_{j=0}^{k} \alpha_j^* y(x_{n+j}) = h \sum_{j=0}^{k-1} \beta_j^* f(x_{n+j}, y(x_{n+j})) + \mathscr{L}^*[y(x_n); h]$$

and (4.7)

$$\tilde{y}_{n+k}^{[0]} + \sum_{j=0}^{k-1} \alpha_j^* y_{n+j}^{[\mu]} = h \sum_{j=0}^{k-1} \beta_j^* f(x_{n+j}, y_{n+j}^{[\mu-t]}).$$

On subtracting and using the localizing assumption and (4.6) we obtain

$$y(x_{n+k}) - \tilde{y}_{n+k}^{[0]} = C_{p^*+1}^* h^{p^*+1} y^{(p^*+1)}(x_n) + O(h^{p^*+2}). \tag{4.8}$$

Since the predictor–corrector pair is being applied in $P(EC)^\mu E^{1-t}$ mode, as defined by (4.4), the equations for the corrector, corresponding to (4.7), are

$$\sum_{j=0}^{k} \alpha_j y(x_{n+j}) = h \sum_{j=0}^{k} \beta_j f(x_{n+j}, y(x_{n+j})) + \mathscr{L}[y(x_n); h]$$

and

$$\tilde{y}_{n+k}^{[\nu+1]} + \sum_{j=0}^{k-1} \alpha_j y_{n+j}^{[\mu]} = h\beta_k f(x_{n+k}, \tilde{y}_{n+k}^{[\nu]}) + h \sum_{j=0}^{k-1} \beta_j f(x_{n+j}, y_{n+j}^{[\mu-t]}), \qquad \nu = 0, 1, \ldots, \mu - 1.$$

On subtracting and using the localizing assumption, we have

$$\begin{aligned} y(x_{n+k}) - \tilde{y}_{n+k}^{[\nu+1]} &= h\beta_k [f(x_{n+k}, y(x_{n+k})) - f(x_{n+k}, \tilde{y}_{n+k}^{[\nu]})] + \mathscr{L}[y(x_n); h] \\ &= h\beta_k \overline{\frac{\partial f}{\partial y}}(x_{n+k}, \eta_\nu)[y(x_{n+k}) - \tilde{y}_{n+k}^{[\nu]}] + C_{p+1} h^{p+1} y^{(p+1)}(x_n) \\ &\quad + O(h^{p+2}), \qquad \nu = 0, 1, \ldots, \mu - 1 \end{aligned} \tag{4.9}$$

(using the notation of §1.3). What follows depends on the relative magnitudes of p^* and p.

First consider the case $p^* \geqslant p$. On substituting (4.8) into (4.9) with $\nu = 0$ we get that

$$y(x_{n+k}) - \tilde{y}_{n+k}^{[1]} = C_{p+1} h^{p+1} y^{(p+1)}(x_n) + O(h^{p+2}).$$

This expression for $y(x_{n+k}) - \tilde{y}_{n+k}^{[1]}$ can now be substituted into (4.9) with $\nu = 1$ to get

$$y(x_{n+k}) - \tilde{y}_{n+k}^{[2]} = C_{p+1} h^{p+1} y^{(p+1)}(x_n) + O(h^{p+2}).$$

Continuing in this manner we find that

$$y(x_{n+k}) - \tilde{y}_{n+k}^{[\mu]} = C_{p+1} h^{p+1} y^{(p+1)}(x_n) + O(h^{p+2}).$$

Thus, if $p^* \geqslant p$, the PLTE of the $P(EC)^\mu E^{1-t}$ mode is, for all $\mu \geqslant 1$, precisely that of the corrector alone.

Now consider the case $p^* = p - 1$. On substituting (4.8) into (4.9) with $\nu = 0$ we now get that

$$y(x_{n+k}) - \tilde{y}^{[1]}_{n+k} = \left[\beta_k \overline{\frac{\partial f}{\partial y}} C^*_p y^{(p)}(x_n) + C_{p+1} y^{(p+1)}(x_n) \right] h^{p+1} + 0(h^{p+2}).$$

Thus if $\mu = 1$, that is if the mode is PECE^{1-t}, the PLTE is not identical with that of the corrector, but the order of the PC method is that of the corrector. However, on successive substitution into (4.9) we find that for $\mu \geqslant 2$,

$$y(x_{n+k}) - \tilde{y}^{[\mu]}_{n+k} = C_{p+1} h^{p+1} y^{(p+1)}(x_n) + 0(h^{p+2}),$$

and the PLTE of the PC method becomes that of the corrector alone.

Now consider the case $p^* = p - 2$. On substituting (4.8) into (4.9) with $\nu = 0$, we get

$$y(x_{n+k}) - \tilde{y}^{[1]}_{n+k} = \beta_k \overline{\frac{\partial f}{\partial y}} C^*_{p-1} h^p y^{(p-1)}(x_n) + 0(h^{p+1}). \tag{4.10}$$

Thus if $\mu = 1$, the order of the PC method is only $p - 1$. Substituting (4.10) into (4.9) with $\nu = 1$ gives

$$y(x_{n+k}) - \tilde{y}^{[2]}_{n+k} = \left[\left(\beta_k \overline{\frac{\partial f}{\partial y}} \right)^2 C^*_{p-1} y^{(p-1)}(x_n) + C_{p+1} y^{(p+1)}(x_n) \right] h^{p+1} + 0(h^{p+2}),$$

and thus for $\mu = 2$ the order of the PC method is that of the corrector, but the two PLTEs are not identical. Further successive substitutions into (4.9) show that for $\mu \geqslant 3$

$$y(x_{n+k}) - \tilde{y}^{[\mu]}_{n+k} = C_{p+1} h^{p+1} y^{(p+1)}(x_n) + 0(h^{p+2}),$$

and the PLTE is that of the corrector alone.

It is now clear that the order and the PLTE of a PC method depend on the gap between p^* and p and on μ, the number of times the corrector is called. Specifically,

(i) if $p^* \geqslant p$ (or if $p^* < p$ and $\mu > p - p^*$), the PC method and the corrector have the same order and the same PLTE,
(ii) if $p^* < p$ and $\mu = p - p^*$, the PC method and the corrector have the same order but different PLTEs, and
(iii) if $p^* < p$ and $\mu \leqslant p - p^* - 1$, the order of the PC method is $p^* + \mu$ ($< p$).

Note that the modes $\text{P(EC)}^\mu\text{E}$ and P(EC)^μ always have the same order and PLTE.

4.3 MILNE'S ESTIMATE FOR THE PLTE; LOCAL EXTRAPOLATION

At the end of §3.7 we noted that attempts to estimate the PLTE of a linear multistep method directly from the formula

$$y(x_{n+k}) - \tilde{y}_{n+k} = C_{p+1} h^{p+1} y^{(p+1)}(x_n)$$

ran up against the difficulty of trying to estimate $y^{(p+1)}(x_n)$ numerically. Predictor–

corrector methods have a substantial advantage over linear multistep methods in that, due to a device due originally to W. E. Milne, it is possible to estimate the PLTE of the former without any need to attempt a direct estimate of $y^{(p+1)}(x_n)$. The device, *which works only if* $p^* = p$, is as close as one ever gets in numerical analysis to getting something for nothing, and is indeed a major motivation for using predictor–corrector methods.

It follows from the preceding section that if $p^* = p$, then

$$C^*_{p+1}h^{p+1}y^{(p+1)}(x_n) = y(x_{n+k}) - \tilde{y}^{[0]}_{n+k} + 0(h^{p+2})$$

and

$$C_{p+1}h^{p+1}y^{(p+1)}(x_n) = y(x_{n+k}) - \tilde{y}^{[\mu]}_{n+k} + 0(h^{p+2}).$$

On subtracting we obtain

$$(C^*_{p+1} - C_{p+1})h^{p+1}y^{(p+1)}(x_n) = \tilde{y}^{[\mu]}_{n+k} - \tilde{y}^{[0]}_{n+k} + 0(h^{p+2}),$$

whence we obtain the Milne estimate for the PLTE

$$\left.\begin{aligned} \text{PLTE} = C_{p+1}h^{p+1}y^{(p+1)}(x_n) &= W(y^{[\mu]}_{n+k} - y^{[0]}_{n+k}) \\ \text{where}\qquad W &:= \frac{C_{p+1}}{C^*_{p+1} - C_{p+1}}. \end{aligned}\right\} \tag{4.11}$$

On the right side of the first of (4.11) we have replaced $\tilde{y}^{[\mu]}_{n+k}$ by $y^{[\mu]}_{n+k}$ since it is no longer necessary to remind ourselves that the localizing assumption is in force; recall the argument in §3.7 that PLTE is an acceptable measure of local accuracy despite the localizing assumption.

The main use made of the estimate (4.11) is the monitoring of steplength, which could be decreased if the norm of the error estimate exceeds a given tolerance, and increased if the norm is less than the tolerance by a given factor. However, as is the case with all error estimates, one is tempted also to add the error estimate to the numerical solution, thereby increasing the accuracy. This addition used to be known as a *modifier*, but is now usually called *local extrapolation*. It is clearly equivalent to raising the order of the method by one. It is common practice in many modern codes to perform local extrapolation at each step and still use the error estimate (4.11) to monitor steplength, the rationale being that if the steplength is chosen so that the error estimate (4.11) for $y^{[\mu]}_{n+k}$ is acceptable, then surely the error in the more accurate locally extrapolated value will also be acceptable. This argument is not altogether sound—higher order methods do not invariably produce smaller errors than do lower order ones—and there is no avoiding the fact that local extrapolation is basically an attempt to eat one's cake and have it! Nevertheless, local extrapolation is an accepted feature of many modern codes. It can be applied in more than one way. We could apply local extrapolation after each call of the corrector; using L as the mnemonic for local extrapolation, the resulting modes can be denoted by $\text{P(ECL)}^{\mu}\text{E}^{1-t}, t = 0$ or 1. Alternatively, we could choose to apply local extrapolation only after the final application of the corrector, resulting in the modes $\text{P(EC)}^{\mu}\text{LE}^{1-t}, t = 0$ or 1; clearly the two families of modes coincide when $\mu = 1$. Noting that, from (4.11), local extrapolation is equivalent to replacing $y^{[\nu]}_n$ by

$$y^{[\nu]}_n + W(y^{[\nu]}_n - y^{[0]}_n),$$

these modes are formally defined as follows:

$P(ECL)^{\mu}E^{1-t}$:

$$\left.\begin{array}{ll}
P: & y^{[0]}_{n+k} + \sum_{j=0}^{k-1} \alpha_j^* y^{[\mu]}_{n+j} = h \sum_{j=0}^{k-1} \beta_j^* f^{[\mu-t]}_{n+j} \\
(ECL)^{\mu}: & \left.\begin{array}{l} f^{[\nu]}_{n+k} = f(x_{n+k}, y^{[\nu]}_{n+k}) \\ \hat{y}^{[\nu+1]}_{n+k} + \sum_{j=0}^{k-1} \alpha_j y^{[\mu]}_{n+j} = h\beta_k f^{[\nu]}_{n+k} + h \sum_{j=0}^{k-1} \beta_j f^{[\mu-t]}_{n+j}, \\ y^{[\nu+1]}_{n+k} = (1+W)\hat{y}^{[\nu+1]}_{n+k} - W y^{[0]}_{n+k} \end{array}\right\} \nu = 0, 1, \dots, \mu - 1 \\
E^{(1-t)}: & f^{[\mu]}_{n+k} = f(x_{n+k}, y^{[\mu]}_{n+k}), \quad \text{if } t = 0.
\end{array}\right\} \tag{4.12}$$

$P(EC)^{\mu}LE$:

$$\left.\begin{array}{ll}
P: & \hat{y}^{[0]}_{n+k} + \sum_{j=0}^{k-1} \alpha_j^* y^{[\mu]}_{n+j} = h \sum_{j=0}^{k-1} \beta_j^* f^{[\mu]}_{n+j} \\
(EC)^{\mu}: & \left.\begin{array}{l} \hat{f}^{[\nu]}_{n+k} = f(x_{n+k}, \hat{y}^{[\nu]}_{n+k}) \\ \hat{y}^{[\nu+1]}_{n+k} + \sum_{j=0}^{k-1} \alpha_j y^{[\mu]}_{n+j} = h\beta_k \hat{f}^{[\nu]}_{n+k} + h \sum_{j=0}^{k-1} \beta_j f^{[\mu]}_{n+j} \end{array}\right\} \nu = 0, 1, \dots, \mu - 1 \\
L: & y^{[\mu]}_{n+k} = (1+W)\hat{y}^{[\mu]}_{n+k} - W \hat{y}^{[0]}_{n+k} \\
E: & f^{[\mu]}_{n+k} = f(x_{n+k}, y^{[\mu]}_{n+k}).
\end{array}\right\} \tag{4.13(i)}$$

$P(EC)^{\mu}L$:

$$\left.\begin{array}{ll}
P: & \hat{y}^{[0]}_{n+k} + \sum_{j=0}^{k-1} \alpha_j^* y^{[\mu]}_{n+j} = h \sum_{j=0}^{k-1} \beta_j^* \hat{f}^{[\mu-1]}_{n+j} \\
(EC)^{\mu}: & \left.\begin{array}{l} \hat{f}^{[\nu]}_{n+k} = f(x_{n+k}, \hat{y}^{[\nu]}_{n+k}) \\ \hat{y}^{[\nu+1]}_{n+k} + \sum_{j=0}^{k-1} \alpha_j y^{[\mu]}_{n+j} = h\beta_k \hat{f}^{[\nu]}_{n+k} + h \sum_{j=0}^{k-1} \beta_j \hat{f}^{[\mu-1]}_{n+j} \end{array}\right\} \nu = 0, 1, \dots, \mu - 1 \\
L: & y^{[\mu]}_{n+k} = (1+W)\hat{y}^{[\mu]}_{n+k} - W \hat{y}^{[0]}_{n+k}.
\end{array}\right\} \tag{4.13(ii)}$$

Exercises

4.3.1. The predictor P and the two correctors $C, \tilde{C}$ are defined by their characteristic polynomials as follows:

$$\begin{array}{lll}
P: & \rho^*(\zeta) = \zeta^4 - 1, & \sigma^*(\zeta) = \tfrac{4}{3}(2\zeta^3 - \zeta^2 + 2\zeta) \\
C: & \rho(\zeta) = \zeta^2 - 1, & \sigma(\zeta) = \tfrac{1}{3}(\zeta^2 + 4\zeta + 1) \\
\tilde{C}: & \rho(\zeta) = \zeta^3 - \tfrac{9}{8}\zeta^2 + \tfrac{1}{8}, & \sigma(\zeta) = \tfrac{3}{8}(\zeta^3 + 2\zeta^2 - \zeta).
\end{array}$$

(Note that C is Simpson's Rule.) Show that Milne's estimate is applicable to the predictor–corrector pairs (P, C) and $(P, \tilde{C})$. Write down the algorithms which use (a) P and C in PECE mode and (b) P and $\tilde{C}$ in PECLE mode. (Algorithm (a) is often known as *Milne's method* and (b) as *Hamming's method*.)

4.3.2*. The scalar initial value problem $y' = -10(y-1)^2$, $y(0) = 2$ has exact solution $y(x) = (2+10x)/(1+10x)$. Knowledge of the exact solution enables us to implement the localizing assumption, and thus compute the actual LTE. Using P and $\tilde{C}$ defined in Exercise 4.3.1 compare the actual LTE with the Milne estimate (a) for the mode of correcting to convergence and (b) in PECE mode. Use exact starting values, take $h = 0.01$ and compute from $x = 0$ to $x = 0.2$.

4.3.3. In (4.12) we wrote the two modes $P(ECL)^{\mu}E^{1-t}$, $t = 0, 1$, in a single statement. Why was this not done in (4.13) for the two modes $P(EC)^{\mu}LE^{1-t}$, $t = 0, 1$?

4.3.4. Let P have order p^* and C have order p. Show that an estimate of the PLTE similar to, but not identical with (4.11) can be constructed when $p^* > p$, but not when $p^* < p$.

4.3.5*. A result of Henrici (1962) shows that Milne's estimate holds *without* the localizing assumption provided $\rho^*(\zeta) \equiv \rho(\zeta)$, the mode of correcting to convergence is employed and the starting errors are $O(h^q)$ with $q > p$, where the PC method has order p. Find a fourth-order predictor which when used with the corrector $\tilde{C}$ defined in Exercise 4.3.1 satisfies the condition $\rho^*(\zeta) \equiv \rho(\zeta)$. Devise and carry out a numerical experiment, using the problem of Exercise 4.3.2*, to test this result.

4.4 PREDICTOR–CORRECTOR METHODS BASED ON ADAMS METHODS IN BACKWARD DIFFERENCE FORM

Almost all modern predictor–corrector codes for non-stiff problems use Adams–Bashforth methods as predictors and Adams–Moulton methods as correctors; such PC methods are consequently sometimes called *ABM methods*, but the phrase 'Adams method' is also somewhat loosely used to mean an ABM method. We saw in §3.9 that the Adams methods, when expressed in backward difference form, had particularly simple and attractive structures; these structures can be fully exploited in the framework of predictor–corrector methods.

Since we shall of course be making use of the Milne estimate for the PLTE, it is necessary that predictor and corrector have the same order. This is achieved by taking the predictor to be a k-step Adams–Bashforth method and the corrector to be a $(k-1)$-step Adams–Moulton; both then have order $p = k$. From (3.96), the k-step kth order ABM pair is thus

$$\left.\begin{aligned} y_{n+1} - y_n &= h\sum_{i=0}^{k-1} \gamma_i^* \nabla^i f_n, \qquad & p^* = k, \quad C_{k+1}^* = \gamma_k^* \\ y_{n+1} - y_n &= h\sum_{i=0}^{k-1} \gamma_i \nabla^i f_{n+1}, \qquad & p = k, \quad C_{k+1} = \gamma_k \end{aligned}\right\} \quad k = 1, 2, \ldots. \tag{4.14}$$

(On setting $k = 1$ in (4.14) we note that the PC pair consisting of Euler's Rule and the Backward Euler method is now considered as an ABM pair of order 1; see the comments towards the end of §3.9.)

We can at once anticipate a notational difficulty. If we envisage (4.14) being applied in $P(EC)^{\mu}E^{1-t}$ mode then, in the second of (4.14), y_{n+1} will be replaced by $y_{n+1}^{[\nu+1]}$, and the *single value* f_{n+1} on the right side by $f_{n+1}^{[\nu]}$, the remaining values f_{n-j} being replaced by $f_{n-j}^{[\mu-t]}$, $j=0,1,\dots,k-1$. We can overcome this difficulty by defining $\nabla_{\nu}^{i}f_{n+1}^{[\mu]}$ to be $\nabla^{i}f_{n+1}^{[\mu]}$ with the single value $f_{n+1}^{[\mu]}$ replaced by $f_{n+1}^{[\nu]}$. That is,

$$\nabla_{\nu}^{i}f_{n+1}^{[\mu]}:=\nabla^{i}f_{n+1}^{[\mu]}+f_{n+1}^{[\nu]}-f_{n+1}^{[\mu]}. \tag{4.15}$$

Thus, for example, $\nabla_{\nu}^{2}f_{n+1}^{[\mu]}=f_{n+1}^{[\nu]}-2f_{n}^{[\mu]}+f_{n-1}^{[\mu]}$. Note that it follows from (4.15) that

$$\nabla_{\nu}^{i}f_{n+1}^{[\mu]}-\nabla_{\omega}^{i}f_{n+1}^{[\mu]}=f_{n+1}^{[\nu]}-f_{n+1}^{[\omega]}. \tag{4.16}$$

We now make use of Property 3 of §3.10, amended to take account of the notation we have just introduced. It is easily seen that the proof of Property 3 is unaffected by the particular value taken by f_{n+1}, so that we may write the result in the form

$$\sum_{i=0}^{k-1}(\gamma_{i}\nabla_{\nu}^{i}f_{n+1}^{[\mu]}-\gamma_{i}^{*}\nabla^{i}f_{n}^{[\mu]})=\gamma_{k-1}^{*}\nabla_{\nu}^{k}f_{n+1}^{[\mu]} \tag{4.17}$$

where the notation is defined by (4.15).

We now apply the pair (4.14) in $P(EC)^{\mu}E^{1-t}$ mode, and use the structure of the Adams methods to develop a form of the ABM method which is computationally convenient and economical. In what follows, those equations which constitute the implementation are enclosed in boxes. The mode is defined by

$$\text{P:}\qquad \boxed{y_{n+1}^{[0]}=y_{n}^{[\mu]}+h\sum_{i=0}^{k-1}\gamma_{i}^{*}\nabla^{i}f_{n}^{[\mu-t]}} \tag{4.18}$$

$$(\text{EC})^{\mu}:\qquad \left.\begin{aligned} f_{n+1}^{[\nu]}&=f(x_{n+1},y_{n+1}^{[\nu]})\\ y_{n+1}^{[\nu+1]}&=y_{n}^{[\mu]}+h\sum_{i=0}^{k-1}\gamma_{i}\nabla_{\nu}^{i}f_{n+1}^{[\mu-t]}\end{aligned}\right\}\ \nu=0,1,\dots,\mu-1 \tag{4.19}$$

$$\text{E}^{1-t}:\qquad f_{n+1}^{[\mu]}=f(x_{n+1},y_{n+1}^{[\mu]})\qquad \text{if } t=0.$$

If we were to apply (4.19) as it stands then we would need to compute and store the differences $\nabla_{\nu}^{i}f_{n+1}^{[\mu-t]}$ for $i=1,2,\dots,k-1$ for each call, $\nu=0,1,\dots,\mu-1$, of the corrector. The computational effort can be reduced, by the following approach, to the computation of just one such difference.

Subtracting (4.18) from (4.19) with $\nu=0$ gives

$$y_{n+1}^{[1]}-y_{n+1}^{[0]}=h\sum_{i=0}^{k-1}(\gamma_{i}\nabla_{0}^{i}f_{n+1}^{[\mu-t]}-\gamma_{i}^{*}\nabla^{i}f_{n}^{[\mu-t]})$$

and on using (4.17) we obtain

$$\boxed{y_{n+1}^{[1]}=y_{n+1}^{[0]}+h\gamma_{k-1}^{*}\nabla_{0}^{k}f_{n+1}^{[\mu-t]}.} \tag{4.20}$$

Now subtract successive equations in (4.19) to get

$$y_{n+1}^{[\nu+1]} - y_{n+1}^{[\nu]} = h\sum_{i=0}^{k-1} \gamma_i(\nabla_\nu^i f_{n+1}^{[\mu-t]} - \nabla_{\nu-1}^i f_{n+1}^{[\mu-t]}), \qquad \nu = 1, 2, \ldots, \mu - 1.$$

But, by (4.16),

$$\nabla_\nu^i f_{n+1}^{[\mu-t]} - \nabla_{\nu-1}^i f_{n+1}^{[\mu-t]} = f_{n+1}^{[\nu]} - f_{n+1}^{[\nu-1]}.$$

Moreover, $\sum_{i=0}^{k-1}\gamma_i = \gamma_{k-1}^*$, by Property 1 of §3.10, and hence we have that

$$\boxed{y_{n+1}^{[\nu+1]} = y_{n+1}^{[\nu]} + h\gamma_{k-1}^*(f_{n+1}^{[\nu]} - f_{n+1}^{[\nu-1]}), \qquad \nu = 1, 2, \ldots, \mu - 1.} \tag{4.21}$$

To apply the Milne estimate, we need to compute $y_{n+1}^{[\mu]} - y_{n+1}^{[0]}$. Subtracting (4.18) from (4.19) with $\nu = \mu - 1$ gives

$$\begin{aligned} y_{n+1}^{[\mu]} - y_{n+1}^{[0]} &= h\sum_{i=0}^{k-1} (\gamma_i \nabla_{\mu-1}^i f_{n+1}^{[\mu-t]} - \gamma_i^* \nabla^i f_n^{[\mu-t]}) \\ &= h\gamma_{k-1}^* \nabla_{\mu-1}^k f_{n+1}^{[\mu-t]}, \end{aligned}$$

by (4.17).

(Alternatively, we could add all the equations in (4.21) and add (4.20) to the result to get

$$\begin{aligned} y_{n+1}^{[\mu]} - y_{n+1}^{[0]} &= h\gamma_{k-1}^*(f_{n+1}^{[\mu-1]} - f_{n+1}^{[0]} + \nabla_0^k f_{n+1}^{[\mu-t]}) \\ &= h\gamma_{k-1}^* \nabla_{\mu-1}^k f_{n+1}^{[\mu-t]}, \end{aligned}$$

by (4.15).) Since $C_{p+1}^* = \gamma_k^*$ and $C_{p+1} = \gamma_k$, the Milne estimate (4.11) for the PLTE at x_{n+1} (which we shall denote by T_{n+1}) is given by

$$T_{n+1} = \frac{C_{p+1}}{C_{p+1}^* - C_{p+1}}(y_{n+1}^{[\mu]} - y_{n+1}^{[0]}) = \frac{\gamma_k}{\gamma_k^* - \gamma_k} h\gamma_{k-1}^* \nabla_{\mu-1}^k f_{n+1}^{[\mu-t]}.$$

But, by Property 2 of §3.10, $\gamma_k^* - \gamma_k = \gamma_{k-1}^*$, whence

$$\boxed{T_{n+1} = h\gamma_k \nabla_{\mu-1}^k f_{n+1}^{[\mu-t]}.} \tag{4.22}$$

Note that in the case $t = 1$, $T_{n+1} = h\gamma_k \nabla^k f_{n+1}^{[\mu-1]}$.

The actual computation takes the form of implementing each of the boxed results (4.18), (4.20), (4.21) and (4.22). Note that the implementation of (4.18) and (4.20) and each call of (4.21) if $t = 0$ (each call except the last if $t = 1$) is followed by a function evaluation, making $\mu + 1 - t$ evaluations in all, which must be the case for a $P(EC)^\mu E^{1-t}$ mode.

It is assumed that the back data need in (4.18), namely the differences $\nabla^i f_n^{[\mu-t]}, i = 0, 1, \ldots, k-1$, have been stored. The difference $\nabla_0^k f_{n+1}^{[\mu-t]}$ needed in (4.20) can be obtained by computing

$$\nabla_0^{i+1} f_{n+1}^{[\mu-t]} = \nabla_0^i f_{n+1}^{[\mu-t]} - \nabla^i f_n^{[\mu-t]}$$

for $i = 0, 1, \ldots, k-1$. Using (4.16), the difference $\nabla_{\mu-1}^k f_{n+1}^{[\mu-t]}$ appearing in (4.22) can be computed from

$$\nabla_{\mu-1}^k f_{n+1}^{[\mu-t]} = \nabla_0^k f_{n+1}^{[\mu-t]} + f_{n+1}^{[\mu-1]} - f_{n+1}^{[0]}.$$

Finally, the back data can be updated by computing

$$\nabla^{i+1} f_{n+1}^{[\mu-t]} = \nabla^i f_{n+1}^{[\mu-t]} - \nabla^i f_n^{[\mu-t]}, \qquad i = 0, 1, \ldots, k-2,$$

or by adding $f_{n+1}^{[\mu-t]} - f_{n+1}^{[0]}$ to $\nabla_0^i f_{n+1}^{[\mu-t]}, i = 1, 2, \ldots, k-1$. We are then ready to compute the next step.

The computation takes a particularly simple form in the case of the PECE^{1-t} mode (a popular choice). Of the four boxed equations, (4.21) no longer applies and the remaining three become

$$\left.\begin{aligned} y_{n+1}^{[0]} &= y_n^{[1]} + h \sum_{i=0}^{k-1} \gamma_i^* \nabla^i f_n^{[1-t]} \\ y_{n+1}^{[1]} &= y_{n+1}^{[0]} + h\gamma_{k-1}^* \nabla_0^k f_{n+1}^{[1-t]} \\ T_{n+1} &= h\gamma_k \nabla_0^k f_{n+1}^{[1-t]} \end{aligned}\right\} \tag{4.23}$$

and we note that the *same* difference now appears on the right sides of the last two equations.

We illustrate the above procedure by considering the case of a third-order ABM method in P(EC)^2 mode. Recall from §3.9 that $\gamma_0^* = 1$, $\gamma_1^* = \frac{1}{2}$, $\gamma_2^* = \frac{5}{12}$ and $\gamma_3 = -\frac{1}{24}$; note that we do not need γ_0, γ_1 or γ_2. We assume that the back data $f_n^{[1]}, \nabla f_n^{[1]}$ and $\nabla^2 f_n^{[1]}$ are available. The sequence of sub-steps for the integration step from x_n to x_{n+1} is

P: $\quad y_{n+1}^{[0]} = y_n^{[2]} + h(f_n^{[1]} + \frac{1}{2}\nabla f_n^{[1]} + \frac{5}{12}\nabla^2 f_n^{[1]})$

E: $\quad f_{n+1}^{[0]} = f(x_{n+1}, y_{n+1}^{[0]})$

$$\nabla_0 f_{n+1}^{[1]} = f_{n+1}^{[0]} - f_n^{[1]}$$

$$\nabla_0^2 f_{n+1}^{[1]} = \nabla_0 f_{n+1}^{[1]} - \nabla f_n^{[1]}$$

$$\nabla_0^3 f_{n+1}^{[1]} = \nabla_0^2 f_{n+1}^{[1]} - \nabla^2 f_n^{[1]}$$

C: $\quad y_{n+1}^{[1]} = y_{n+1}^{[0]} + \frac{5}{12} h \nabla_0^3 f_{n+1}^{[1]}$

E: $\quad f_{n+1}^{[1]} = f(x_{n+1}, y_{n+1}^{[1]})$

C: $\quad y_{n+1}^{[2]} = y_{n+1}^{[1]} + \frac{5}{12} h(f_{n+1}^{[1]} - f_{n+1}^{[0]})$

$$\nabla_1^3 f_{n+1}^{[1]} = \nabla_0^3 f_{n+1}^{[1]} + f_{n+1}^{[1]} - f_{n+1}^{[0]}$$

Error $\quad T_{n+1} = -\frac{1}{24} h \nabla^3 f_{n+1}^{[1]}$

Update $\quad \nabla f_{n+1}^{[1]} = f_{n+1}^{[1]} - f_n^{[1]}$

$$\nabla^2 f_{n+1}^{[1]} = \nabla f_{n+1}^{[1]} - \nabla f_n^{[1]}.$$

On expanding the differences in terms of function values, we easily find that

$$y_{n+1}^{[0]} = y_n^{[2]} + \frac{h}{12}(23f_n^{[1]} - 16f_{n-1}^{[1]} + 5f_{n-2}^{[1]})$$

$$\tfrac{5}{12}h\nabla_0^3 f_{n+1}^{[1]} = \frac{5h}{12}(f_{n+1}^{[0]} - 3f_n^{[1]} + 3f_{n-1}^{[1]} - f_{n-2}^{[1]})$$

whence

$$y_{n+1}^{[1]} = y_n^{[2]} + \frac{h}{12}(5f_{n+1}^{[0]} + 8f_n^{[1]} - f_{n-1}^{[1]})$$

and

$$y_{n+1}^{[2]} = y_n^{[2]} + \frac{h}{12}(5f_{n+1}^{[1]} + 8f_n^{[1]} - f_{n-1}^{[1]}).$$

These equations will be recognized as the third-order Adams–Bashforth and Adams–Moulton methods, now in standard linear multistep form (see Table 3.2, §3.11), implemented in $P(EC)^2$ mode. Further,

$$\begin{aligned} T_{n+1} &= -\tfrac{1}{24}h(f_{n+1}^{[1]} - 3f_n^{[1]} + 3f_{n-1}^{[1]} + f_{n-2}^{[1]}) \\ &= -\tfrac{1}{24}h^4 \frac{d^3}{dx^3} f(x_n, y(x_n)) + 0(h^5) \\ &= -\tfrac{1}{24}h^4 y^{(4)}(x_n) + 0(h^5), \end{aligned}$$

the expected result, since the error constant of the third-order Adams–Moulton method is $-\frac{1}{24}$. We have thus demonstrated that the implementation developed in this section is equivalent to the predictor–corrector method in standard form.

Finally, we consider the effect of incorporating local extrapolation into an ABM method. From (4.19) with $v = \mu - 1$ we have that

$$y_{n+1}^{[\mu]} = y_n^{[\mu]} + h\sum_{i=0}^{k-1} \gamma_i \nabla_{\mu-1}^i f_{n+1}^{[\mu-t]}$$

while, from (4.22),

$$T_{n+1} = h\gamma_k \nabla_{\mu-1}^k f_{n+1}^{[\mu-t]}.$$

Thus local extrapolation, that is, adding T_{n+1} to $y_{n+1}^{[\mu]}$ has the same effect as replacing k by $k+1$. In other words, local extrapolation is equivalent to replacing the kth-order Adams–Moulton corrector by the $(k+1)$th-order Adams–Moulton corrector *throughout* the argument leading to the boxed equations; note that the equivalence is thus to the $P(ECL)^\mu E^{1-t}$ mode, *not* to the $P(EC)^\mu LE^{1-t}$ mode, a result which will be corroborated when we consider, in §4.6, stability regions for predictor–corrector methods.

Some additional notation enables us to express this result more succinctly. Let the kth-order Adams–Bashforth and the kth-order Adams–Moulton methods be denoted by $P_{(k)}$ and $C_{(k)}$ respectively, where, as in §3.10, the subscript k denotes order and not stepnumber. (The brackets round the subscript are necessary to avoid confusion with the error constant, C_{k+1}.) Using the notation defined by (3.97) of §3.10 and noting that in an ABM method the corrector is right-shifted by one steplength, $P_{(k)}$ and $C_{(k)}$ can be defined in terms of the polynomials $\rho_k^*(r), \sigma_k^*(r), \rho_k(r)$ and $\sigma_k(r)$ as follows:

$$\left.\begin{aligned} &P_{(k)}:[\rho_k^*(r),\sigma_k^*(r)], \qquad C_{(k)}:[r\rho_k(r), r\sigma_k(r)] \\ &\rho_k^*(r)=r^k-r^{k-1}, \qquad \rho_k(r)=r^{k-1}-r^{k-2}, \end{aligned}\right\} \tag{4.24}$$
$$\sigma_k^*(r) \text{ and } \sigma_k(r) \text{ have degree } k-1$$

by (3.97). The result we have established above may now be written as

$$P_{(k)}(EC_{(k)}L)^{\mu}E^{1-t} \equiv P_{(k)}(EC_{(k+1)})^{\mu}E^{1-t}. \tag{4.25}$$

If we replace k by $k+1$ in the corrector (4.19) (but not in the predictor (4.18)) and repeat the previous working, we easily find (by way of Property 2 of §3.10) that (4.18) holds without change and that (4.20) and (4.21) are replaced by

$$y_{n+1}^{[1]} = y_{n+1}^{[0]} + h\gamma_k^* \nabla_0^k f_{n+1}^{[\mu-t]}$$

and

$$y_{n+1}^{[\nu+1]} = y_{n+1}^{[\nu]} + h\gamma_k^*(f_{n+1}^{[\nu]} - f_{n+1}^{[\nu-1]}), \qquad \nu = 1,2,\dots,\mu-1$$

respectively.

Exercises

4.4.1. In the above section we illustrated the application of ABM methods in backward difference form by examining the case of a third-order ABM method in $P(EC)^2$ mode, and showed that the process was equivalent to that obtained by applying the methods in standard form. Do likewise for the fourth-order ABM method in PECE mode.

4.4.2*. Devise and carry out a numerical experiment, using the problem of Exercise 4.3.2, to test the accuracy of the local error estimate given by the fourth-order ABM method in PECE mode. (Note the result of Henrici stated in Exercise 4.3.5*.)

4.5 PREDICTOR–CORRECTOR METHODS BASED ON GENERAL LINEAR MULTISTEP METHODS IN BACKWARD DIFFERENCE FORM

In §3.11, we saw that it was possible to express, in backward difference form, the general linear k-step method with $k-1$ free parameters, the order of such methods being k if explicit and $k+1$ if implicit. We recall that such formulations involved expressing the linear multistep methods as linear combinations of s-Adams methods. It is natural to ask whether predictor–corrector methods based on these forms can be implemented in a manner similar to that discussed for Adams methods in the preceding section. It turns out that this is indeed possible, *provided that the predictor and the corrector have the same left side*, that is provided that $\rho^* \equiv \rho$. Of course, we must also have $p^* = p$ in order for the Milne estimate to be available. We may assume $k \geqslant 2$ (see comment after (3.108)).

From (3.110), we may write the implicit linear k-step method with $k-1$ free parameters as

$$\sum_{s=0}^{k-1} A_s \nabla y_{n+1-s} = h \sum_{s=0}^{k-1} A_s \sum_{i=0}^{k} \gamma_i^{s+1} \nabla^i f_{n+1}, \tag{4.26}$$

where $A_0 = 1$. This method has order $k+1$ and error constant $C_{k+2} = \sum_{s=0}^{k-1} A_s \gamma_{k+1}^{s+1}$. Since, with a k-step predictor of similar structure, we can achieve only order k, it is necessary to reduce the order of (4.26) to k, that is to replace k by $k-1$ in (4.26), yielding

$$\sum_{s=0}^{k-2} A_s \nabla y_{n+1-s} = h \sum_{s=0}^{k-2} A_s \sum_{i=0}^{k-1} \gamma_i^{s+1} \nabla^i f_{n+1}, \qquad A_0 = 1. \tag{4.27}$$

Note that this is the usual situation where, in order to achieve parity of order, the corrector has to have a smaller stepnumber than the predictor. We shall use (4.27) as corrector; note that it has $k-2$ free parameters, order k and error constant

$$C_{k+1} = \sum_{s=0}^{k-2} A_s \gamma_k^{s+1}. \tag{4.28}$$

The explicit linear k-step method with $k-1$ free parameters can, by (3.111), be written as

$$\sum_{s=0}^{k-1} A_s \nabla y_{n+1-s} = h \sum_{s=0}^{k-1} A_s \sum_{i=0}^{k-1} \gamma_i^{s} \nabla^i f_n, \qquad A_0 = 1. \tag{4.29}$$

It has order k and error constant $C_{k+1}^* = \sum_{s=0}^{k-1} A_s \gamma_k^s$. In order that the left sides of predictor and corrector be identical, we must set $A_{k-1} = 0$ in (4.29), yielding

$$\sum_{s=0}^{k-2} A_s \nabla y_{n+1-s} = h \sum_{s=0}^{k-2} A_s \sum_{i=0}^{k-1} \gamma_i^{s} \nabla^i f_n, \qquad A_0 = 1. \tag{4.30}$$

We shall use (4.30) as predictor; like the corrector (4.27), it has $k-2$ free parameters and order k, but its error constant is

$$C_{k+1}^* = \sum_{s=0}^{k-2} A_s \gamma_k^s. \tag{4.31}$$

Note that when $k=2$, (4.27) and (4.30) become the second-order Adams–Bashforth and Adans–Moulton methods respectively.

We need the following natural generalization of Property 3 (§3.10) of the Adams coefficients; its proof, which is almost identical with that of Property 3, uses the fact that $\gamma_j^{s+1} - \gamma_j^s = -\gamma_{j-1}^s$, which follows from (3.104).

$$\sum_{i=0}^{k-1} (\gamma_i^{s+1} \nabla^i f_{n+1} - \gamma_i^s \nabla^i f_n) = \gamma_{k-1}^s \nabla^k f_{n+1}, \qquad s = 0, 1, 2, \ldots,$$

where the γ_i^s, $s = 0, 1, 2, \ldots$ are defined by (3.10). This result can be amended, as in the preceding section, to read

$$\sum_{i=0}^{k-1} (\gamma_i^{s+1}\nabla_v^i f_{n+1}^{[\mu]} - \gamma_i^s \nabla^i f_n^{[\mu]}) = \gamma_{k-1}^s \nabla_v^k f_{n+1}^{[\mu]}, \quad s = 0, 1, 2, \ldots, \tag{4.32}$$

where the notation is defined by (4.15).

We apply the k-step kth-order pair (4.30) and (4.27) in $P(EC)^\mu E^{1-t}$ mode and, by an argument which is exactly parallel to that used in the preceding section and which uses (4.32), we obtain the following sequence of sub-steps:

$$\left.\begin{aligned}
y_{n+1}^{[0]} &= y_n^{[\mu]} - \sum_{s=1}^{k-2} A_s \nabla y_{n+1-s}^{[\mu]} + h\sum_{s=0}^{k-2} A_s \sum_{i=0}^{k-1} \gamma_i^s \nabla^i f_n^{[\mu-t]}\\
y_{n+1}^{[1]} &= y_{n+1}^{[0]} + h\left(\sum_{s=0}^{k-2} A_s \gamma_{k-1}^s\right)\nabla_0^k f_{n+1}^{[\mu-t]}\\
y_{n+1}^{[v+1]} &= y_{n+1}^{[v]} + h\left(\sum_{s=0}^{k-2} A_s \gamma_{k-1}^s\right)(f_{n+1}^{[v]} - f_{n+1}^{[v-1]}), \qquad v = 1, 2, \ldots, \mu - 1\\
T_{n+1} &= h\left(\sum_{s=0}^{k-2} A_s \gamma_k^{s+1}\right)\nabla_{\mu-1}^k f_{n+1}^{[\mu-t]}
\end{aligned}\right\} \tag{4.33}$$

where $A_0 = 1$ throughout. On comparing these equations with the boxed equations (4.18), (4.20), (4.21) and (4.22) of the preceding section, we see that, despite the somewhat formidable notation, the extension is really quite straightforward. The structure is preserved and the same differences have to be computed and updated (apart from the trivial addition of the differences $\nabla y_{n+1-s}^{[\mu]}$ in the first equation). The remaning equations differ from their counterparts in the preceding section only inasmuch as γ_{k-1}^* is replaced by $\sum_{s=0}^{k-2} A_s \gamma_{k-1}^s$ and γ_k by $\sum_{s=0}^{k-2} A_s \gamma_k^{s+1}$.

It should be noted that there is no possibility of choosing the free parameters A_s, $s = 1, 2, \ldots, k-2$ so as to increase the order of the method (although of course local extrapolation can still be performed). Were we to choose the A_s to force the corrector to have order $k+1$, then we would also have to ensure that the order of the predictor was also $k+1$, since otherwise the Milne estimate would not be applicable; but there exist no explicit linear k-step methods of order $k+1$ which satisfy the root condition and, since $\rho^* \equiv \rho$, if the predictor fails to satisfy the root condition then so too does the corrector and the PC method becomes zero-unstable. The free parameters could, however, be used to attempt to improve the regions of absolute stability of the PC method or to reduce its PLTE.

Again as in the preceding section, we find that applying local extrapolation to the mode defined by (4.33) (in $P(ECL)^\mu E^{1-t}$ mode) is equivalent to replacing the kth-order corrector (4.27) by the $(k+1)$th-order corrector given by (4.26) with $A_{k-1} = 0$.

4.6 LINEAR STABILITY THEORY FOR PREDICTOR–CORRECTOR METHODS

Recall that in §3.8 we considered the application of a linear multistep method to the test equation $y' = Ay$, where the eigenvalues λ_t, $t = 0, 1, \ldots, m$ of A are distinct and have negative real parts, and deemed the method to be absolutely stable if all solutions $\{y_n\}$ of the difference system resulting from applying the method to the test equation tended to zero as n tended to infinity. By means of the transformation $y_n = Qz_n$, where $Q^{-1}AQ = \mathrm{diag}[\lambda_1, \lambda_2, \ldots, \lambda_m]$, we saw that it was enough to apply the method to the scalar test equation $y' = \lambda y$, where λ, a complex scalar, represented any eigenvalue of A. The characteristic polynomial of the resulting scalar linear constant coefficient difference equation, which we called the stability polynomial, $\pi(r, \hat{h})$ where $\hat{h} = h\lambda$, determined the region of absolute stability of the method.

A similar aproach can be applied to predictor–corrector methods. It is not difficult to show that the same diagonalizing transformation $y_n = Qz_n$ uncouples the general predictor–corrector method so that, once again, it is enough to consider the scalar test equation $y' = \lambda y$. Our first task is to find the stability polynomial for the $\mathrm{P(EC)^{\mu}E^{1-t}}$ mode defined by (4.4) or (4.5). Note that this will be the characteristic polynomial of the difference equation for the *final* value $y_n^{[\mu]}$.

It is highly advantageous to define the mode in operator notation as in (4.5). Applying this to the test equation $y' = \lambda y$, we obtain

$$E^k y_n^{[0]} + [\rho^*(E) - E^k] y_n^{[\mu]} = \hat{h}\sigma^*(E) y_n^{[\mu - t]} \tag{4.34}$$

$$E^k y_n^{[\nu+1]} + [\rho(E) - E^k] y_n^{[\mu]} = \hat{h}\beta_k E^k y_n^{[\nu]} + \hat{h}[\sigma(E) - \beta_k E^k] y_n^{[\mu - t]}, \quad \nu = 0, 1, \ldots \mu - 1, \tag{4.35}$$

where, as in §3.8, we have written $\hat{h}$ for $h\lambda$. On defining

$$H := \hat{h}\beta_k \tag{4.36}$$

and subtracting successive equations in (4.35) we obtain

$$E^k(y_n^{[\nu+1]} - y_n^{[\nu]}) = HE^k(y_n^{[\nu]} - y_n^{[\nu-1]}), \qquad \nu = 1, 2, \ldots, \mu - 1,$$

which can be re-written as

$$y_n^{[\nu+1]} - (1 + H) y_n^{[\nu]} + H y_n^{[\nu-1]} = 0.$$

Regard this as a difference equation in $\{y_n^{[\nu]}, \nu = 0, 1, \ldots, \mu\}$. The equation is linear with constant coefficients, and its characteristic polynomial is $s^2 - (1 + H)s + H$ which has roots 1 and H. It follows from §1.7 that the solution for y_n takes the form

$$y_n^{[\nu]} = A_{n\mu} + B_{n\mu} H^{\nu}, \qquad \nu = 0, 1, \ldots, \mu - 1$$

where $A_{n\mu}$ and $B_{n\mu}$ are independent of ν. These constants can be evaluated if we regard any two of the $y_n^{[\nu]}$, $\nu = 0, 1, \ldots, \mu$ as being given boundary values. Choosing the boundary values to be $y_n^{[0]}$ and $y_n^{[\mu]}$, the resulting solution for $y_n^{[\nu]}$ turns out to be given by

$$(1 - H^{\mu}) y_n^{[\nu]} = y_n^{[\mu]} - H^{\mu} y_n^{[0]} + H^{\nu}(y_n^{[0]} - y_n^{[\mu]}), \qquad \nu = 0, 1, \ldots, \mu. \tag{4.37}$$

(It is easily checked that this solution has the required form and takes the chosen boundary values.) Now put $\nu = \mu - 1$ to get

$$(1 - H^{\mu}) y_n^{[\mu-1]} = H^{\mu-1}(1 - H) y_n^{[0]} + (1 - H^{\mu-1}) y_n^{[\mu]},$$

which, on defining

$$M_\mu(H) := \frac{H^\mu(1-H)}{1-H^\mu}, \tag{4.38}$$

can be rewritten in the form

$$Hy_n^{[\mu-1]} = M_\mu(H)y_n^{[0]} + [H - M_\mu(H)]y_n^{[\mu]}. \tag{4.39}$$

On eliminating $y_n^{[0]}$ between (4.34) and (4.39) we obtain, after a little rearranging,

$$HE^k y_n^{[\mu-1]} = [HE^k - M_\mu(H)\rho^*(E)]y_n^{[\mu]} + M_\mu(H)\hat{h}\sigma^*(E)y_n^{[\mu-t]}. \tag{4.40}$$

Since $t = 0$ or 1, this is an equation involving $y_n^{[\mu]}$ and $y_n^{[\mu-1]}$ only. Another such equation is afforded by setting $\nu = \mu - 1$ in (4.35):

$$\rho(E)y_n^{[\mu]} = HE^k y_n^{[\mu-1]} + [\hat{h}\sigma(E) - HE^k]y_n^{[\mu-t]}. \tag{4.41}$$

The final stage is to eliminate $y_n^{[\mu-1]}$ between (4.40) and (4.41) to obtain a difference equation in $y_n^{[\mu]}$.

Case $t = 0$ The elimination gives

$$[\rho(E) - \hat{h}\sigma(E) + HE^k]y_n^{[\mu]} = [HE^k - M_\mu(H)\rho^*(E) + M_\mu(H)\hat{h}\sigma^*(E)]y_n^{[\mu]}$$

or

$$\{\rho(E) - \hat{h}\sigma(E) + M_\mu(H)[\rho^*(E) - \hat{h}\sigma^*(E)]\}y_n^{[\mu]} = 0.$$

The stability polynomial, which is the characteristic polynomial of this difference equation, is therefore

$$\pi_{\mathrm{P(EC)^\mu E}}(r, \hat{h}) = \rho(r) - \hat{h}\sigma(r) + M_\mu(H)[\rho^*(r) - \hat{h}\sigma^*(r)], \tag{4.42}$$

where $M_\mu(H)$ and H are given by (4.38) and (4.36).

Case $t = 1$ Equations (4.41) and (4.40) now become

$$\rho(E)y_n^{[\mu]} = \hat{h}\sigma(E)y_n^{[\mu-1]}$$

and

$$[HE^k - M_\mu(H)\rho^*(E)]y_n^{[\mu]} = [HE^k - M_\mu(H)\hat{h}\sigma^*(E)]y_n^{[\mu-1]}.$$

On eliminating $y_n^{[\mu-1]}$ between these two equations we obtain

$$\{HE^k[\rho(E) - \hat{h}\sigma(E)] + \hat{h}M_\mu(H)[\rho^*(E)\sigma(E) - \rho(E)\sigma(E)]\}y_n^{[\mu]} = 0,$$

whence, by (4.36), we find the stability polynomial

$$\pi_{\mathrm{P(EC)^\mu}}(r, \hat{h}) = \beta_k r^k[\rho(r) - \hat{h}\sigma(r)] + M_\mu(H)[\rho^*(r)\sigma(r) - \rho(r)\sigma^*(r)]. \tag{4.43}$$

Our first observation is that, whereas the principal local truncation error of the $\mathrm{P(EC)^\mu E^{1-t}}$ mode is, in normal circumstances, that of the corrector alone, the linear stability characteristics of the $\mathrm{P(EC)^\mu E^{1-t}}$ mode are not those of the corrector and, moreover, differ markedly in the cases $t = 0$ and $t = 1$. Recalling from §3.8 that the stability polynomial of the linear multistep method defined by the polynomials $\rho(r)$ and $\sigma(r)$ is $\pi(r, \hat{h}) = \rho(r) - \hat{h}\sigma(r)$, we see from (4.42), (4.43) and (4.38) that the stability polynomial of the $P(EC)^{\mu}E^{1-t}$ mode is essentially an $O(\hat{h}^{\mu})$ perturbation of the stability polynomial of the corrector; $\pi_{P(EC)^{\mu}E}(r, \hat{h})$ has the simpler structure, and is a linear combination of the stability polynomials of the predictor and of the corrector.

From (4.36), $H = \hat{h}\beta_k = h\lambda\beta_k$, whence

$$|H| = h|\lambda|\ |\beta_k| \leqslant h\|A\|\ |\beta_k|$$

since λ represents any eigenvalue of A, and any norm of A is greater than the spectral radius of A. Further, we may take $\|A\|$ to be the Lipschitz constant L of the system $y' = Ay$, and we have that

$$|H| \leqslant hL|\beta_k| < 1$$

by (3.8). It then follows from (4.38) that $M_{\mu}(H) \to 0$ as $\mu \to \infty$. Thus, as we would expect, the stability polynomial of $P(EC)^{\mu}E^{1-t}$ tends (essentially) to that of C as $\mu \to \infty$. (The factor r^k in $\pi_{P(EC)^{\mu}}(r, \hat{h})$ has no effect in the limit as $\mu \to \infty$.)

As for linear multistep methods, a predictor–corrector method is said to be absolutely stable for given $\hat{h}$ if for that $\hat{h}$ all the roots r_s of the stability polynomial satisfy $|r_s| < 1$; the region $\mathscr{R}_A$ of the complex $\hat{h}$-plane in which the method is absolutely stable is the region of absolute stability. Such regions are found using the boundary locus technique described in §3.8. However, for predictor–corrector methods the polynomial $\pi(r, \hat{h})$ is nonlinear in $\hat{h}$, and we can no longer solve explicitly for $\hat{h}$ the equation $\pi(\exp(i\theta), \hat{h}) = 0$ which defines the boundary $\partial\mathscr{R}_A$ of the region $\mathscr{R}_A$. We can, however, solve numerically (by Newton iteration, for example) for a range of values of θ, and thus obtain a plot of $\partial\mathscr{R}_A$. There is a single exception to this, namely the PEC mode. From (4.43) and (4.38) it follows that

$$\frac{1}{\beta_k}\pi_{PEC}(r, \hat{h}) = r^k[\rho(r) - \hat{h}\sigma(r)] + \hat{h}[\rho^*(r)\sigma(r) - \rho(r)\sigma^*(r)] \tag{4.44}$$

which is linear in $\hat{h}$.

For linear multistep methods we showed that the root r_1 of the stability polynomial satisfied $r_1 = \exp(\hat{h}) + O(\hat{h}^{p+1})$ (see (3.72). The proof of this result hinged on the fact that $\pi(\exp(\hat{h}), \hat{h}) = O(\hat{h}^{p+1})$ (see (3.70)). Now if the predictor and corrector both have order p, then by the argument which led to (3.70), we have that

$$\rho^*(\exp(\hat{h})) - \hat{h}\sigma^*(\exp(\hat{h})) = O(\hat{h}^{p+1}), \qquad \rho(\exp(\hat{h})) - \hat{h}\sigma(\exp(\hat{h})) = O(\hat{h}^{p+1}),$$

and it follows that

$$\rho^*(\exp(\hat{h}))\sigma(\exp(\hat{h})) - \rho(\exp(\hat{h}))\sigma^*(\exp(\hat{h})) = O(\hat{h}^{p+1}).$$

From (4.42) and (4.43) it follows that $\pi_{P(EC)^{\mu}E^{1-t}}(\exp(\hat{h}), \hat{h}) = O(\hat{h}^{p+1})$ and consequently $r_1 = \exp(\hat{h}) + O(\hat{h}^{p+1})$. Thus, for predictor–corrector methods, just as for linear multistep methods, the region of absolute stability cannot contain the positive real axis in the neighbourhood of the origin.

Let us now consider the degree of the stability polynomials defined by (4.42) and (4.43). In general, $\pi_{P(EC)^{\mu}E}(r,\hat{h})$ has degree k whilst $\pi_{P(EC)^{\mu}}(r,\hat{h})$ has degree $2k$. Thus, to achieve absolute stability, twice as many roots have to be controlled for the P(EC)$^{\mu}$ mode as for the P(EC)$^{\mu}$E mode, suggesting that the latter class of methods will have the larger regions of absolute stability, a conjecture that we can go some way towards substantiating by the following observations. From (4.38) we have that for $\mu = 1, 2, \ldots$

$$\frac{HM_{\mu}(H)}{1+M_{\mu}(H)} = \frac{H^{\mu+1}(1-H)}{1-H^{\mu}+H^{\mu}(1-H)} = \frac{H^{\mu+1}(1-H)}{1-H^{\mu+1}} = M_{\mu+1}(H). \tag{4.45}$$

Now consider the mode P(EC)$^{\mu+1}$; by (4.43), its stability polynomial is

$$\pi_{P(EC)^{\mu+1}}(r,\hat{h}) = \beta_k r^k[\rho(r) - \hat{h}\sigma(r)] + M_{\mu+1}(H)[\rho^*(r)\sigma(r) - \rho(r)\sigma^*(r)].$$

By (4.45) and (4.36)

$$\begin{aligned}\frac{1+M_{\mu}(H)}{\beta_k}&\pi_{P(EC)^{\mu+1}}(r,\hat{h})\\ &= r^k[1+M_{\mu}(H)][\rho(r)-\hat{h}\sigma(r)] + \hat{h}M_{\mu}(H)[\rho^*(r)\sigma(r)-\rho(r)\sigma^*(r)]\\ &= r^k[\rho(r)-\hat{h}\sigma(r)] + M_{\mu}(H)\{r^k\rho(r) - \hat{h}[(r^k-\rho^*(r))\sigma(r)+\rho(r)\sigma^*(r)]\}.\end{aligned} \tag{4.46}$$

The form of the right side suggests that we make the following definitions:

$$\left.\begin{aligned}&\tilde{\rho}(r) := r^k\rho(r), \qquad \tilde{\sigma}(r) := r^k\sigma(r)\\ &\tilde{\rho}^*(r) := r^k\rho(r), \qquad \tilde{\sigma}^*(r) := [r^k - \rho^*(r)]\sigma(r) + \rho(r)\sigma^*(r).\end{aligned}\right\} \tag{4.47}$$

Now $\rho(r)$, $\sigma(r)$ and $\rho^*(r)$ all have degree k, while $\sigma^*(r)$ has degree $k-1$. It follows from (4.47) that $\tilde{\rho}(r)$, $\tilde{\sigma}(r)$ and $\tilde{\rho}^*(r)$ have degree $2k$, but $\tilde{\sigma}^*(r)$ has degree at $2k-1$ at most (since $\alpha_k^* = 1$). The linear multistep method with first and second characterstic polynomials $\tilde{\rho}^*(r)$, $\tilde{\sigma}^*(r)$ is therefore explicit, and we shall accordingly denote it by $\tilde{P}$. The linear multistep method similarly defined by the polynomials $\tilde{\rho}(r)$, $\tilde{\sigma}(r)$ is implicit, and we shall denote it by $\tilde{C}$; note that $\tilde{C}$ is just C right-shifted by k steplengths. With this notation, (4.46) can be written as

$$\frac{1+M_{\mu}(H)}{\beta_k}\pi_{P(EC)^{\mu+1}}(r,\hat{h}) = \pi_{\tilde{P}(E\tilde{C})^{\mu}E}(r,\hat{h}), \qquad \mu = 1, 2, \ldots. \tag{4.48}$$

It follows directly from (4.44) that a similar result holds in the case $\mu = 0$, namely

$$\frac{1}{\beta_k}\pi_{PEC}(r,\hat{h}) = \pi_{\tilde{P}}(r,\hat{h}), \tag{4.49}$$

where $\pi_{\tilde{P}}(r,\hat{h})$ is the stability polynomial of the explicit linear multistep method $\tilde{P}$. The factors $[1+M_{\mu}(H)]/\beta_k$ in (4.48) and $1/\beta_k$ in (4.49) being independent of r, we conclude that

$$\mathscr{R}_A[P(EC)^{\mu+1}] = \mathscr{R}_A[\tilde{P}(E\tilde{C})^{\mu}E], \qquad \mu = 0, 1, \ldots, \tag{4.50}$$

with the obvious interpretation that, in the case $\mu = 0$, $\tilde{P}(E\tilde{C})^{\mu}E = \tilde{P}E = \tilde{P}$.

It can be shown that $\tilde{P}$ has the same order and error constant as P. The mode $P(EC)^{\mu+1}$ is not, however, computationally equivalent to $\tilde{P}(E\tilde{C})^{\mu}E$, although there does exist a relationship between the solutions computed by the two modes; see Lambert (1971) for fuller details. Clearly, $\tilde{C}$ has the same order and error constant as C; moreover, the Milne estimate for the PLTE will be the same for the two modes $P(EC)^{\mu+1}$ and $\tilde{P}(E\tilde{C})^{\mu}E$. Thus, to any $P(EC)^{\mu+1}$ mode there corresponds a $\tilde{P}(E\tilde{C})^{\mu}E$ mode with the same PLTE and Milne estimate, and with identical region of absolute stability. The converse is not true, and it follows that if we look for the predictor–corrector method costing $\mu+1$ evaluations per step which has the 'best' stability region, then we ought to search the class of $P(EC)^{\mu}E$ methods. Such a 'best' method may or may not be replaceable by a $P(EC)^{\mu+1}$ method with the same PLTE and stability region; when such a replacement is possible, the stepnumber will be reduced.

The relationships (4.47) linking $P(EC)^{\mu+1}$ to $\tilde{P}(E\tilde{C})^{\mu}E$ take a particularly simple form in the case of ABM methods. For a kth-order ABM, we have from (4.24) that the polynomials $[\rho^*, \sigma^*]$, $[\rho, \sigma]$ defining P and C respectively are given by

$$\rho^*(r) = \rho(r) = r^{k-1}(r-1), \qquad \sigma^*(r) = \sigma_k^*(r), \qquad \sigma(r) = r\sigma_k(r).$$

Substituting these in (4.47) gives

$$\tilde{\rho}(r) = \tilde{\rho}^*(r) = r^{2k-1}(r-1),$$
$$\tilde{\sigma}(r) = r^{k+1}\sigma_k(r), \qquad \tilde{\sigma}^*(r) = r^{k-1}[r\sigma_k(r) + (r-1)\sigma_k^*(r)].$$

All four polynomials $\tilde{\rho}(r), \tilde{\rho}^*(r), \tilde{\sigma}(r)$ and $\tilde{\sigma}^*(r)$ now have a common factor r^{k-1} which can be disregarded, since zero roots of the stability polynomial do not have any effect on the region of absolute stability. We may therefore replace (4.47) by

$$\tilde{\rho}(r) = \tilde{\rho}^*(r) = r^k(r-1), \qquad \tilde{\sigma}(r) = r^2\sigma_k(r),$$
$$\tilde{\sigma}^*(r) = r\sigma_k(r) + (r-1)\sigma_k^*(r)$$

and the $\tilde{P}(E\tilde{C})^{\mu}E$ method now has stepnumber $k+1$ rather than $2k$. Consider, for example, the second-order ABM for which

$$\rho_2^*(r) = r^2 - r, \quad \sigma_2^*(r) = (3r-1)/2, \quad \rho_2(r) = r-1, \quad \sigma_2(r) = (r+1)/2$$

and we obtain

$$\tilde{P}: \tilde{\rho}^*(r) = r^2(r-1), \qquad \tilde{\sigma}^*(r) = (4r^2 - 3r + 1)/2$$

$$\tilde{C}: \tilde{\rho}(r) = r^2(r-1), \qquad \tilde{\sigma}(r) = r^2(r+1)/2.$$

We note that the stepnumber is indeed 3 and that $\tilde{P}$ is explicit.

The regions of absolute stability of the kth-order ABM methods, $k = 1, 2, 3, 4$, in PEC, PECE and $P(EC)^2$ modes are shown in Figure 4.1.

The stability region of the kth-order Adams–Bashforth method is also included and is labelled PE. As was the case for linear multistep methods, all these regions are symmetric about the real axis, and Figure 4.1 shows the regions only in the half-plane $\text{Im}(\hat{h}) > 0$; note that the scale in Figure 4.1 is larger than in Figures 3.2 and 3.3 of §3.8. We have had occasion to remark previously that $k = 1$ gives a somewhat anomalous

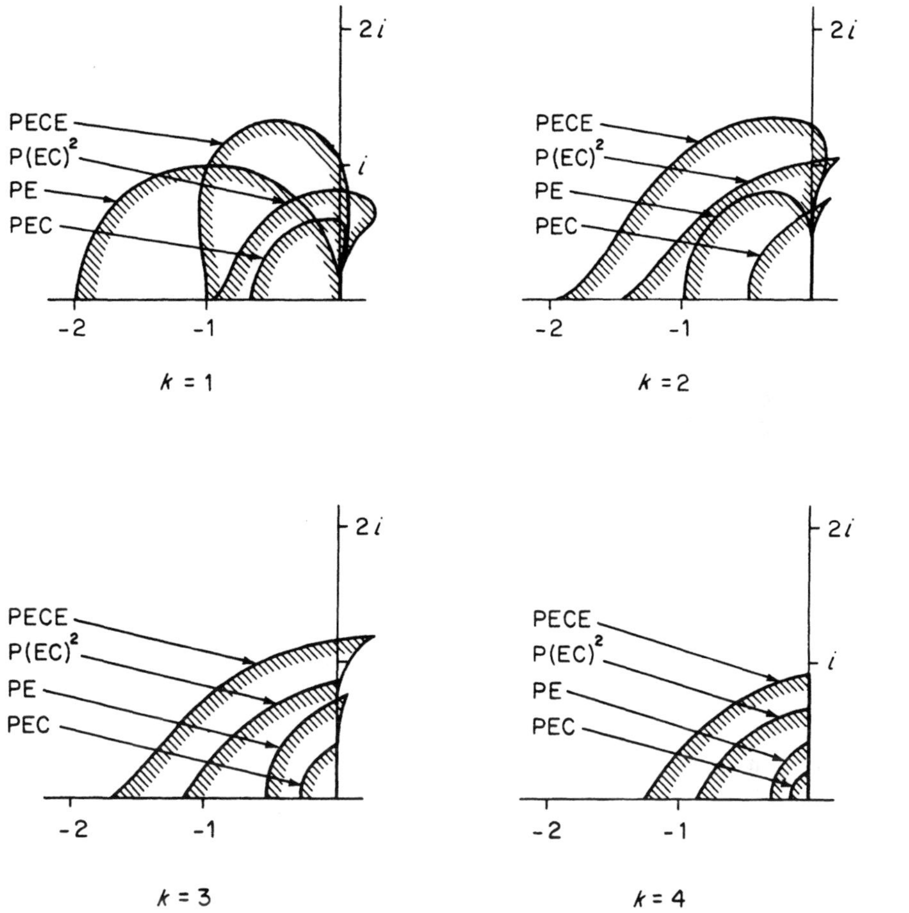

Figure 4.1 Regions of absolute stability for *k*th-order ABM methods.

member of the ABM family, and that is again the case in Figure 4.1. For the cases $k = 2, 3, 4$ there is a clear pecking order in terms of size of stability region, namely (in descending order) PECE, $P(EC)^2$, *PE*, *PEC*. That PEC has poorer stability than PE (the straight Adams–Bashforth method) is not surprising; we have already seen that *PEC* has the same stability region as an explicit method $\tilde{P}$, which turns out to have poorer stability than the Adams–Bashforth method. Comparison with Figure 3.2 of §3.8 shows that the stability region for the Adams–Moulton method in the mode of correcting to convergence is (not surprisingly) in every case greater than that for the modes displayed in Figure 4.1. With the exception of the anomalous case $k = 1$, the regions for each mode become smaller as the order increases.

From (4.42) we note that the stability polynomial of a $P(EC)^{\mu}E$ mode is a linear combination of the stability polynomials of the predictor and of the corrector. However, the region of absolute stability is a highly nonlinear function of the stability polynomial, so we cannot infer that if the predictor and the corrector separately have good stability regions then the $P(EC)^{\mu}E$ will also have a good stability region. (The same holds true *a fortiori* for the P(EC)$^\mu$ mode.) By way of illustration, let us construct a family of predictor–corrector methods in which the correctors are the BDF described in §3.12 (not, we add, the usual application of the BDF—see Chapter 6 for that.) The kth-order corrector is then given by (3.119), and a suitable kth-order predictor is

$$\tau_k \sum_{i=1}^{k} \delta_i \nabla^i y_{n+k} = h\tau_k(1 - \nabla^k) f_{n+k} \tag{4.51}$$

with the notation of §3.12. In the case $k = 1$, this predictor–corrector pair coincides with the first-order ABM, but for $k = 2, 3, 4$, the above predictor has a larger region of absolute stability than has the kth-order Adams–Bashforth. The BDF correctors have, as we have seen in Figure 3.4 of §3.12, infinite stability regions and are greatly superior to the Adams–Moulton methods in this respect. Yet the constructed BDF predictor–corrector pair in PECE mode has, for $k = 2, 3, 4$, stability regions which differ only slightly from those of the ABM.

Let us now consider the effect on absolute stability of local extrapolation, in the two classes of modes P(ECL)$^\mu$E^{1-t} and P(EC)$^\mu$LE^{1-t} defined by (4.12) and (4.13). By an argument similar to that which produced the stability polynomials (4.42) and (4.43) for the P(EC)$^\mu$E^{1-t} modes, the following stability polynomials can be derived:

$$\pi_{\mathrm{P(ECL)^\mu E}}(r, \hat{h}) = (1 + W)[\rho(r) - \hat{h}\sigma(r)] + [M_\mu(H + WH) - W][\rho^*(r) - \hat{h}\sigma^*(r)] \tag{4.52}$$

$$\begin{aligned}\pi_{\mathrm{P(ECL)^\mu}}(r, \hat{h}) = \beta_k r^k\{(1 + W)[\rho(r) - \hat{h}\sigma(r)] - W[\rho^*(r) - \hat{h}\sigma^*(r)]\}\\ + M_\mu(H + WH)[\rho^*(r)\sigma(r) - \rho(r)\sigma^*(r)]\end{aligned} \tag{4.53}$$

$$\pi_{\mathrm{P(EC)^\mu LE}}(r, \hat{h}) = (1 + W)[\rho(r) - \hat{h}\sigma(r)] + [M_\mu(H) + (H - 1)W][\rho^*(r) - \hat{h}\sigma^*(r)] \tag{4.54}$$

$$\begin{aligned}\pi_{\mathrm{P(EC)^\mu L}}(r, \hat{h}) = \beta_k r^k\{(1 + W)[\rho(r) - \hat{h}\sigma(r)] - W[\rho^*(r) - \hat{h}\sigma^*(r)]\}\\ + [M_\mu(H) + HW][\rho^*(r)\sigma(r) - \rho(r)\sigma^*(r)]\end{aligned} \tag{4.55}$$

where, as before,

$$H = \beta_k \hat{h}, \qquad M_\mu(H) = \frac{H^\mu(1 - H)}{1 - H^\mu}, \qquad W = \frac{C_{p+1}}{C^*_{p+1} - C_{p+1}} \tag{4.56}$$

(see (4.36), (4.38) and (4.11)).

Note that on putting $W = 0$, that is, *not* performing local extrapolation, (4.52) and (4.54) both revert to $\pi_{\mathrm{P(EC)^\mu E}}(r, \hat{h})$ given by (4.42), and (4.53) and (4.55) both revert to $\pi_{\mathrm{P(EC)^\mu}}(r, \hat{h})$ given by (4.43). Note also that in the case $\mu = 1$, when P(ECL)$^\mu$E$^{1-t} \equiv$ P(EC)$^\mu$LE^{1-t}, we have that $M_1(H) = H$ and it follows that

$$M_1(H + WH) - W = H + WH - W = H + (H - 1)W = M_1(H) + (H - 1)W,$$
$$M_1(H + WH) = H + WH = M_1(H) + WH,$$

and (4.52) and (4.54) coincide as do (4.53) and (4.55).

Let us now consider the case when the PC pair consists of a kth-order ABM method in P(ECL)$^\mu$E mode. By (4.25) we know that $\mathrm{P_{(k)}(EC_{(k)}L)^\mu E^{1-t}} \equiv \mathrm{P_{(k)}(EC_{(k+1)})^\mu E^{1-t}}$ so that the stability polynomials given by (4.52) and (4.53) with $\mathrm{P} = \mathrm{P}_{(k)}$, $C = C_{(k)}$ must coincide respectively with those given by (4.42) and (4.43) with $P = P_{(k)}$, $C = C_{(k+1)}$. To show that this is indeed the case takes a little work. By (4.24) $P_{(k)}$ and $C_{(k)}$ are defined by

$$\rho^*(r) = \rho_k^*(r), \qquad \sigma^*(r) = \sigma_k^*(r), \qquad \rho(r) = r\rho_k(r), \qquad \sigma(r) = r\sigma_k(r),$$

where $\rho_k^*(r) = r\rho_k(r) = r^k - r^{k-1}$. Substituting in (4.52) gives

$$\begin{aligned}\pi_{P_{(k)}(EC_{(k)}L)^{\mu}E}(r, \hat{h}) &= (1 + W)r[\rho_k(r) - \hat{h}\sigma_k(r)] + [M_\mu(H + WH) - W][\rho_k^*(r) - \hat{h}\sigma_k^*(r)]\\ &= r\rho_k(r) - \hat{h}[(1 + W)r\sigma_k(r) - W\sigma_k^*(r)] + M_\mu(H + WH)[\rho_k^*(r) - \hat{h}\sigma_k^*(r)].\end{aligned} \tag{4.57}$$

Now $H\ (= \hat{h}\beta_k)$ depends, through β_k, on k, and it is necessary in this argument to make that clear by writing H_k for H. By Property 4 of §3.10 we have

$$H_k = \hat{h}\gamma_{k-1}^*. \tag{4.58}$$

Further, by (3.96) of §3.9, the error constants of $P_{(k)}$ and $C_{(k)}$ are given by $C_{k+1}^* = \gamma_k^*$, $C_{k+1} = \gamma_k$, whence, by (4.56)

$$W = \gamma_k/(\gamma_k^* - \gamma_k) = \gamma_k/\gamma_{k-1}^*, \qquad 1 + W = \gamma_k^*/\gamma_{k-1}^* \tag{4.59}$$

where we have used Property 2 of §3.10. The terms on the right side of (4.57) can now be simplified. By (4.59),

$$[(1 + W)r\sigma_k(r) - W\sigma_k^*(r)] = [\gamma_k^* r\sigma_k(r) - \gamma_k\sigma_k^*(r)]/\gamma_{k-1}^* = \sigma_{k+1}(r), \qquad k \geqslant 2,$$

by Property 6 of §3.10. Further, by (4.58) and (4.59),

$$H_k + WH_k = (\gamma_k^*/\gamma_{k-1}^*)\hat{h}\gamma_{k-1}^* = \hat{h}\gamma_k^* = H_{k+1}.$$

Hence (4.57) now reads

$$\begin{aligned}\pi_{P_{(k)}(EC_{(k)}L)^{\mu}E}(r, \hat{h}) &= r\rho_k(r) - \hat{h}\sigma_{k+1}(r) + M_\mu(H_{k+1})[\rho_k^*(r) - \hat{h}\sigma_k^*(r)]\\ &= \pi_{P_{(k)}(EC_{(k+1)})^{\mu}E}(r, \hat{h}), \qquad k \geqslant 2\end{aligned} \tag{4.60}$$

by (4.42). Note that when a kth-order predictor is combined with a $(k+1)$th-order corrector in an ABM method, it is no longer necessary to right shift the corrector; thus $\rho(r) = r\rho_k(r), \sigma(r) = \sigma_{k+1}(r)$ properly define $C_{(k+1)}$ in this context. It is readily established by direct substitution that (4.60) also holds for $k = 1$. A similar argument shows that

$$\pi_{P_{(k)}(EC_{(k)}L)^{\mu}}(r, \hat{h}) = \pi_{P_{(k)}(EC_{(k+1)})^{\mu}}(r, \hat{h})$$

and we have demonstrated the required result.

There is one further point of interest to be extracted from the ABM case. Consider the mode $P_{(k)}(EC_{(k+1)})^{\mu+1}$; it is defined by setting

$$\rho^*(r) = \rho_k^*(r), \qquad \sigma^*(r) = \sigma_k^*(r), \qquad \rho(r) = \rho_{k+1}(r), \qquad \sigma(r) = \sigma_{k+1}(r),$$

where $\rho_k^*(r) = \rho_{k+1}(r) = r^k - r^{k-1}$. (See note following (4.60).) Now let us apply the equations (4.47) to get

$$\tilde{\rho}(r) = r^k\rho_{k+1}(r), \quad \tilde{\sigma}(r) = r^k\sigma_{k+1}(r), \quad \tilde{\rho}^*(r) = r^k\rho_k^*(r)$$

and

$$\begin{aligned}\tilde{\sigma}^*(r) &= (r^k - r^k + r^{k-1})\sigma_{k+1}(r) + (r^k - r^{k-1})\sigma_k^*(r)\\ &= r^{k-1}[\sigma_{k+1}(r) + (r-1)\sigma_k^*(r)]\\ &= r^{k-1}\sigma_{k+1}^*(r), \qquad k \geq 1\end{aligned}$$

by Property 7 of §3.10. On dividing out the common factor r^{k-1}, we have

$$\begin{aligned}&\tilde{\rho}(r) = r\rho_{k+1}(r), \quad \tilde{\sigma}(r) = r\sigma_{k+1}(r),\\ &\tilde{\rho}^*(r) = r\rho_k^*(r) = \rho_{k+1}^*(r), \quad \sigma^*(r) = \sigma_{k+1}^*(r)\end{aligned}$$

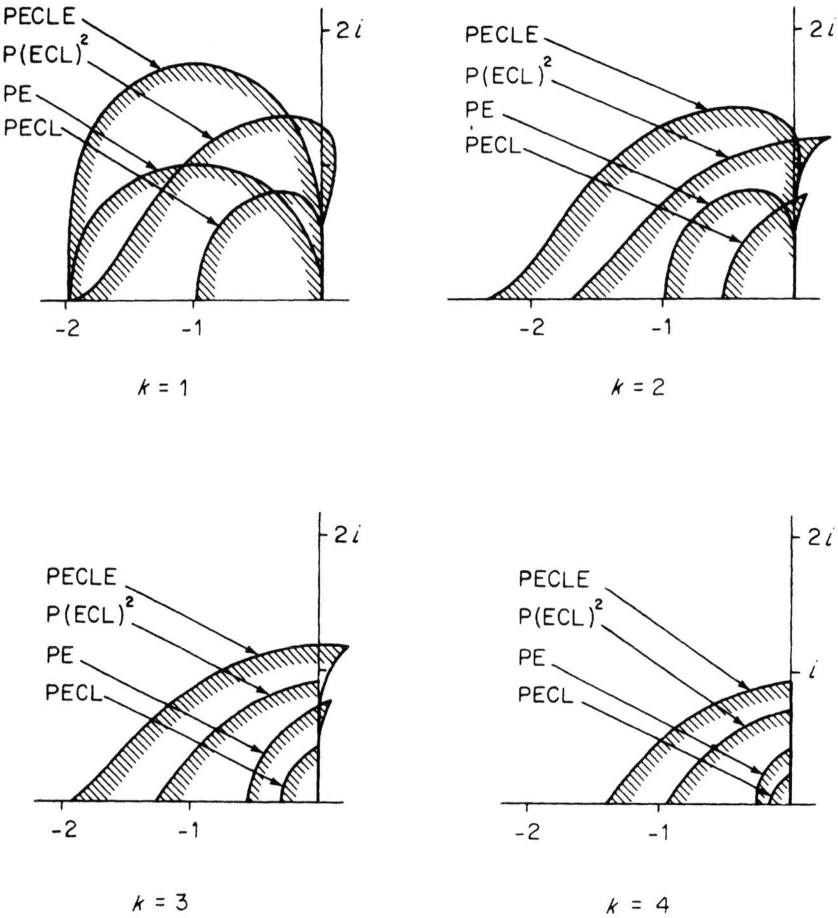

Figure 4.2 Regions of absolute stability for kth-order ABM methods with local extrapolation.

whence $\tilde{P} = P_{(k+1)}$, $\tilde{C} = C_{(k+1)}$, and it follows from (4.50) that

$$\mathscr{R}_A[\mathrm{P}_{(k)}(\mathrm{EC}_{(k+1)})^{\mu+1}] = \mathscr{R}_A[\mathrm{P}_{(k+1)}(\mathrm{EC}_{(k+1)})^{\mu}\mathrm{E}].$$

Combining this with the result (4.25) we have

$$\mathscr{R}_A[\mathrm{P}_{(k)}(\mathrm{EC}_{(k)}\mathrm{L})^{\mu+1}] = \mathscr{R}_A[\mathrm{P}_{(k+1)}(\mathrm{EC}_{(k+1)})^{\mu}\mathrm{E}]. \tag{4.61}$$

Thus, a kth-order ABM method applied in the mode $\mathrm{P(ECL)}^{\mu+1}$ (i.e. with local extrapolation applied at every call of the iteration) has the same stability region as a $(k+1)$th-order ABM applied in the mode $\mathrm{P(EC)}^{\mu}\mathrm{E}$ (i.e. without local extrapolation).

The regions of absolute stability for the kth-order ABM methods, $k = 1, 2, 3, 4$ in the modes PECL, PECLE and $\mathrm{P(ECL)}^2$ modes are shown in Figure 4.2. As in Figure 4.1, the stability region for the kth-order Adams–Bashforth method is also included and is labelled PE. Again, the case $k = 1$ is anomalous, but for $k = 2, 3, 4$ we observe that although local extrapolation has the effect of enlarging the stability regions, the pecking order of PECLE, $\mathrm{P(ECL)}^2$, PE, PECL is unaltered. The result (4.61) is well illustrated in Figure 4.2; for example, the region for $\mathrm{P(ECL)}^2$, $k = 2$ is the same as for PECLE, $k = 3$, and the region for PECL, $k = 2$ coincides with that for PE, $k = 3$.

Exercises

4.6.1. Use the Routh–Hurwitz criterion to find the intervals of absolute stability of the predictor (4.51) with $k = 2$ and of the PECE algorithm constructed from the same predictor and the 2-step BDF as corrector. Compare these intervals with the corresponding ones for the 2-step Adams–Bashforth method and the 2-step ABM. (See Figure 4.1 and Figure 3.3 of §3.8.) Hence corroborate the conclusions drawn in the above section.

4.6.2. The predictor $y_{n+2} + 4y_{n+1} - 5y_n = h(4f_{n+1} + 2f_n)$ is combined with Simpson's Rule as corrector in PECE mode. Show that if the roots of the stability polynomial are r_1 and r_2 (r_1 being the perturbation of the principal root of ρ) then for all real $\bar{h}$, the method is relatively stable according to the criterion $|r_2| \leqslant |r_1|$ (which is not quite Criterion A of §3.8). Show, however, that the method is not relatively stable (in the above sense) for small imaginary $\bar{h}$.

4.6.3. The result (4.49) shows that a PEC method has the same region of absolute stability as the explicit method $\tilde{\mathrm{P}}$, defined by the characteristic polynomials $\tilde{\rho}^*(r)$, $\tilde{\sigma}^*(r)$ given in (4.47). Let $k = 2$, and consider the following five equations which arise from the application of the PEC method at a number of consecutive steps:

$$y^{[0]}_{n+q+2} + \alpha_1^* y^{[1]}_{n+q+1} + \alpha_0^* y^{[1]}_{n+q} = h(\beta_1^* f^{[0]}_{n+q+1} + \beta_0^* f^{[0]}_{n+q}), \qquad q = 0, 1, 2$$
$$y^{[1]}_{n+q+2} + \alpha_1 y^{[1]}_{n+q+1} + \alpha_0 y^{[1]}_{n+q} = h(\beta_2 f^{[0]}_{n+q+2} + \beta_1 f^{[0]}_{n+q+1} + \beta_0 f^{[0]}_{n+q}), \qquad q = 0, 1.$$

Hence show that the *predicted* values $\{y^{[0]}_n\}$ generated by the PEC method satisfy the explicit method $\tilde{\mathrm{P}}$. Deduce that the numerical results produced by the PEC method could alternatively be obtained by using $\tilde{\mathrm{P}}$ to calculate the sequence $\{y^{[0]}_n\}$ for the whole range of integration, and then modifying $y^{[0]}_n$ to $y^{[1]}_n$ by applying the 3-term recurrence relation

$$y^{[1]}_{n+2} + \alpha_1 y^{[1]}_{n+1} + \alpha_0 y^{[1]}_n = h(\beta_2 f^{[0]}_{n+2} + \beta_1 f^{[0]}_{n+1} + \beta_0 f^{[0]}_n).$$

Explain why this interpretation of the PEC method corroborates the result (4.49).

4.6.4. Verify the result (4.61) by calculating the stability polynomials for $P_{(2)}(EC_{(2)}L)^2$ and $P_{(3)}EC_{(3)}E$.

4.7 CHANGING THE STEPLENGTH IN AN ABM ALGORITHM

As we have seen, predictor–corrector methods possess many advantages, notably the facility for monitoring the local truncation error cheaply and efficiently. However, there is a balancing disadvantage, shared by all multistep methods, namely the difficulties encountered in implementing a change of steplength. In the remainder of this chapter we shall be discussing ways in which predictor–corrector methods are implemented in modern codes, and inevitably heuristic arguments based on computational experience will play a significant role. Such codes are almost always based on ABM methods, so we shall restrict our discussion to that family of methods.

Suppose that we have used a kth-order ABM method (which will have stepnumber k) to compute y_n, but before going on to compute y_{n+1} we want to change the steplength from h to αh. In order to apply the method to compute an approximation to y at $x_n + \alpha h$, we need back data at x_n, which we have, and at $x_n - \alpha h, x_n - 2\alpha h, \ldots, x_n - (k-1)\alpha h$, which we do not have. (The codes referred to above use a range of ABM methods of orders up to 13, so that quite a lot of new back data may have to be generated on change of steplength.) Many different ways of tackling this problem have been proposed, but we shall discuss only those that have found to be the most successful; a good reference on this topic is Krogh (1973). The available techniques can be categorized into two different groups. The first, known as *interpolatory techniques*, use polynomial interpolation of the existing back data in order to approximate the missing back data; there are several ways of doing this. In the second group, the ABM methods themselves are replaced by ABM-like methods which assume that the data is unevenly spaced, and whose coefficients therefore vary as the steplength varies. Stepchanging techniques based on such methods are usually known as *variable step techniques*, a name which the author finds unsatisfactory; algorithms which have the facility for changing both steplength and order are widely known as 'variable step variable order' or VSVO algorithms, whether they use interpolatory or 'variable step' techniques to implement a change of steplength, and there is clearly a clash in the nomenclature. Accordingly, we prefer to call this second group of techniques *variable coefficient techniques*.

With interpolatory techniques, ABM methods have an advantage over other predictor–corrector methods, in that, since $\rho^*(r) = \rho(r) = r^k - r^{k-1}$, we never need to generate missing back values of y, only those of f. The interpolation can be done wholly in the $x - f$ space, thus avoiding any call for additional function evaluations. The key piece of information we work from is the unique polynomial of degree $k-1$ which interpolates the available back data $(x_{n-\tau}, f_{n-\tau})$, $\tau = 0, 1, \ldots, k-1$. Note that, by (1.31) of §1.10, the errors in this interpolation will be $O(h^k)$; since, in an ABM implementation the back values of f (or their differences) are multiplied by h, the error in y due to interpolation errors in the back data will be $O(h^{k+1})$, that is, of the same order as the LTE. Now the interpolating polynomial can be defined (and stored) in a number of different ways, thus giving rise to a number of different interpolatory techniques We could simply work out its coefficients (which would not be a very efficient way to proceed), we could specify it by the data $f_{n-\tau}, \tau = 0, 1, \ldots, k-1$, or by the backward differences $\nabla^i f_n, i = 0, \ldots, k-1$, or we could fix the polynomial by specifying its value and that of its first $k-1$ derivatives at the point x_n. In §4.8 and §4.9 we shall consider in some detail two interpolatory techniques; in the first of these the interpolating polynomial is specified by the backward differences of f, and in the second the values of the interpolant and its derivatives are specified at x_n.

Variable coefficient techniques essentially consist of Adams methods in backward difference form, as in §4.4, but derived under the assumption that the solution has been computed at unevenly spaced values of x. We shall carry out such a re-derivation in §4.10. Computation of the variable coefficients at each step becomes the major computational effort in using these techniques. When the steplength is held constant for a number of steps, then the coefficients naturally become constant, and the methods become equivalent to standard ABM methods.

It is not a straightforward matter to compare the computational effort of interpolatory and variable coefficient techniques. With interpolatory techniques, the amount of computation involved clearly increases as the dimension m of the initial value problem increases, whereas the effort of computing the coefficients in a variable coefficient technique remains independent of m. Thus variable coefficient techniques become more attractive if the system is large. Nevertheless, it is still generally true that algorithms employing variable coefficient techniques are computationally more expensive than those using interpolatory techniques. On the other hand, they are more flexible in handling very frequent changes of steplength and are, in practice, rather more robust since, unlike interpolatory techniques, they always use computed and not interpolated back data.

4.8 CHANGING THE STEPLENGTH; DOUBLING AND HALVING

In the preceding section, we listed a number of different ways in which, in an interpolatory step-changing technique, the interpolant of the available back data could be specified. If we are implementing an ABM method in $P(EC)^{\mu}E^{1-t}$ mode in backward difference form as described in §4.4 then, of the options listed, that of defining the polynomial by specifying the backward differences is clearly the most natural. From (4.18) of §4.4, the back data that have been stored on conclusion of the step from x_{n-1} to x_n are $\nabla^i f_n^{[\mu-t]}, i = 0, 1, \ldots, k-1$. In what follows we shall drop the superscript $[\mu - t]$, which is to be taken as read. Ideally, what we would like is an algorithm in which the input consists of these differences and the output is the corresponding differences of the interpolated values at $x_n - \tau\alpha h, \tau = 0, 1, \ldots, k-1$. A remarkably simple algorithm, due to Krogh (1973), does just this, in the case when stepchanging is restricted to doubling or halving the current steplength. One might think that no such algorithm is necessary in the case of doubling the steplength, since the previously computed values f_{n-2}, $f_{n-4}, \ldots, f_{n-2k+2}$ could be used as the new back data. However, Krogh (1973) reports that such a technique is consistently less accurate in practice than the algorithm we are about to describe. This is not altogether surprising, since the technique of using every other value of f uses information which is further away from the current step than that used in an interpolatory technique (recall that k can be large), and the solution may have locally changed in character; indeed it is often such a change that creates the need to alter the steplength in the first place.

Let $I(x)$ be the unique polynomial (with coefficients in $\mathbb{R}^m$) of degree $k-1$ passing through the k points $(x_{n-\tau}, f_{n-\tau}), \tau = 0, 1, \dots, k-1$. We then have that

$$\nabla^i I(x_n) = \nabla^i f_n, \qquad i = 0, 1, \dots, k-1 \tag{4.62}$$

where $\nabla I(x) := I(x) - I(x-h), \nabla^i I(x) := \nabla^{i-1} I(x) - \nabla^{i-1} I(x-h), i = 2, 3, \dots$. Since the pth differences of a polynomial of degree p are constant, it follows that

$$\nabla^i I(x_n) = 0, \qquad i = k, k+1, \dots. \tag{4.63}$$

In the step-doubling case, we wish to generate a set of differences of the data $I(x_n)$, $I(x_n - 2h), \dots, I(x_n - 2(k-1)h)$. Let us denote such differences by $\nabla^i_{(D)} I(x_n), i = 0, 1, \dots, k-1$, defined by $\nabla_{(D)} I(x) := I(x) - I(x-2h), \nabla^i_{(D)} I(x) := \nabla^{i-1}_{(D)} I(x) - \nabla^{i-1}_{(D)} I(x-2h)$, $i = 2, 3, \dots, k-1$. (The subscript D is bracketed to avoid any possible confusion with the notation introduced in (4.15) of §4.4.) Now,

$$\nabla I(x_n) = I(x_n) - I(x_n - h),$$

$$\nabla^2 I(x_n) = I(x_n) - 2I(x_n - h) + I(x_n - 2h),$$

whence

$$(2\nabla - \nabla^2) I(x_n) = I(x_n) - I(x_n - 2h) = \nabla_{(D)} I(x_n).$$

Thus we have the following identity:

$$\nabla_{(D)} \equiv 2\nabla - \nabla^2. \tag{4.64}$$

From equations (4.62), (4.63) and (4.64) we are able to generate $\nabla^i_{(D)} I(x_n)$ in terms of $\nabla^i f_n, i = 1, 2, \dots, k-1$. (The case $i = 0$ is trivial.) We illustrate the procedure in the case $k = 5$.

$$\left.\begin{aligned}
\nabla_{(D)} I(x_n) &= \nabla(2-\nabla) I(x_n) = 2\nabla f_n - \nabla^2 f_n \\
\nabla^2_{(D)} I(x_n) &= \nabla^2 (2-\nabla)^2 I(x_n) = (4\nabla^2 - 4\nabla^3 + \nabla^4) I(x_n) \\
&= 4\nabla^2 f_n - 4\nabla^3 f_n + \nabla^4 f_n \\
\nabla^3_{(D)} I(x_n) &= \nabla^3 (2-\nabla)^3 I(x_n) = (8\nabla^3 - 12\nabla^4 + 6\nabla^5 - \nabla^6) I(x_n) \\
&= (8\nabla^3 - 12\nabla^4) I(x_n) \qquad \text{(by (4.63))} \\
&= 8\nabla^3 f_n - 12\nabla^4 f_n \\
\nabla^4_{(D)} I(x_n) &= \nabla^4 (2-\nabla)^4 I(x_n) = (16\nabla^4 + \cdots) I(x_n) \\
&= 16\nabla^4 I(x_n) = 16\nabla^4 f_n.
\end{aligned}\right\} \tag{4.65}$$

For general k, the above procedure is neatly accomplished by a segment of code due to Krogh (1973). (For ease of exposition, the segment is written for the case $f \in \mathbb{R}^1$; adaptation to the case $f \in \mathbb{R}^m$ is straightforward.) Let $A_i (= A[i]) := \nabla^i f_n, i = 1, 2, \dots, k-1$, and consider the following few lines of code:

```
for j:= 1 to k - 2 do
  begin
    for i:= j to k - 2 do A[i]:= 2.0*A[i] - A[i + 1];
    A[k - 1]:= 2.0*A[k - 1];
  end;
A[k - 1]:= 2.0*A[k - 1];
```
(4.66)

Applying this in the case $k = 5$ gives

	A_1	A_2	A_3	A_4
	∇f_n	$\nabla^2 f_n$	$\nabla^3 f_n$	$\nabla^4 f_n$
$j=1$	$(2\nabla - \nabla^2)f_n$	$(2\nabla^2 - \nabla^3)f_n$	$(2\nabla^3 - \nabla^4)f_n$	$2\nabla^4 f_n$
$j=2$		$(4\nabla^2 - 4\nabla^3 + \nabla^4)f_n$	$(4\nabla^3 - 4\nabla^4)f_n$	$4\nabla^4 f_n$
$j=3$			$(8\nabla^3 - 12\nabla^4)f_n$	$8\nabla^4 f_n$
				$16\nabla^4 f_n$

and, on comparing with (4.65), we see that the code segment (4.66) has transformed the vector $[\nabla f_n, \nabla^2 f_n, \nabla^3 f_n, \nabla^4 f_n]^T$ into $[\nabla_{(D)} f_n, \nabla^2_{(D)} f_n, \nabla^3_{(D)} f_n, \nabla^4_{(D)} f_n]^T$.

A similar procedure for halving the steplength can be deduced from the above. We now wish to generate a set of differences from the data $I(x_n), I(x_n - h/2), \ldots, I(x_n - (k-1)h/2)$. Let us denote such differences by $\nabla_{(H)} I(x_n), i = 0, 1, \ldots, k-1$, defined by $\nabla_{(H)} I(x) := I(x) - I(x - h/2)$, $\nabla^i_{(H)} I(x) := \nabla^{i-1}_{(H)} I(x) - \nabla^{i-1}_{(H)} I(x - h/2), i = 2, 3, \ldots, k-1$. Now, deducing $\nabla_{(D)}$ from ∇ is obviously the same process as deducing ∇ from $\nabla_{(H)}$, and from (4.64) we can thus write

$$\nabla \equiv 2\nabla_{(H)} - \nabla^2_{(H)}.$$

We can no longer express $\nabla_{(H)}$ explicitly in terms of ∇ but, by analogy with (4.65), we see that the transformation which takes $[(\nabla_{(H)} f_n)^T, (\nabla^2_{(H)} f_n)^T, (\nabla^3_{(H)} f_n)^T, (\nabla^4_{(H)} f_n)^T]^T$ into $[(\nabla f_n)^T, (\nabla^2 f_n)^T, (\nabla^3 f_n)^T, (\nabla^4 f_n)^T]^T$ is a linear one and, moreover, the transforming matrix is triangular, making the inversion of the transformation simple. Specifically, in the case $k = 5$, we have by analogy with (4.65)

$$\begin{aligned}
\nabla f_n &= (2\nabla_{(H)} - \nabla^2_{(H)}) I(x_n) \\
\nabla^2 f_n &= (4\nabla^2_{(H)} - 4\nabla^3_{(H)} + \nabla^4_{(H)}) I(x_n) \\
\nabla^3 f_n &= (8\nabla^3_{(H)} - 12\nabla^4_{(H)}) I(x_n) \\
\nabla^4 f_n &= 16\nabla^4_{(H)} I(x_n).
\end{aligned}$$

This linear system is readily solved to give

$$\left.\begin{aligned}
\nabla^4_{(H)} I(x_n) &= \tfrac{1}{16}\nabla^4 f_n \\
\nabla^3_{(H)} I(x_n) &= \tfrac{1}{8}\nabla^3 f_n + \tfrac{3}{32}\nabla^4 f_n \\
\nabla^2_{(H)} I(x_n) &= \tfrac{1}{4}\nabla^2 f_n + \tfrac{1}{8}\nabla^3 f_n + \tfrac{5}{64}\nabla^4 f_n \\
\nabla_{(H)} I(x_n) &= \tfrac{1}{2}\nabla f_n + \tfrac{1}{8}\nabla^2 f_n + \tfrac{1}{16}\nabla^3 f_n + \tfrac{5}{128}\nabla^4 f_n.
\end{aligned}\right\} \qquad (4.67)$$

Once again, this procedure can be carried out, for general k, by a segment of code which is, in effect, (4.66) applied backwards. As before, let $A_i(=A[i]) := \nabla^i f_n, i = 1, 2, \ldots, k-1$; then the segment of code (again written for the case $f \in \mathbb{R}^1$) is

```
A[k - 1] := A[k - 1]/2.0;
for j := k - 2 downto 1 do
  begin
    A[k - 1] := A[k - 1]/2.0;
    for i := k - 2 downto j do A[i] := (A[i] + A[i + 1])/2.0;
  end;
```
(4.68)

Applying this in the case $k = 5$ gives

	A_4	A_3	A_2	A_1
	$\nabla^4 f_n$	$\nabla^3 f_n$	$\nabla^2 f_n$	∇f_n
	$\frac{1}{2}\nabla^4 f_n$			
$j=3$	$\frac{1}{4}\nabla^4 f_n$	$(\frac{1}{2}\nabla^3 + \frac{1}{8}\nabla^4) f_n$		
$j=2$	$\frac{1}{8}\nabla^4 f_n$	$(\frac{1}{4}\nabla^3 + \frac{1}{8}\nabla^4) f_n$	$(\frac{1}{2}\nabla^2 + \frac{1}{8}\nabla^3 + \frac{1}{16}\nabla^4) f_n$	
$j=1$	$\frac{1}{16}\nabla^4 f_n$	$(\frac{1}{8}\nabla^3 + \frac{3}{32}\nabla^4) f_n$	$(\frac{1}{4}\nabla^2 + \frac{1}{8}\nabla^3 + \frac{5}{64}\nabla^4) f_n$	$(\frac{1}{2}\nabla + \frac{1}{8}\nabla^2 + \frac{1}{16}\nabla^3 + \frac{5}{128}\nabla^4) f_n$

and, on comparing with (4.67), we see that (4.68) has achieved the desired result.

In practice, using (4.66) to implement step-doubling works very satisfactorily, but using (4.68) to implement step-halving can run into difficulties over adverse accumulation of error when k is large. Krogh (1973) reports cases where using (4.68) to halve the step-length in an ABM method can result in the error estimate *increasing*! Problems of this sort, encountered when reducing steplength, are not confined to the use of (4.68); they can arise with any interpolatory technique, and are essentially due to the fact that the underlying polynomial interpolant which the ABM method uses to advance the solution does not, after a step change, pass through previously computed points, but through an interpolant of these points. There exists a modification (applicable in the case of an ABM method in PECE mode with step-having by (4.68)) which successfully overcomes this difficulty; it is rather too elaborate to quote here, and the reader is referred to Krogh (1973) for details.

Exercise

4.8.1. Corroborate (4.65) by the following calculations: Let $I(x) = x^4 + 4x^3 + 3x^2 + 2x + 1$; evaluate $I(x)$ for $x = 0, -\frac{1}{2}, -1, -\frac{3}{2}, -2$, and construct a table of backward differences $\nabla^i I(0)$, $i = 0, 1, 2, 3, 4$. Apply (4.65) to find the corresponding differences $\nabla^i_{(D)} I(0)$. Now evaluate $I(x)$ at $x = 0, -1, -2, -3, -4$ and check that the differences generated by these values coincide with the $\nabla^i_{(D)} I(0)$. Why do they coincide *exactly*? Carry out a similar calculation to check (4.67).

4.9 CHANGING THE STEPLENGTH; THE NORDSIECK VECTOR AND GEAR'S IMPLEMENTATION

We now look at another interpolatory technique in which we use a different option for identifying the vector polynomial $I(x)$ of degree $k-1$ which interpolates the known back values $f_{n-\tau}, \tau = 0, 1, \ldots, k-1$, namely that of fixing $I(x)$ and its first $k-1$ derivatives at the point x_n. Let $F(x_n) \in \mathbb{R}^{mk}$ be defined by

$$\begin{aligned} F(x_n) &:= [(f_n)^T, (\nabla f_n)^T, \ldots, (\nabla^{k-1} f_n)^T]^T \\ &= [(I(x_n))^T, (\nabla I(x_n))]^T, \ldots, (\nabla^{k-1} I(x_n))^T]^T, \end{aligned}$$

by (4.62). Observing that $\nabla^i I(x_n) = 0(h^i)$, it seems appropriate, when defining a vector whose m-block components are $I(x)$ and its first $k-1$ derivatives evaluated at x_n, to scale the ith derivative by h^i, and define $G(x_n) \in \mathbb{R}^{mk}$ by

$$G(x_n) := [(I(x_n))^T, h(I^{(1)}(x_n))^T, \ldots, h^{k-1}(I^{(k-1)}(x_n))^T]^T. \tag{4.69}$$

The technique of storing back data in terms of an interpolant and its derivatives evaluated at a single point was first proposed by Nordsieck (1962), and it is appropriate to refer to $G(x)$ as a *Nordsieck vector*. The result of the scaling by powers of h is that we can obtain $G(x_n)$ in terms of $F(x_n)$ by means of a linear transformation

$$G(x_n) = AF(x_n) \tag{4.70}$$

where the matrix A is *independent of* h. If we now wish to replace h by αh, all we have to do to the vector $G(x_n)$ is multiply the ith m-block component by α^{i-1} (see (4.69)); the corresponding vector of differences of $I(x_n)$ evaluated at x_n-$\tau\alpha h$, $\tau = 0, 1, \ldots, k-1$, is then obtained by inverting the transformation (4.70).

Consider for example the case $k = 5$. Then $I(x) = I(x_n + rh) =: P(r)$, where

$$\begin{aligned} P(r) = f_n + r\nabla f_n + \tfrac{1}{2}r(r+1)\nabla^2 f_n + \tfrac{1}{6}r(r+1)(r+2)\nabla^3 f_n \\ + \tfrac{1}{24}r(r+1)(r+2)(r+3)\nabla^4 f_n. \end{aligned}$$

Since $h^i I^{(i)}(x_n) = P^{(i)}(r)|_{r=0}, i = 1, 2, 3, 4$, we find that

$$\begin{aligned} hI^{(1)}(x_n) &= \nabla f_n + \tfrac{1}{2}\nabla^2 f_n + \tfrac{1}{3}\nabla^3 f_n + \tfrac{1}{4}\nabla^4 f_n \\ h^2 I^{(2)}(x_n) &= \nabla^2 f_n + \nabla^3 f_n + \tfrac{11}{12}\nabla^4 f_n \\ h^3 I^{(3)}(x_n) &= \nabla^3 f_n + \tfrac{3}{2}\nabla^4 f_n \\ h^4 I^{(4)}(x_n) &= \nabla^4 f_n, \end{aligned}$$

whence (4.70) holds with

$$A = \begin{bmatrix} I & 0 & 0 & 0 & 0 \\ 0 & I & \tfrac{1}{2}I & \tfrac{1}{3}I & \tfrac{1}{4}I \\ 0 & 0 & I & I & \tfrac{11}{12}I \\ 0 & 0 & 0 & I & \tfrac{3}{2}I \\ 0 & 0 & 0 & 0 & I \end{bmatrix} \tag{4.71}$$

where I is the $m \times m$ unit matrix and 0 the $m \times m$ null matrix.

Suppose that we wish to double the steplength; this is equivalent to multiplying $G(x_n)$ by the block diagonal matrix

$$D = \text{diag}[I, \quad 2I, \quad 4I, \quad 8I, \quad 16I].$$

If we define

$$F_{(D)}(x_n) := [(I(x_n))^{\mathrm{T}}, (\nabla_{(D)} I(x_n))^{\mathrm{T}}, (\nabla^2_{(D)} I(x_n))^{\mathrm{T}}, (\nabla^3_{(D)} I(x_n))^{\mathrm{T}}, (\nabla^4_{(D)} I(x_n))^{\mathrm{T}}]^{\mathrm{T}},$$

then $F_{(D)}(x_n)$ is given in terms of $F(x_n)$ by

$$F_{(D)}(x_n) = A^{-1} D A F(x_n).$$

On performing the matrix arithmetic we find that

$$F_D(x_n) = \begin{bmatrix} I & 0 & 0 & 0 & 0 \\ 0 & 2I & -I & 0 & 0 \\ 0 & 0 & 4I & -4I & I \\ 0 & 0 & 0 & 8I & -12I \\ 0 & 0 & 0 & 0 & 16I \end{bmatrix} F(x_n) \tag{4.72}$$

thus reproducing the result we obtained in (4.65).

This is all by way of introduction, and we would not seriously propose that the above procedure be used as a step-changing technique. What this illustration does is highlight the advantages and disadvantages of the approach. The advantage is clear—the ability to change steplength by an arbitrary factor. The disadvantage is that the computation of the matrix A (which is different for different values of k) and of its inverse and the matrix multiplications together represent a quite unacceptably large amount of computation.

A development due to Gear (1967, 1971a), in which the ABM method is implemented in standard—not backward difference—form, makes ingenious use of a Nordsieck vector. The device, which successfully overcomes the disadvantage noted above, is best seen when the ABM method is applied in P(EC)$^\mu$ mode. In both the original Nordsieck methods and in the Gear development the interpolant to be stored in terms of derivatives is not that which interpolates the back data $f_{n-\tau}$, $\tau = 1, 2, \dots, k-1$, but the Hermite interpolant $P(x)$ which, in the sense of §3.3, is equivalent to the Adams–Bashforth predictor in the ABM method. If the ABM method, applied in P(EC)$^\mu$ mode, has order k, then $P(x)$ has degree k and satisfies

$$\begin{aligned} P(x_{n+k}) &= y^{[0]}_{n+k}, \quad P(x_{n+k-1}) = y^{[\mu]}_{n+k-1} \\ P'(x_{n+j}) &= f^{[\mu-1]}_{n+j}, \quad j = 0, 1, \dots, k-1. \end{aligned} \tag{4.73}$$

($P(x)$ has $k+1$ m-vector coefficients; as in §3.3, the eliminant of these $k+1$ coefficients between the $k+2$ conditions (4.73) is the kth-order Adams–Bashforth method.) The back data used by the predictor can be lined up to define a back vector $Y^{[\mu]}_{n+k} \in \mathbb{R}^{m(k+1)}$ given by

$$Y^{[\mu]}_{n+k-1} := [(y^{[\mu]}_{n-k-1})^{\mathrm{T}}, h(f^{[\mu-1]}_{n-k-1})^{\mathrm{T}}, \dots, h(f^{[\mu-1]}_{n})^{\mathrm{T}}]^{\mathrm{T}}. \tag{4.74}$$

Clearly $Y^{[\mu]}_{n+k-1}$ determines $P(x)$ uniquely; so does the vector of $P(x)$ and its first k derivatives evaluated at x_{n+k-1}. For the reasons discussed earlier, it is appropriate to scale these derivatives by powers of h and it turns out to be helpful also to scale them by factorials. We therefore define the Nordsieck vector $Z^{[\mu]}_{n+k-1} \in \mathbb{R}^{m(k+1)}$ by

$$Z^{[\mu]}_{n+k-1} := \left[(P(x_{n+k-1}))^{\mathrm{T}}, h(P^{(1)}(x_{n+k-1}))^{\mathrm{T}}, \ldots, \frac{h^k}{k!}(P^{(k)}(x_{n+k-1}))^{\mathrm{T}} \right]^{\mathrm{T}}. \tag{4.75}$$

The transformation from $Y^{[\mu]}_{n+k-1}$ to $Z^{[\mu]}_{n+k-1}$ is a linear one,

$$Z^{[\mu]}_{n+k-1} = QY^{[\mu]}_{n+k-1} \tag{4.76}$$

and, as in our earlier illustration, the scaling of the derivatives of $P(x)$ by powers of h results in Q being independent of h. The elements of Q are thus constants which depend only on the coefficients of the kth-order Adams–Bashforth method. However, from (4.74), (4.75) and (4.73) we see that the first m-block components of $Y^{[\mu]}_{n+k-1}$ and $Z^{[\mu]}_{n+k-1}$ are identical, as also are the second m-block components. Thus Q must have the form

$$Q = \begin{bmatrix} I & 0 & 0 & 0 & \cdots & 0 \\ 0 & I & 0 & 0 & \cdots & 0 \\ * & * & * & * & \cdots & * \\ \vdots & \vdots & \vdots & \vdots & & \vdots \\ * & * & * & * & \cdots & * \end{bmatrix}. \tag{4.77}$$

The kth-order ABM in $\mathrm{P(EC)}^\mu$ mode written in standard form is, by (4.4),

$$\left.\begin{aligned} &\mathrm{P:}\quad y^{[0]}_{n+k} = y^{[\mu]}_{n+k-1} + h\sum_{j=0}^{k-1}\beta^*_j f^{[\mu-1]}_{n+j} \\ &\mathrm{C:}\quad y^{[\nu+1]}_{n+k} = y^{[\mu]}_{n+k-1} + h\beta_k f^{[\nu]}_{n+k} + h\sum_{j=0}^{k-1}\beta_j f^{[\mu-1]}_{n+j}, \qquad \nu = 0,1,\ldots,\mu-1 \end{aligned}\right\} \tag{4.78}$$

where, since the corrector has stepnumber one less than that of the corrector, $\beta_0 = 0$.

The Gear approach amounts to twisting the arm of (4.78) to make it *look* like a one-step method (which it will not be) and then applying the transformation (4.76) (in a more general context) whereupon the method genuinely becomes one-step. It follows from (4.78) that

$$y^{[\nu+1]}_{n+k} - y^{[\nu]}_{n+k} = h\beta_k(f^{[\nu]}_{n+k} - f^{[\nu-1]}_{n+k}), \qquad \nu = 1,2,\ldots,\mu-1 \tag{4.79}$$

and

$$y^{[1]}_{n+k} - y^{[0]}_{n+k} = h\beta_k\left\{ f^{[0]}_{n+k} - \sum_{j=0}^{k-1}[(\beta^*_j - \beta_j)/\beta_k]f^{[\mu-1]}_{n+j} \right\}. \tag{4.80}$$

Now introduce the simplifying notation

$$\delta^*_j := (\beta^*_j - \beta_j)/\beta_k, \qquad j = 0,1,\ldots,k-1; \qquad d_{n+k} = \sum_{j=0}^{k-1}\delta^*_j f^{[\mu-1]}_{n+j}. \tag{4.81}$$

The coefficients δ^*_j turn out to be very simple functions of k. Consider the polynomial

$$\sum_{j=0}^{k}\delta^*_j r^j := \left[\sum_{j=0}^{k}(\beta^*_j - \beta_j)r^j\right] \Big/ \beta_k,$$

where $\beta_k^* = 0$, $\beta_0 = 0$. Now, in the notation of §3.10, $\sigma_k^*(r)$ and $\sigma_k(r)$ are the second characteristic polynomials of the Adams–Bashforth and Adams–Moulton methods, respectively, of *order* k, and both have degree $k-1$ (see (3.97)). Recalling that in the present context we have right-shifted the corrector by one steplength, we may write

$$\sum_{j=0}^{k} \beta_j^* r^j = \sigma_k^*(r), \quad \sum_{j=0}^{k} \beta_j r^j = r\sigma_k(r),$$

whence

$$\sum_{j=0}^{k} \delta_j^* r^j = [\sigma_k^*(r) - r\sigma_k(r)]/\beta_k.$$

It follows from Properties 8 and 4 of §3.10 that

$$\sum_{j=0}^{k} \delta_j r^j = -(r-1)^k,$$

whence

$$\delta_j^* = (-1)^{k+j+1}\binom{k}{j}, \qquad j = 0, 1, \ldots, k. \tag{4.82}$$

(Note also that, from (4.81), $d_{n+k} = [1 + (-1)^{k+1}\nabla^k] f_{n+k}^{[\mu-1]}$.)

We can now write (4.80) as

$$y_{n+k}^{[1]} - y_{n+k}^{[0]} = h\beta_k(f_{n+k}^{[0]} - d_{n+k}).$$

Now define the vector $Y_{n+k}^{[\nu]} \in \mathbb{R}^{m(k+1)}$ by

$$Y_{n+k}^{[\nu]} := \begin{cases} [(y_{n+k}^{[0]})^T, h(d_{n+k})^T, h(f_{n+k-1}^{[\mu-1]})^T, h(f_{n+k-2}^{[\mu-1]})^T, \ldots, h(f_{n+1}^{[\mu-1]})^T]^T & \text{if } \nu = 0 \\ [(y_{n+k}^{[\nu]})^T, h(f_{n+k}^{[\nu-1]})^T, h(f_{n+k-1}^{[\mu-1]})^T, h(f_{n+k-2}^{[\mu-1]})^T, \ldots, h(f_{n+1}^{[\mu-1]})^T]^T & \text{if } \nu = 1, 2, \ldots, \mu. \end{cases} \tag{4.83}$$

Note that we have used the same notation in (4.74) and (4.83), but there is no contradiction; on putting $\nu = \mu$ and replacing n by $n-1$ in (4.83), we recover (4.74).

By (4.83), (4.81), (4.79) and (4.80) we can now write (4.78) in the following form:

$$\left.\begin{aligned} &\text{P:} \quad Y_{n+k}^{[0]} = BY_{n+k-1}^{[\mu]} \\ &\text{C:} \quad Y_{n+k}^{[\nu+1]} = Y_{n+k}^{[\nu]} + GF(Y_{n+k}^{[\nu]}), \qquad \nu = 0, 1, \ldots, \mu - 1 \end{aligned}\right\} \tag{4.84}$$

where B is an $m(k+1) \times m(k+1)$ matrix, G is an $m(k+1) \times m$ matrix and F, an m-vector, is a function of an $m(k+1)$ vector argument, given by

$$B = \begin{bmatrix} I & \beta_{k-1}^* I & \beta_{k-2}^* I & \cdots & \beta_1^* I & \beta_0^* I \\ 0 & \delta_{k-1}^* I & \delta_{k-2}^* I & \cdots & \delta_1^* I & \delta_0^* I \\ 0 & I & 0 & \cdots & 0 & 0 \\ 0 & 0 & I & \cdots & 0 & 0 \\ \vdots & \vdots & \vdots & & \vdots & \vdots \\ 0 & 0 & 0 & \cdots & I & 0 \end{bmatrix}, \qquad G = \begin{bmatrix} \beta_k I \\ I \\ 0 \\ 0 \\ \vdots \\ 0 \end{bmatrix} \tag{4.85}$$

and

$$F(Y_{n+k}^{[\nu]}) = \begin{cases} h(f_{n+k}^{[0]} - d_{n+k}), & \text{if } \nu = 0 \\ (f_{n+k}^{[\nu]} - f_{n+k}^{[\nu-1]}), & \text{if } \nu = 1, 2, \ldots, \mu - 1. \end{cases} \tag{4.86}$$

Note that, since $f_{n+k}^{[\nu]} = f(x_{n+k}, y_{n+k}^{[\nu]})$, F is indeed a function of $Y_{n+k}^{[\nu]}$. Note also that B and G depend only on the coefficients in the ABM method and are independent of h. In (4.84) we have achieved an apparently one-step form, but this is of course illusory, since the vector of back values, $Y_{n+k-1}^{[\mu]}$, depends on values of f at $x_{n+k-1}, x_{n+k-2}, \ldots, x_n$. However, let us now apply the transformation

$$Z_{n+k}^{[\nu]} = QY_{n+k}^{[\nu]}, \qquad \nu = 0, 1, \ldots, \mu, \tag{4.87}$$

where Q is defined by (4.76). Note that from (4.86) $F(Y_{n+k}^{[\nu]})$ is a function of the first two block components only of $Y_{n+k}^{[\nu]}$ and that, from the structure of Q given by (4.77), the first two block components of $Y_{n+k}^{[\nu]}$ will be identical with those of $Z_{n+k}^{[\nu]}$ $(= QY_{n+k}^{[\nu]})$; it follows that $F(Y_{n+k}^{[\nu]}) \equiv F(Z_{n+k}^{[\nu]})$. Hence, on applying the transformation (4.87), we can write (4.84) wholly in terms of $Z_{n+k}^{[\nu]}, Z_{n+k-1}^{[\nu]}$ as

$$\begin{aligned} &\text{P:} \quad Z_{n+k}^{[0]} = QBQ^{-1}Z_{n+k-1}^{[\mu]} \\ &\text{C:} \quad Z_{n+k}^{[\nu+1]} = Z_{n+k}^{[\nu]} + \tilde{G}F(Z_{n+k}^{[\nu]}), \qquad \nu = 0, 1, \ldots, \mu - 1 \end{aligned} \tag{4.88}$$

where $\tilde{G} = QG$. Now (4.88) is genuinely a one-step method since, by (4.75), the vector of back values, $Z_{n+k-1}^{[\mu]}$, depends only an information at the point x_{n+k-1}. We can change steplength from h to αh simply by multiplying the ith block component of $Z_{n+k-1}^{[\mu]}$ by α^{i-1}, $i = 1, 2, \ldots, k+1$, and the advantage of the Nordsieck approach is realized. It would appear that the disadvantage—the excessive computational effort—still remains, particularly in the prediction step. However, it turns out that the product QBQ^{-1} is precisely the $m(k+1) \times m(k+1)$ block *Pascal matrix* Π, defined by

$$\Pi := \begin{bmatrix} I & I & I & I & \cdots & I \\ 0 & I & 2I & 3I & \cdots & \binom{k}{1}I \\ 0 & 0 & I & 3I & \cdots & \binom{k}{2}I \\ 0 & 0 & 0 & I & \cdots & \binom{k}{3}I \\ \vdots & \vdots & \vdots & \vdots & & \vdots \\ 0 & 0 & 0 & 0 & & I \end{bmatrix}. \tag{4.89}$$

This interesting result can be established in a number of ways. The following is an outline of a direct proof.

The matrix Q^{-1} is easier to handle than is Q. Define $\xi := x_{n+k-1}$; from (4.73), (4.74), (4.75) and (4.76), we see that Q^{-1} maps

$$[(P(\xi))^{\mathrm{T}}, h(P^{(1)}(\xi))^{\mathrm{T}}, h^2(P^{(2)}(\xi))^{\mathrm{T}}/2!, \ldots, h^k(P^{(k)}(\xi))^{\mathrm{T}}/k!]^{\mathrm{T}}$$

into

$$[(P(\xi))^T, h(P^{(1)}(\xi))^T, h(P^{(1)}(\xi-h))^T, \dots, h(P^{(1)}(\xi-(k-1)h))^T]^T.$$

From Taylor expansions and the fact that, since $P(x)$ is a polynomial of degree $k, P^{(s)}(x) \equiv 0$ for $s > k$, we have that for $i = 1, 2, \dots, k-1$

$$hP^{(1)}(\xi - ih) = hP^{(1)}(\xi) - ih^2P^{(2)}(\xi) + \cdots + (-i)^{k-1}h^kP^{(k)}(\xi)/(k-1)!$$

whence

$$Q^{-1} = \begin{bmatrix} I & 0 & 0 & 0 & \cdots & 0 \\ 0 & I & 0 & 0 & \cdots & 0 \\ 0 & I & -2I & 3I & \cdots & k(-1)^{k-1}I \\ 0 & I & -2.2I & 3.2^2I & \cdots & k(-2)^{k-1}I \\ \vdots & \vdots & \vdots & \vdots & & \vdots \\ 0 & I & -2(k-1)I & 3(k-1)^2I & \cdots & k(-k+1)^{k-1}I \end{bmatrix}$$

and

$$Q^{-1}\Pi = \begin{bmatrix} I & I & I & I & \cdots & I \\ 0 & I & 2I & 3I & \cdots & kI \\ 0 & I & 0 & 0 & \cdots & 0 \\ 0 & I & -2I & 3I & \cdots & k(-1)^{k-1}I \\ 0 & I & -2.2I & 3.2^2I & \cdots & k(-2)^{k-1}I \\ \vdots & \vdots & \vdots & \vdots & & \vdots \\ 0 & I & -2(k-2)I & 3(k-2)^2I & \cdots & k(-k+2)^{k-1}I \end{bmatrix} \quad (4.90)$$

Now, from (4.85) and (4.90) it is clear that for $i \geqslant 2$ the ith block row of BQ^{-1} is identical with the ith block row of $Q^{-1}\Pi$. That the first and second block rows of the two products are also identical can be established by expanding (about ξ) the linear difference operators associated with the predictor and corrector and using the fact that both are of order k. We thus show that $BQ^{-1} = Q^{-1}\Pi$, or $\Pi = QBQ^{-1}$.

That $QBQ^{-1} = \Pi$ is much more than just a pretty result. In the first of (4.88) we no longer need to compute (for each k) the matrices B, Q and QBQ^{-1}. Moreover, multiplication of a vector by a Pascal matrix can be achieved extremely cheaply. It is left to the reader to verify that the following segment of code (again written for the case $m = 1$) computes the product Πa, where $a = [a_0, a_1, \dots, a_k]^T$:

```
for i:= 0 to k − 1 do
for j:= k downto i + 1 do
a[j − 1]:= a[j − 1] + a[j];
```

(Note that no multiplications are involved.)

4.10 CHANGING THE STEPLENGTH; VARIABLE COEFFICIENT TECHNIQUES

An early example of a variable coefficient technique was afforded by Ceschino (1961) who derived variable coefficient formulae of Adams type for orders up to 4. Although there are now much more efficient ways of implementing variable coefficient techniques, Ceschino's formulae serve to illustrate the sort of problems encountered with such techniques. Suppose we have used a third-order ABM method to compute an acceptable numerical solution at x_n, but before proceeding further we want to change the steplength from h to αh. Let $y_{n+j\alpha}$ denote the numerical solution at $x_{n+j\alpha} := x_n + j\alpha h, j = 1, 2$, and let $f_{n+j\alpha} = f(x_{n+j\alpha}, y_{n+j\alpha})$. Then, using Ceschino's third-order formulae, the solution is advanced for x_n to $x_{n+\alpha}$ by the predictor–corrector pair

$$\left.\begin{aligned} y_{n+\alpha} &= y_n + \frac{\alpha h}{12}[(12 + 9\alpha + 2\alpha^2)f_n - 4\alpha(3+\alpha)f_{n-1} + \alpha(3+2\alpha)f_{n-2}] \\ y_{n+\alpha} &= y_n + \frac{\alpha h}{6(1+\alpha)}[(3+2\alpha)f_{n+\alpha} + (3+\alpha)(1+\alpha)f_n - \alpha^2 f_{n-1}]. \end{aligned}\right\} \quad (4.91(\text{i}))$$

However, to advance the solution one further step to $x_{n+2\alpha}$ we need another special predictor together with the standard third-order Adams–Moluton corrector:

$$\left.\begin{aligned} y_{n+2\alpha} &= y_{n+\alpha} + \frac{\alpha h}{6(1+\alpha)}[(9+14\alpha)f_{n+\alpha} - (3+5\alpha)(1+\alpha)f_n + 5\alpha^2 f_{n-1}] \\ y_{n+2\alpha} &= y_{n+\alpha} + \frac{\alpha h}{12}(5f_{n+2\alpha} + 8f_{n+\alpha} - f_n). \end{aligned}\right\} \quad (4.91(\text{ii}))$$

(Note that the above formulae all revert to standard ABM formulae when $\alpha = 1$.) The difficulties are now apparent. For a third-order method we needed to compute (and store) three special formulae; for a kth-order method we would need $2k - 3$ such special formulae, and if, as in modern codes, we wish to operate with k ranging from 1 to 13, say, then the grand total of special methods needed would be 144. But there is an even more serious drawback, namely that, if the order is k, then it takes $k - 1$ steps to complete a change to steplength. During these $k - 1$ steps, there may arise a need to change the steplength again (and perhaps more than once); we leave it to the reader to contemplate the ensuing complications!

A more constructive approach is to assume from the outset that the back data is already unevenly spaced. The development we describe here is essentially due to Krogh (1974) but we shall (partially) adopt an approach due to Hall (1976) which is notationally easier to follow. We restrict ourselves to the case of Adams-like methods applied in PECE mode; adaptation to other modes is straightforward. Let us refresh our memory about how we developed Adams–Bashforth methods in backward difference form in §3.9. To derive a k-step Adams–Bashforth method (of order k), we started from the identity

$$y(x_{n+1}) - y(x_n) = \int_{x_n}^{x_{n+1}} y'(x)\,dx,$$

replaced $y'(x)$ by $f(x, y(x))$ and approximated the integrand by the Newton–Gregory interpolant (3.88) which interpolated the data

$$(x_n, f_n), (x_{n-1}, f_{n-1}), \ldots, (x_{n-k+1}, f_{n-k+1})$$

which was, of course, assumed evenly spaced. Now that we are dealing with unevenly spaced data, the appropriate interpolating polynomial of degree $k-1$ is given by the Newton divided difference interpolation formula (see (1.34) of §1.10):

$$I^*_{k-1}(x) = f^1[n] + (x - x_n)f^1[n, n-1] + \cdots + (x - x_n)(x - x_{n-1})\cdots(x - x_{n-k+2})f^1[n, n-1, \ldots, n-k+1]. \quad (4.92)$$

Here, the superscript $*$ indicates, as always, that we are dealing with a predictor, and the notation $f^1[\cdots]$ indicates that, since we are setting up a PECE mode, the function values used to construct the divided differences are $f^{[1]}_{n-i}, i = 0, 1, \ldots, k-1$.

On integrating this interpolant and arguing as in §3.9, we obtain the k-step kth-order predictor

$$y^{[0]}_{n+1} - y^{[1]}_n = \sum_{i=0}^{k-1} g^*_i f^1[n, n-1, \ldots, n-i], \quad (4.93)$$

where

$$g^*_i = \begin{cases} \displaystyle\int_{x_n}^{x_{n+1}} \mathrm{d}x = x_{n+1} - x_n & \text{if } i = 0 \\ \displaystyle\int_{x_n}^{x_{n+1}} (x - x_n)(x - x_{n-1})\cdots(x - x_{n-i+1})\,\mathrm{d}x & \text{if } i = 1, 2, \ldots, . \end{cases} \quad (4.94)$$

(We shall return later to the question of how to compute the coefficients g^*_i efficiently.)

Recalling from §4.4 that the efficient way to implement the corrector in a constant steplength ABM method is to express the corrected value as an update of the predicted value (see (4.23)), it is natural to try to develop the interpolant for the corrector from that for the predictor. The result (1.37) of §1.10 allows us to do just that, and tells us that the polynomial of degree k which interpolates the data

$$(x_{n+1}, f_{n+1}), (x_n, f_n), \ldots, (x_{n-k+1}, f_{n-k+1})$$

is

$$I_k(x) = I^*_{k-1}(x) + (x - x_n)(x - x_{n-1})\cdots(x - x_{n-k+1})f^1_0[n+1, n, \ldots, n-k+1] \quad (4.95)$$

where the notation f^1_0 indicates that, when evaluating the divided difference, the single function value $f^{[1]}_{n+1}$ is replaced by $f^{[0]}_{n+1}$ (consistent with the notation defined by (4.15)).

Note that $I_k(x)$, being of degree k, will generate a corrector of order $k+1$ which, together with the kth order predictor (4.93), will be equivalent to a kth order PECE algorithm with local extrapolation. Let us pursue that option for a moment. Integrating $I_k(x)$ from x_n to x_{n+1} will yield a formula for $y^{[1]}_{n+1} - y^{[1]}_n$, just as doing likewise to $I^*_{k-1}(x)$ yielded the formula (4.93) for $y^{[0]}_{n+1} - y^{[1]}_n$. It follows from (4.95) and (4.94) that the corrector stage (with local extrapolation) can be written as

$$y^{[1]}_{n+1} = y^{[0]}_{n+1} + g^*_k f^1_0[n+1, n, \ldots, n-k+1]. \quad (4.96)$$

Equations (4.93) and (4.96) define the kth-order PECE method with local extrapolation.

To get the kth-order PECE method with no local extrapolation, we obtain from (4.95), with k replaced by $k-1$, the polynomial $I_{k-1}(x)$ of degree $k-1$,

$$I_{k-1}(x) = I^*_{k-2}(x) + (x - x_n)(x - x_{n-1})\cdots(x - x_{n-k+2})f^1_0[n+1, n, \ldots, n-k+2] \quad (4.97)$$

which interpolates the data

$$(x_{n+1}, f^{[0]}_{n+1}), (x_n, f^{[1]}_n), \ldots, (x_{n-k+2}, f^{[1]}_{n-k+2}).$$

Recalling that in a kth-order ABM method the corrector is right-shifted by one steplength relative to the predictor, this is the appropriate data set for the corrector. Further, from (4.92), the polynomial $I^*_{k-2}(x)$, which now appears in (4.97), can be written in the form

$$I^*_{k-2}(x) = I^*_{k-1}(x) - (x - x_n)(x - x_{n-1})\cdots(x - x_{n-k+2}) f^1[n, n-1, \ldots, n-k+1].$$

The interpolant representing the kth-order corrector is then, by (4.97),

$$\begin{aligned} I_{k-1}(x) &= I^*_{k-1}(x) + (x - x_n)(x - x_{n-1})\cdots(x - x_{n-k+2}) \\ &\quad \times \{f^1_0[n+1, n, \ldots, n-k+2] - f^1[n, n-1, \ldots, n-k+1]\} \\ &= I^*_{k-1}(x) + (x_{n+1} - x_{n-k+1})\{(x - x_n)\cdots(x - x_{n-k+2}) f^1_0[n+1, n, \ldots, n-k+1]\} \end{aligned}$$

by the definition of divided differences (see (1.33)). Following the same argument as before and using (4.94), we have that the corrector step can be written as

$$y^{[1]}_{n+1} - y^{[0]}_{n+1} = (x_{n+1} - x_{n-k+1}) g^*_{k-1} f^1_0[n+1, n, \ldots, n-k+1], \tag{4.98}$$

which, together with (4.93), defines the k-step kth-order PECE method.

We can obtain an error estimate by comparing the value of $y^{[1]}_{n+1}$ given by the above kth-order corrector with that given by the $(k+1)$th-order corrector (4.96); thus, on subtracting (4.98) from (4.96) we have

$$T_{n+1} = \{g^*_k - (x_{n+1} - x_{n-k+1}) g^*_{k-1}\} f^1_0[n+1, n, \ldots, n-k+1]. \tag{4.99}$$

It is instructive to compare the prediction, correction and error estimation stages, given by (4.93), (4.98) and (4.99) respectively, with their equal-spacing-case counterparts (4.23) of §4.3. In the case when the data is equally spaced, we find from (4.94) and (1.38) of §1.10 that

$$g^*_i = i!h^{i+1}\gamma^*_i, \qquad \left.\begin{aligned} f^1[n, n-1, \ldots, n-i+1] &= \nabla^i f^{[1]}_n/(i!h^i) \\ f^1_0[n+1, n, \ldots, n-k+1] &= \nabla^k_0 f^{[1]}_{n+1}/(k!h^k) \end{aligned}\right\} \tag{4.100}$$

from which it is straightforward to show that the formulae given above do revert to (4.23).

We return to the problem of how best to compute the coefficients g^*_i defined by (4.94). Let us evaluate the first few g^*_i directly:

$$\left.\begin{aligned} g^*_0 &= x_{n+1} - x_n \\ g^*_1 &= \int_{x_n}^{x_{n+1}} (x - x_n)\,dx = \tfrac{1}{2}(x_{n+1} - x_n)^2 \\ g^*_2 &= \int_{x_n}^{x_{n+1}} (x - x_n)(x - x_{n-1})\,dx = (x_{n+1} - x_{n-1})(x_{n+1} - x_n)^2/2 - (x_{n+1} - x_n)^3/6 \end{aligned}\right\} \tag{4.101}$$

(integrating by parts in the last of these). Clearly, we need to find a way of generating the g_i^* that is suitable for automatic computation. The way to do it turns out to be to embed the g_i^* in a more extended definition. Let us use the notation

$$\underbrace{\int_{x_n}^{x_{n+1}}\int_{x_n}^{x}\cdots\int_{x_n}^{x}}_{j\text{-times}} F(x)\,\mathrm{d}x$$

to denote the j-fold integral

$$\underbrace{\int_{x_n}^{x_{n+1}}\left[\int_{x_n}^{x}\left[\cdots\left[\int_{x_n}^{x}}_{j\text{-times}} F(x)\,\mathrm{d}x\right]\cdots\right]\mathrm{d}x\right]\mathrm{d}x,$$

and define g_{ij} for $j = 1, 2, \ldots$ by

$$g_{ij} = \begin{cases} \underbrace{\int_{x_n}^{x_{n+1}}\int_{x_n}^{x}\cdots\int_{x_n}^{x}}_{j\text{-times}} \mathrm{d}x = (x_{n+1} - x_n)^j/j! & \text{if } i = 0 \\ \underbrace{\int_{x_n}^{x_{n+1}}\int_{x_n}^{x}\cdots\int_{x_n}^{x}}_{j\text{-times}} (x - x_n)(x - x_{n-1})\cdots(x - x_{n-i+1})\,\mathrm{d}x & \text{if } i = 1, 2, \ldots \end{cases} \tag{4.102}$$

Clearly, from (4.94), we have that $g_i^* = g_{i1}, i = 0, 1, 2, \ldots$. The point of introducing the coefficients g_{ij} is that, by repeated use of integration by parts, it is possible to establish the following recurrence relation:

$$g_{ij} = (x_{n+1} - x_{n-i+1})g_{i-1,j} - jg_{i-1,j+1}, \qquad i = 1, 2, \ldots,\ j = 1, 2, \ldots, \tag{4.103}$$

from which it is possible to build up the following triangular array which generates the $g_{i,j}$ and hence the g_i^*:

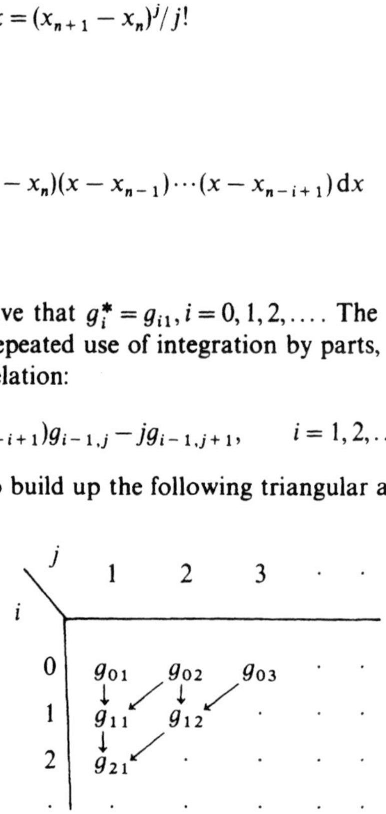

The entries in the first row are given directly by (4.102), and the arrows indicate the dependence of elements in later rows on those in the preceding row. Thus we get

$$g_{01} = x_{n+1} - x_n, \qquad g_{02} = (x_{n+1} - x_n)^2/2, \qquad g_{03} = (x_{n+1} - x_n)^3/6$$
$$g_{11} = (x_{n+1} - x_n)g_{01} - g_{02} = (x_{n+1} - x_n)^2/2$$
$$g_{12} = (x_{n+1} - x_n)g_{02} - 2g_{03} = (x_{n+1} - x_n)^3/6$$
$$g_{21} = (x_{n+1} - x_{n-1})g_{11} - g_{12} = (x_{n+1} - x_{n-1})(x_{n+1} - x_n)^2/2 - (x_{n+1} - x_n)^3/6.$$

We see that the generated values for $g_i^* = g_{i1}, i = 0, 1, 2$, agree with those found in (4.101).

If the PECE algorithm described above is run with the steplength held constant for a number of steps, then the coefficients g_i^* become constant, and need not be re-computed at each step. However, as we have seen in (4.100), they do not become the standard Adams–Bashforth coefficients γ_i^*, just as the divided differences do not revert to standard backward differences. Equations (4.100) indicate that in the general case of unequally spaced data points, the coefficients g_i^* will be very small if the current steplength is small (and i reasonably large), while the divided differences will be very large. Multiplication of a very small number into a very large one is a process likely to exacerbate the effect of round-off error, and it would be attractive to find an alternative formulation in which the coefficients *and* the 'divided differences' separately revert to standard Adams coefficients and backward differences respectively when the steplength is held constant. Such an alternative has been developed by Krogh (1974), who replaces the divided differences $f[n, n-1, \ldots, n-i]$ by *modified divided differences* $\hat{f}[n, n-1, \ldots, n-i]$, defined by

$$\hat{f}[n, n-1, \ldots, n-i] = (x_n - x_{n-1})(x_n - x_{n-2}) \cdots (x_n - x_{n-i}) f[n, n-1, \ldots, n-i].$$

It is readily seen that $\hat{f}[n, n-1, \ldots, n-i]$ does revert to $\nabla^i f_n$ when the data are equally spaced. The development in terms of these modified divided differences follows the general pattern of our development for unmodified divided differences, but is considerably more complicated. The reader is referred to the paper by Krogh or to a particularly readable account in Shampine and Gordon (1975). Both of these references describe several devices which increase the efficiency with which the method is implemented, the error estimated and the modified divided differences updated. The resulting method is more than just a technique one calls up when a change of steplength is required; it is used throughout the computation, since it automatically reverts to the standard ABM method in backward difference form whenever the steplength is held constant. Indeed, the recurrence relation (similar to (4.103)) for the coefficients which replace the g_i^* (and which will revert to the γ_i^* when the steplength is constant) turns out to be one of the best ways of generating the coefficients γ_i^*.

4.11 THE STRUCTURE OF VSVO ALGORITHMS

One fact that has clearly emerged from the extensive computational experience that has accumulated over the years is that the key to high efficiency in predictor–corrector algorithms is the capacity to vary automatically not only the steplength, but also the order (and hence the stepnumber) of the methods employed. Algorithms with such a capability are known as *variable step, variable order*, or VSVO, algorithms. It is not our purpose here to advocate, far less to study in detail, any particular VSVO algorithm, but rather to describe, in a general way, how they work. It is emphasized that we deal here only with algorithms for non-stiff initial value problems, although several existing VSVO codes include options for dealing with stiff systems, options which we shall ignore in this section.

The essential components of VSVO algorithm are:

(i) a **family of methods**,
(ii) a **starting procedure**,
(iii) a local **error estimator**,
(iv) a **strategy** for deciding when to change steplength and/or order, and
(v) a **technique** for changing the steplength and/or order.

We consider each of these components in turn. Note that we have not included in this list any mention of linear stability. Algorithms do not normally test to see whether the condition of absolute stability is satisfied for a given steplength (much too expensive!), and rely on the fact that if such a condition is not satisfied then the error estimate will grow sharply, and the algorithm will then take appropriate action.

Family of methods This is almost always the family of ABM methods, of orders ranging from 1 to around 13. In some algorithms the low-order ABM are replaced by predictor–corrector methods with better regions of absolute stability, since, at low order, it is usually lack of stability rather than accuracy that limits the steplength. Various modes can be employed, but PECE with local extrapolation is probably the most popular.

Starting procedure This is simplicity itself. The algorithm always starts with the one-step ABM pair (or its alternate) which does not require any additional starting values, and allows the steplength/order-changing strategy to take over. This usually results in the order building up rapidly over the initial few steps.

Error estimator This is always afforded by some form of Milne's device which, as we have seen in §4.4, takes a particularly simple and efficient form for ABM methods.

Strategy The account here essentially follows that given by Hall (1976). Suppose that the algorithm is currently working with a kth-order method; let E_k be the norm of the local error estimate at x_{n+1} and let τ be a user-defined tolerance. Then an obvious criterion for acceptance of the step from x_n to x_{n+1} is that

$$E_k \leqslant \tau, \tag{4.104}$$

the so-called 'error-per-step' criterion. It can, however, be argued that since the user is really interested in the accumulated and not the local error, it is inconsistent to worry only about the size of the local error and not about how many steps the error is committed at between two given values of x. This gives rise to an alternative criterion, $E_k \leqslant h\tau$, the so-called 'error-per-unit-step' criterion. Arguments can be put for both criteria. We shall develop a strategy based on the criterion (4.104), but adaptation to the other criterion will be obvious.

Recall from §4.4 that the norm of the error estimate for a kth-order ABM method is (see (4.22))

$$E_k = \| T_{n+1} \| = \| h\gamma_k \nabla^k_{\mu-1} f^{[\mu-t]}_{n+1} \|. \tag{4.105}$$

This estimate is available after the final application of the corrector in a $P(EC)^\mu E^{1-t}$ algorithm, and before the final evaluation (in the case $t = 0$); thus if the error estimate does not satisfy (4.104) and the step aborted, that final evaluation need not be made.

At first sight, we now seem to have an impossible task; we have only one criterion from which to deduce two pieces of information, namely what should be the steplength *and* the order at the next step. However, there is a very common situation in elementary calculus where we ask for one piece of information and invariably get two! It is when we ask what is the maximum of a function, when, on the way to getting that maximum, we always find the point at which the maximum is taken. The very same mechanism comes to our rescue here.

Suppose that on completing the step from x_n to x_{n+1} the criterion (4.104) is satisfied, and the computed value for $y^{[\mu]}_{n+1}$ consequently accepted; suppose further that the final evaluation has been made (in the case of a $P(EC)^\mu E$ mode) and that the backward differences have been updated, so that we are ready to take the next step. Before doing that, we ask what would have been the maximum steplength we could have used on the just completed step from x_n to x_{n+1} using ABM methods of orders $k-1, k$ and $k+1$. Whichever of these three steplengths turns out to be the greatest will be the steplength we shall use for the new step, and the value of k which produces that maximum steplength will be the order we shall use on the new step.

Let the steplength used with the kth-order method on the step from x_n to x_{n+1} be h_k. Since that step was successful it must have produced an error estimate which satisfied (4.104); that is, we must have that

$$\begin{aligned} E_k &= \beta\tau, \qquad 0 \leqslant \beta \leqslant 1 \\ &\approx K h_k^{k+1}, \end{aligned} \tag{4.106}$$

(where K is a constant) since the method has order k. The maximum steplength $\bar{h}_k$ we could have taken with this kth-order method would have produced an error estimate $\bar{E}_k$ satisfying

$$\bar{E}_k = \tau \approx K \bar{h}_k^{k+1}. \tag{4.107}$$

It follows from (4.106) and (4.107) that $\beta \approx (h_k/\bar{h}_k)^{k+1}$, whence

$$\bar{h}_k \approx h_k \left[\frac{\tau}{E_k} \right]^{1/(k+1)}, \tag{4.108}$$

an estimate we can compute, using (4.105). Suppose now that we had computed from x_n to x_{n+1}, using ABM methods of orders $k-1$ and $k+1$, and steplengths h_{k-1} and h_{k+1} respectively. Then it can be shown that the formulae

$$E_{k-1} = \| h_{k-1}\gamma_{k-1} \nabla^{k-1}_{\mu-1} f^{[\mu-t]}_{n+1} \|, \qquad E_{k+1} = \| h_{k+1}\gamma_{k+1} \nabla^{k+1}_{\mu-1} f^{[\mu-t]}_{n+1} \| \tag{4.109}$$

also hold, so that E_{k+i}, $i = -1, 0, 1$ can be computed. (There is an argument here in favour of local extrapolation, balancing our arguments against it in §4.3; Hall (1976) shows that the second of (4.109) is strictly valid only if local extrapolation is performed.) The argument used above to derive (4.107) can then be repeated to yield

$$\bar{h}_{k+i} \approx h_{k+i}\left[\frac{\tau}{E_{k+i}}\right]^{1/(k+1+i)}, \qquad i = -1, 0, 1.$$

The order we use for the next step is $k + \bar{\imath}$, where

$$\bar{h}_{k+\bar{\imath}} = \max_{i \in \{-1,0,1\}} \bar{h}_{k+i}$$

and the maximum steplength we can use on the next step is $\bar{h}_{k+\bar{\imath}}$. Note that one possible outcome is that we retain the same order at the next step, but change the steplength.

In this area of implementation there is, not surprisingly, a large heuristic element. Thus, most algorithms would multiply the maximum steplength advocated by the above argument by a heuristically chosen factor less than but close to one, so as to reduce the chances of a marginal rejection at the next step. Algorithms may also bias decisions in favour of retaining order, and may (particularly in the case of those using interpolatory step-changing techniques) include an embargo against changing the steplength too frequently.

Technique In §4.7–§4.10 we have already considered in detail techniques for step-changing. The technique for changing the order is much more straightforward. If the technique for changing the steplength is either variable coefficient or interpolatory using backward differences, then the reduction of the order from k to $k - 1$ is achieved by throwing away the kth difference; an increase of order from k to $k + 1$ is accomplished by forming the $(k + 1)$th difference, which is simply achieved by retaining all of the back data at the completion of the step rather than throwing away the data at x_{n+k-1}, as one normally would do. In the case when step-changing is done via a Nordsieck vector, an order reduction of one is achieved by ignoring the last component of the Nordsieck vector (though, as shown by Hall (1976), the resulting method is no longer equivalent to an Adams method). An order increase of one can be achieved by estimating an extra derivative in the Nordsieck vector by differencing the current last component.

It is natural to ask whether the convergence properties of predictor–corrector methods, previously established on the assumption of constant steplength and constant order, still hold in a VSVO formulation. Results by Gear and Tu (1974) and Gear and Watanabe (1974) show that a VSVO algorithm based on ABM methods with step-changing achieved by a variable coefficient technique is always convergent (as the maximum steplength employed in the interval of integration tends to zero). If an interpolatory technique is used then convergence is assured if the step/order-changing technique is such that there exists a constant N such that in any N consecutive steps there are always k steps of constant length taken by the same kth-order ABM method, for some value of k. These results emphasize yet again that variable coefficient techniques, though usually more expensive to implement, are basically sounder than interpolatory techniques.

Finally, we mention a few of the better known VSVO codes and indicate which step-changing techniques they use. The first two VSVO implementations appeared simultaneously, when Gear (1969) and Krogh (1969) gave concurrent presentations at the same conference in 1968. The corresponding codes are DIFSUB (Gear, 1971a), which uses the Nordsieck vector, and DVDQ (Krogh, 1969), which uses step doubling or halving. GEAR (Hindmarsh, 1974) is a much modified and extended version of DIFSUB. A widely used code, EPISODE (Byrne and Hindmarsh, 1975), uses the variable coefficient technique, as does DE/STEP, which is described in considerable detail in the book by Shampine and Gordon (1975). Initial value problems for systems of ordinary differential equations can be very lively mathematical creatures and, on occasion, are capable of upsetting even the most sophisticated of codes; striking examples of this can be found in Shampine (1980).

5 Runge–Kutta Methods

5.1 INTRODUCTION

The simplest of all numerical methods is Euler's Rule,

$$y_{n+1} = y_n + hf_n.$$

It is linear in y_n and f_n and, being a one-step method, presents no difficulty when we want to change the steplength; but of course it has very low accuracy. Linear multistep methods achieve higher accuracy by retaining linearity with respect to y_{n+j} and f_{n+j}, $j = 0, 1, \dots, k$, but sacrificing the one-step format. The result of retaining the linearity is that the local error has a relatively simple structure, which is why we are able to estimate it so easily via Milne's device; the cost of moving to a multistep format is, as we have seen, the considerable difficulties encountered when we want to change steplength.

Runge–Kutta methods develop from Euler's Rule in exactly the opposite direction; higher order is achieved by retaining the one-step form but sacrificing the linearity. The result is that there is no difficulty in changing the steplength, but the structure of the local error is much more complicated, and there exists no easy and cheap error estimate comparable with Milne's device. We are rather in a Catch-22 situation; with linear multistep methods it is easy to tell when we ought to change steplength but hard to change it, while with Runge–Kutta Methods it is hard to tell when to change steplength but easy to change it!

The general *s-stage Runge–Kutta method* for the problem

$$y' = f(x, y), \quad y(a) = \eta, \quad f: \mathbb{R} \times \mathbb{R}^m \to \mathbb{R}^m \tag{5.1}$$

is defined by

$$\left.\begin{aligned} &y_{n+1} = y_n + h\sum_{i=1}^{s} b_i k_i \\ &\text{where} \\ &k_i = f\left(x_n + c_i h,\, y_n + h\sum_{j=1}^{s} a_{ij} k_j\right), \qquad i = 1, 2, \dots, s \end{aligned}\right\} \tag{5.2}$$

We shall always assume that the following (the *row-sum condition*) holds:

$$c_i = \sum_{j=1}^{s} a_{ij}, \qquad i = 1, 2, \dots, s. \tag{5.3}$$

It is convenient to display the coefficients occurring in (5.2) in the following form, known as a *Butcher array*:

$$\begin{array}{c|cccc} c_1 & a_{11} & a_{12} & \cdots & a_{1s} \\ c_2 & a_{21} & a_{22} & \cdots & a_{2s} \\ \vdots & \vdots & & & \vdots \\ c_s & a_{s1} & a_{s2} & \cdots & a_{ss} \\ \hline & b_1 & b_2 & \cdots & b_s \end{array} \tag{5.4}$$

We define the s-dimensional vectors c and b and the $s \times s$ matrix A by

$$c = [c_1, c_2, \dots, c_s]^\mathsf{T}, \quad b = [b_1, b_2, \dots, b_s]^\mathsf{T}, \quad A = [a_{ij}]. \tag{5.5}$$

Note that, by (5.3), the components of c are the row sums of A. Clearly an s-stage Runge–Kutta method is completely specified by its Butcher array

$$\begin{array}{c|c} c & A \\ \hline & b^\mathsf{T}. \end{array}$$

An alternative form of (5.2), which in some contexts is more convenient, is

$$\left.\begin{aligned} y_{n+1} &= y_n + h \sum_{i=1}^{s} b_i f(x_n + c_i h, Y_i), \\ \text{where} \quad & \\ Y_i &= y_n + h \sum_{j=1}^{s} a_{ij} f(x_n + c_j h, Y_j), \qquad i = 1, 2, \dots, s. \end{aligned}\right\} \tag{5.6}$$

The forms (5.6) and (5.2) are seen to be equivalent if we make the interpretation

$$k_i = f(x_n + c_i h, Y_i), \qquad i = 1, 2, \dots, s. \tag{5.7}$$

If in (5.2) we have that $a_{ij} = 0$ for $j \geqslant i$, $i = 1, 2, \dots, s$, then each of the k_i is given explicitly in terms of previously computed k_j, $j = 1, 2, \dots, i-1$, and the method is then an *explicit* or *classical* Runge–Kutta method. If this is not the case then the method is *implicit* and, in general, it is necessary to solve at each step of the computation an implicit system for the k_i. Note that this system has dimension ms, so that implicitness in a Runge–Kutta method poses an even more daunting computational problem than does implicitness in a linear multistep method. There is a sort of half-way house; if it happens that $a_{ij} = 0$ for $j > i$, $i = 1, 2, \dots, s$, then each k_i is individually defined by

$$k_i = f\left(x_n + c_i h, y_n + \sum_{j=1}^{i} a_{ij} k_j \right), \qquad i = 1, 2, \dots, s$$

and instead of having to solve at each step a nonlinear system of dimension ms, we have to solve s uncoupled systems each of dimension m; this is less demanding, but still more so than in the case of an implicit linear multistep method. Such methods are called *semi-implicit*. Summarizing, we have:

Explicit method:

$$a_{ij} = 0, j \geqslant i, j = 1, 2, \dots, s \Leftrightarrow A \text{ strictly lower triangular.}$$

Semi-implicit method:

$$a_{ij} = 0, j > i, j = 1, 2, \ldots, s \Leftrightarrow A \text{ lower triangular.}$$

Implicit method:

$$a_{ij} \neq 0 \text{ for some } j > i \Leftrightarrow A \text{ not lower triangular.}$$

A remark that can be made about Runge–Kutta methods (and one which seldom appears to be made) is that they constitute a clever and sensible idea. The unique solution of a well-posed initial value problem can be thought of as a single integral curve in $\mathbb{R}^{m+1}$; but, due to truncation and round-off error, any numerical slution is, in effect, going to wander off that integral curve, and the numerical solution is inevitably going to be affected by the behaviour of neighbouring integral curves. Thus it is the behaviour of the *family* of integral curves, and not just that of the unique solution curve, that is of importance, a point we shall return to when discussing stiffness later in this book. Runge–Kutta methods deliberately try to gather information about this family of curves. Such an interpretation is most easily seen in the case of explicit Runge–Kutta methods, where, from (5.2) and (5.3), we have

$$\begin{aligned}
k_1 &= f(x_n, y_n),\\
k_2 &= f(x_n + c_2 h, y_n + c_2 h k_1),\\
k_3 &= f(x_n + c_3 h, y_n + (c_3 - a_{32}) h k_1 + h a_{32} k_2)\\
\vdots\ & \quad \vdots
\end{aligned}$$

Start from (x_n, y_n), take one step of Euler's Rule of length $c_2 h$ and evaluate the derivative vector at the point so reached; the result is k_2. We now have two samples for the derivative, k_1 and k_2, so let us use a weighted mean of k_1 and k_2 as the initial slope in another Euler step (from (x_n, y_n)) of length $c_3 h$, and evaluate the derivative at the point so reached; the result is k_3. Continuing in this manner, we obtain a set k_i, $i = 1, 2, \ldots, s$ of samples of the derivative. The final step is yet another Euler step

$$y_{n+1} = y_n + h \sum_{i=1}^{s} b_i k_i$$

from (x_n, y_n) to (x_{n+1}, y_{n+1}), using as initial slope a weighted mean of the samples $k_1, k_2, \ldots, k_s$. Thus, an explicit Runge–Kutta method sends out feelers into the solution space, to gather samples of the derivative, before deciding in which direction to take an Euler step, an eminently sensible idea.

5.2 CONSISTENCY, LOCAL TRUNCATION ERROR, ORDER AND CONVERGENCE

In Chapter 2 we considered the general class of methods (2.4),

$$\sum_{j=0}^{k} \alpha_j y_{n+j} = h \phi_f(y_{n+k}, y_{n+k-1}, \ldots, y_n, x_n; h). \tag{5.8}$$

On putting

$$\left.\begin{aligned} &k = 1, \qquad \alpha_1 = 1, \qquad \alpha_0 = -1, \qquad \phi_f(y_n, x_n; h) = \sum_{i=1}^{s} b_i k_i \\ &k_i = f\left(x_n + c_i h, y_n + h\sum_{j=1}^{s} a_{ij}k_j\right), \qquad i = 1, 2, \dots, s \end{aligned}\right\} \tag{5.9}$$

we see that the class (5.8) contains the class of Runge–Kutta methods.

Recall from equations (2.9) of §2.4 that the necessary and sufficient conditions for (5.8) to be consistent are

$$\sum_{j=0}^{k} \alpha_j = 0, \quad \phi_f(y(x_n), y(x_n), \dots, y(x_n), x_n; 0)\Big/\left(\sum_{j=0}^{k} j\alpha_j\right) = f(x_n, y(x_n)).$$

Applying these conditions in the case when (5.9) holds, we see that the necessary and sufficient condition for a general Runge–Kutta method to be consistent is

$$\phi_f(y(x_n), x_n; 0) = f(x_n, y(x_n)) \Leftrightarrow \sum_{i=1}^{s} b_i = 1. \tag{5.10}$$

Following the discussion in §3.5, we define the local truncation error T_{n+1} of (5.2) at x_{n+1} to be the residual when y_{n+j} is replaced by $y(x_{n+j})$, $j = 0, 1$; that is,

$$T_{n+1} = y(x_{n+1}) - y(x_n) - h\phi_f(y(x_n), x_n; h), \tag{5.11}$$

where ϕ_f is defined by (5.9). If p is the largest integer such that $T_{n+1} = 0(h^{p+1})$, we say that the method has *order p*. If, as in §3.5, we denote by $\tilde{y}_{n+1}$ the value at x_{n+1} generated by the Runge–Kutta method when the localizing assumption that $y_n = y(x_n)$ is made, then, since

$$\tilde{y}_{n+1} = y_n + \phi_f(y_n, x_n; h),$$

we have from (5.11) that

$$y(x_{n+1}) - \tilde{y}_{n+1} = T_{n+1}. \tag{5.12}$$

(Compare with (3.25) of §3.5.)

If the method is consistent then it follows from (5.10) that

$$\begin{aligned} y(x_{n+1}) - y(x_n) - \phi_f(y(x_n), x_n; h) &= hy'(x_n) - hf(x_n, y(x_n)) + 0(h^2) \\ &= 0(h^2), \end{aligned}$$

since $y'(x) = f(x, y(x))$. Thus, from (5.11), a consistent method has order at least 1, a result in line with our definition of order for linear multistep methods. Note also that Euler's Rule, which is both a Runge–Kutta and a linear multistep method, has order 1 whichever definition of order one uses.

It is obvious from (5.9) that Runge–Kutta methods always satisfy the root condition of §2.5 and hence, by Theorems 2.1 and 2.2, are convergent provided only that the consistency condition (5.10) is satisfied.

It is possible to establish bounds for the local and global truncation errors for Runge–Kutta methods, but these turn out to be, in practice, as useless as the corresponding bounds for linear multistep methods. Once again, if the method has order p, the local truncation error is $0(h^{p+1})$ and the global truncation error is $0(h^p)$.

5.3 DERIVATION OF EXPLICIT RUNGE–KUTTA METHODS FOR SCALAR PROBLEMS

Runge–Kutta methods first appeared in 1895, and up to the 1960s only explicit methods were considered. Moreover, the derivation of such methods invariably assumed a scalar problem, and it was tacitly (and wrongly, as it turns out) assumed that nothing of significance would change when these methods were applied to systems. The technique for deriving the order conditions consisted of matching the expansion of the solution generated by one step of the Runge–Kutta method with the Taylor expansion of the exact solution, the terms in the expansions being calculated essentially by brute force. Such calculations are notoriously heavy and tedious, particularly if high order is sought. In this section we derive in this way explicit Runge–Kutta methods with up to three stages, this being adjudged sufficient to persuade the reader (without exposing him to too much tedium) that some better approach is needed.

By (5.2) and (5.3) the 3-stage explicit Runge–Kutta method can be written as

$$\left.\begin{aligned} y_{n+1} &= y_n + h(b_1k_1 + b_2k_2 + b_3k_3)\\ k_1 &= f(x_n, y_n)\\ k_2 &= f(x_n + hc_2, y_n + hc_2k_1)\\ k_3 &= f(x_n + hc_3, y_n + h(c_3 - a_{32})k_1 + ha_{32}k_2). \end{aligned}\right\} \tag{5.13}$$

We assume that $f(x, y)$ is sufficiently smooth, and introduce the shortened notation

$$f := f(x, y), \quad f_x := \frac{\partial f(x, y)}{\partial x}, \quad f_{xx} := \frac{\partial^2 f(x, y)}{\partial x^2}, \quad f_{xy}(\equiv f_{yx}) := \frac{\partial^2 f(x, y)}{\partial x \partial y}$$

etc. all evaluated at the point $(x_n, y(x_n))$. Then, on expanding $y(x_{n+1})$ about x_n as a Taylor series, we have

$$y(x_{n+1}) = y(x_n) + hy^{(1)}(x_n) + \tfrac{1}{2}h^2y^{(2)}(x_n) + \tfrac{1}{6}h^3y^{(3)}(x_n) + 0(h^4).$$

Now,

$$\left.\begin{aligned} y^{(1)}(x_n) &= f,\\ y^{(2)}(x_n) &= f_x + f_yy' = f_x + ff_y\\ y^{(3)}(x_n) &= f_{xx} + f_{xy}f + f(f_{yx} + f_{yy}f) + f_y(f_x + ff_y)\\ &= f_{xx} + 2ff_{xy} + f^2f_{yy} + f_y(f_x + ff_y). \end{aligned}\right\} \tag{5.14}$$

Let us shorten the notation again by defining

$$F := f_x + ff_y, \quad G := f_{xx} + 2ff_{xy} + f^2f_{yy}, \tag{5.15}$$

so that we can write the expansion for $y(x_{n+1})$ as

$$y(x_{n+1}) = y(x_n) + hf + \tfrac{1}{2}h^2F + \tfrac{1}{6}h^3(Ff_y + G) + 0(h^4). \tag{5.16}$$

In order to use (5.12), we need a similar expansion for $\tilde{y}_{n+1}$. Expanding the k_i given by (5.13) we have $k_1 = f$ and

$$k_2 = f + hc_2(f_x + k_1 f_y) + \tfrac{1}{2}h^2c_2^2(f_{xx} + 2k_1 f_{xy} + k_1^2 f_{yy}) + 0(h^3).$$

On substituting f for k_1, and using the notation in (5.15), we get

$$k_2 = f + hc_2F + \tfrac{1}{2}h^2c_2^2G + 0(h^3)]. \tag{5.17}$$

We can treat k_3 similarly (but now the tedium sets in!).

$$\begin{aligned} k_3 = f &+ h\{c_3f_x + [(c_3 - a_{32})k_1 + a_{32}k_2]f_y\} \\ &+ \tfrac{1}{2}h^2\{c_3^2f_{xx} + 2c_3[(c_3 - a_{32})k_1 + a_{32}k_2]f_{xy} \\ &+ [(c_3 - a_{32})k_1 + a_{32}k_2]^2 f_{yy}\} + 0(h^3). \end{aligned}$$

Now write f for k_1, substitute for k_2 from (5.17) and retain terms up to $0(h^2)$ to get

$$k_3 = f + hc_3F + h^2(c_2a_{32}Ff_y + \tfrac{1}{2}c_3^2G) + 0(h^3). \tag{5.18}$$

On substituting from (5.17) and (5.18) into (5.13) and using the localizing assumption (and the attendant notation introduced in §3.5), we obtain the following expansion for $\tilde{y}_{n+1}$:

$$\begin{aligned} \tilde{y}_{n+1} = y(x_n) &+ h(b_1 + b_2 + b_3)f + h^2(b_2c_2 + b_3c_3)F \\ &\tfrac{1}{2}h^3[2b_3c_2a_{32}Ff_y + (b_2c_2^2 + b_3c_3^2)G] + 0(h^4). \end{aligned} \tag{5.19}$$

It is now a question of trying to match the expansions (5.16) and (5.19). Let us see what can be achieved with one, two and three stages. (For more than three stages, there would be a term k_4 in (5.13) which would contribute additional terms.)

One-stage The method (5.13) becomes 1-stage if we set $b_2 = b_3 = 0$. Then (5.19) reduces to

$$\tilde{y}_{n+1} = y(x_n) + hb_1 f + 0(h^4).$$

From (5.12) and (5.16) we see that the best we can do is set $b_1 = 1$, whence $T_{n+1} = 0(h^2)$. Thus there exists only one explicit one-stage Runge–Kutta method of order 1, namely Euler's Rule.

Two-stage The method becomes two-stage if we set $b_3 = 0$, when (5.19) becomes

$$\tilde{y}_{n+1} = y(x_n) + h(b_1 + b_2)f + h^2b_2c_2F + \tfrac{1}{2}h^3b_2c_2^2G + 0(h^4).$$

On comparing with (5.16) we see that order 2 can be achieved by choosing

$$b_1 + b_2 = 1, \qquad b_2c_2 = \tfrac{1}{2}. \tag{5.20}$$

This is a pair of equations in three unknowns, so there exists a singly infinite family of explicit two-stage Runge–Kutta methods of order 2. It is evident from (5.16) that no member of this family can achieve order higher than 2. Two particular solutions of (5.20) yield well-known methods:

(i) *The modified Euler* or *improved polygon method* is given by $b_1 = 0$, $b_2 = 1$, $c_2 = \frac{1}{2}$. Its Butcher array is

$$\begin{array}{c|cc} 0 & & \\ \frac{1}{2} & \frac{1}{2} & \\ \hline & 0 & 1 \end{array}$$

(ii) The *improved Euler method* is given by $b_1 = b_2 = \frac{1}{2}$, $c_2 = 1$. Its Butcher array is

$$\begin{array}{c|cc} 0 & & \\ 1 & 1 & \\ \hline & \frac{1}{2} & \frac{1}{2} \end{array}$$

Three-stage We can achieve order 3 if we can satisfy the following conditions

$$\begin{aligned} b_1 + b_2 + b_3 &= 1 \\ b_2 c_2 + b_3 c_3 &= \tfrac{1}{2} \\ b_2 c_2^2 + b_3 c_3^2 &= \tfrac{1}{3} \\ b_3 c_2 a_{32} &= \tfrac{1}{6}. \end{aligned}$$

There are now four equations in six unknowns and there exists a doubly infinite family of solutions; consideration of the h^4 term, which we ignored in the derivation, confirms that none of these solutions leads to a method or order greater than three. Two particular solutions lead to well-known methods:

(i) *Heun's third-order formula* with Butcher array

$$\begin{array}{c|ccc} 0 & & & \\ \frac{1}{3} & \frac{1}{3} & & \\ \frac{2}{3} & 0 & \frac{2}{3} & \\ \hline & \frac{1}{4} & 0 & \frac{3}{4} \end{array}$$

(ii) *Kutta's third-order formula* with Butcher array

$$\begin{array}{c|ccc} 0 & & & \\ \frac{1}{2} & \frac{1}{2} & & \\ 1 & -1 & 2 & \\ \hline & \frac{1}{6} & \frac{2}{3} & \frac{1}{6} \end{array}$$

By a similar approach it is possible to show that there exists a doubly infinite family of explicit four-stage Runge–Kutta methods of order 4, none of which has order greater than 4. By far the best known of these is the *classical Runge–Kutta method* which has Butcher array

$$\begin{array}{c|cccc} 0 & & & & \\ \frac{1}{2} & \frac{1}{2} & & & \\ \frac{1}{2} & 0 & \frac{1}{2} & & \\ 1 & 0 & 0 & 1 & \\ \hline & \frac{1}{6} & \frac{1}{3} & \frac{1}{3} & \frac{1}{6} \end{array} \qquad (5.21)$$

So popular is this method that, when one sees a reference to a problem having been solved by 'the Runge–Kutta method', it is almost certainly this method that has been used. As we shall see later, there turn out to be good reasons for choosing a four-stage fourth-order method, but (5.21) does not perform notably differently from other fourth-order Runge–Kutta methods. Of course, the presence of three zeros in A is attractive, but the author suggests another reason for its historical popularity. In the pre-computer days, computations were performed on purely mechanical devices like the 'signal-box' Brunsviga, now to be found only in museums. Multiplication or division was a tiresome business on such machines, involving a great deal of handle-turning. As always, the main effort was in the function evaluations needed to produce the k_i. That the c_i and the a_{ij} in (5.21) are always either 1 or $\frac{1}{2}$ (as opposed to $\frac{1}{3}$, for example) increased the chances of any divisions in the evaluations of f terminating quickly!

This section has thrown up several points of interest. There exists a single explicit one-stage Runge–Kutta method of order 1, a singly-infinite family of two-stage methods of order 2, a doubly-infinite family of three-stage methods of order 3, but a doubly- (not a triply-) infinite family of four-stage methods of order 4; some sort of anomaly is indicated. Secondly, the contrast with the order conditions for linear multistep methods is interesting; in the latter we worked naturally with total derivatives of the exact solution, and local truncation errors were multiples of $h^{p+1}y^{(p+1)}(x_n)$. From (5.16) and (5.19) we see that the natural building blocks of Runge–Kutta methods are not the total derivatives of the exact solution, but rather certain functions, such as F and G (see (5.15)) of the partial derivatives of f. Thirdly, in the case of linear multistep methods, the dimension of the system was unimportant, and the development for a scalar problem would hardly have differed from our development for an m-dimensional system. In contrast, the development given above for a scalar problem does not extend in any obvious way to an m-dimensional problem; for example, how would we interpret (5.14) if $y, f \in \mathbb{R}^m$? This last comment is our starting point for the next section.

Exercises

5.3.1. Find a solution of the third-order conditions for which $c_2 = c_3$ and $b_2 = b_3$; the resulting explicit method is known as *Nyström's third-order method*.

5.3.2. Show that the predictor–corrector method consisting of Euler's Rule and the Trapezoidal

Rule in PECE mode is equivalent to the improved Euler method. Find the three-stage explicit Runge–Kutta method which is equivalent to the same predictor–corrector pair in $P(EC)^2E$ mode.

5.3.3. Show that if $f(x, y) \equiv g(x)$ the improved Euler method reduces to the Trapezoidal Rule for quadrature and that Kutta's third-order rule and the popular fourth-order method (5.21) both reduce to Simpson's Rule for quadrature. Show that Heun's third-order formula reduces to the two-point Radau quadrature formula $\int_{-1}^{+1} F(x)dx = \frac{1}{2}F(-1) + \frac{3}{2}F(\frac{1}{3})$ applied to $\int_{x_h}^{x_n+h} g(x)dx$.

5.4 THE BUTCHER THEORY; INTRODUCTION

The ideas we are about to describe can be traced back to a paper by Merson (1957), but their development into a major theory is due to J. C. Butcher, in a long series of papers starting in the mid-1960s. The reader who wishes to see a full account of the theory is referred to the book by Butcher (1987) which, incidentally, contains the most comprehensive available bibliography on the subject of numerical methods for ordinary differential equations, listing some 2000 entries. In the following sections we shall present only a simplified version of the theory, aimed at enabling the reader to use the theory to establish the order conditions (and the structure of the local truncation error) and to appreciate some of the consequences. In particular, we shall not give any proofs of the theorems involved, and our treatment of the algebraic aspects will be non-rigorous (possibly to the point of offending some readers!).

Recall from §1.4 that, while there is a loss of generality in assuming that the *scalar* initial value problem is autonomous, there is no such loss in the case of a problem involving a system of ordinary differential equations. We are now dealing with systems, and a useful simplification is achieved by assuming the autonomous form

$$y' = f(y), \quad y(a) = \eta, \qquad f: \mathbb{R}^m \to \mathbb{R}^m, \quad m > 1 \tag{5.22}$$

Let us start by seeing what dificulties we get into by trying to adapt the approach of the preceding section to the systems case. The first problem we hit is the counterpart of (5.14), where we need to express the total derivatives of y in terms of f and its partial derivatives. For the *scalar* autonomous case (5.14) reads

$$y^{(1)} = f, \quad y^{(2)} = f_y f, \quad y^{(3)} = f_{yy} f^2 + f_y^2 f. \tag{5.23}$$

What are the corresponding expressions when $y, f \in \mathbb{R}^m$? To keep things simple, let us consider the case $m = 2$, so that $y = [{}^1y, {}^2y]^T$, $f = [{}^1f, {}^2f]^T$. Introducing the notation

$${}^if_j := \frac{\partial({}^if)}{\partial({}^jy)}, \qquad {}^if_{jk} := \frac{\partial^2({}^if)}{\partial({}^jy)\partial({}^ky)} \quad \text{etc.}$$

we have ${}^1y^{(1)} = {}^1f$, ${}^2y^{(1)} = 2f$, whence we find on differentiation that

$${}^1y^{(2)} = {}^1f_1({}^1f) + {}^1f_2({}^2f), \quad {}^2y^{(2)} = {}^2f_1({}^1f) + {}^2f_2({}^2f)$$

or

$$y^{(2)} = \begin{pmatrix} {}^1f_1 & {}^1f_2 \\ {}^2f_1 & {}^2f_2 \end{pmatrix} f \tag{5.24}$$

and

$$\left.\begin{aligned}{}^1y^{(3)} &= {}^1f_{11}({}^1f)^2 + {}^1f_{12}({}^1f)({}^2f) + {}^1f_1[{}^1f_1({}^1f) + {}^1f_2({}^2f)]\\ &\quad + {}^1f_{21}({}^2f)({}^1f) + {}^1f_{22}({}^2f)^2 + {}^1f_2[{}^2f_1({}^1f) + {}^2f_2({}^2f)]\\ {}^2y^{(3)} &= {}^2f_{11}({}^1f)^2 + {}^2f_{12}({}^1f)({}^2f) + {}^2f_1[{}^1f_1({}^1f) + {}^1f_2({}^2f)]\\ &\quad + {}^2f_{21}({}^2f)({}^1f) + {}^2f_{22}({}^2f)^2 + {}^2f_2[{}^2f_1({}^1f) + {}^2f_2].\end{aligned}\right\} \tag{5.25}$$

Note that (5.24) can be written as $y^{(2)} = \partial f/(\partial y)f$, where $\partial f/\partial y$ is the Jacobian matrix, a natural generalization of the second of (5.23). In contrast, (5.25) cannot be written in matrix/vector form, and it does not look at all like a generalization of the last of (5.23).

Clearly we cannot go on like this to evaluate even higher derivatives (and we have not yet considered the case of general m). A better notation is essential, and indeed more than a notation is needed; we need to discover a *structure* for these higher derivatives. (On the question of notation, the reader should be warned that in the following sections we shall continually be changing (and simplifying) notation; the guideline will be that as soon as we become familiar with a particular notation we will try to simplify it further.)

5.5 THE Mth FRECHET DERIVATIVE; ELEMENTARY DIFFERENTIALS

Following Butcher (1972), we introduce a key definition, that of the *Mth Frechet derivative*:

Definition *Let* $z, f(z) \in \mathbb{R}^m$. *The* M**th Frechet derivative** *of* f, *denoted by* $f^{(M)}(z)$, *is an* **operator** *on* $\mathbb{R}^m \times \mathbb{R}^m \times \cdots \times \mathbb{R}^m$ *(M times), linear in each of its operands, with value*

$$\underbrace{f^{(M)}(z)}_{\text{(argument)}}\underbrace{(K_1, K_2, \ldots, K_M)}_{\text{(operands)}} = \sum_{i=1}^{m}\sum_{j_1=1}^{m}\sum_{j_2=1}^{m}, \ldots, \sum_{j_M=1}^{m} {}^if_{j_1 j_2 \cdots j_M}\, {}^{j_1}K_1\, {}^{j_2}K_2, \ldots, {}^{j_M}K_M e_i \tag{5.26}$$

where

$$K_t = [{}^1K_t, {}^2K_t, \ldots, {}^mK_t]^{\mathsf T} \in \mathbb{R}^m, \qquad t = 1, 2, \ldots, M,$$

$${}^if_{j_1 j_2, \ldots, j_M} = \frac{\partial^M}{\partial({}^{j_1}z)\partial({}^{j_2}z), \ldots, \partial({}^{j_M}z)}\, {}^if(z) \tag{5.27}$$

and

$$e_i = [0, 0, \ldots, 0, \underset{\substack{\downarrow \\ (i\text{th component})}}{1}, 0, \ldots, 0]^{\mathsf T} \in \mathbb{R}^m$$

This somewhat daunting definition becomes perhaps a little less so if we make the following comments:

(1) The value of $f^{(M)}(z)(\cdots)$ is a vector in $\mathbb{R}^m$, but it is typographically difficult to display it as a vector. The vector e_i in (5.26) is introduced merely as a notational device to overcome this; the expression between the first summation and the vector e_i is the ith component of the vector.
(2) Repeated subscripts are permitted in (5.26) so that all possible partial derivatives of order M are involved. Thus, if $M = 3$, $m = 2$, the following partial derivatives will appear:

$$ {}^if_{111} = \frac{\partial^3({}^if)}{\partial({}^1z)^3}, \quad {}^if_{112} = {}^if_{121} = {}^if_{211} = \frac{\partial^3({}^if)}{\partial({}^1z)^2\partial({}^2z)} $$

$$ {}^if_{122} = {}^if_{212} = {}^if_{221} = \frac{\partial^3({}^if)}{\partial({}^1z)\partial({}^2z)^2}, \qquad {}^if_{222} = \frac{\partial^3({}^if)}{\partial({}^2z)^3}, \qquad i = 1, 2. $$

(3) The argument z simply denotes the vector with respect to whose component we are performing the partial differentiations.
(4) **An *M*th Frechet derivative has *M* operands**; this is the key property to note.

Let us now put $m = 2$, and see if we can interpret the results (5.24) and (5.25) of the preceding section in terms of Frechet derivatives.

Case $M = 1$

$$ f^{(1)}(z)(K_1) = \sum_{i=1}^{2} \sum_{j_1=1}^{2} {}^if_{j_1}({}^{j_1}K_1)e_i $$

$$ = \begin{bmatrix} {}^1f_1({}^1K_1) + {}^1f_2({}^2K_1) \\ {}^2f_1({}^1K_1) + {}^2f_2({}^2K_1) \end{bmatrix} \tag{5.28} $$

where

$$ {}^if_1 = \frac{\partial({}^if)}{\partial({}^1z)}, \qquad {}^if_2 = \frac{\partial({}^if)}{\partial({}^2z)}, \qquad i = 1, 2. $$

Now replace z by y, and K_1 by f (noting that all four are 2-dimensional vectors). Equation (5.28) now reads

$$ f^{(1)}(y)(f(y)) = \begin{bmatrix} {}^1f_1({}^1f) + {}^1f_2({}^2f) \\ {}^2f_1({}^1f) + {}^2f_2({}^2f) \end{bmatrix} = y^{(2)} \tag{5.29} $$

by (5.24). In the context of deriving Runge–Kutta methods, we do not really need to be told that partial derivatives of f are to be taken with respect to the component of y (what else is there?), so that we can shorten the notation in (5.29) to

$$ y^{(2)} = f^{(1)}(f). \tag{5.30} $$

Thus, the second derivative of y is the first Frechet derivative of f operating on f; from

(5.24) we see that this is equivalent to saying that the second derivative of y is the Jacobian of f operating on f.

Case $M = 2$

$$f^{(2)}(z)(K_1, K_2) = \sum_{i=1}^{2} \sum_{j_1=1}^{2} \sum_{j_2=1}^{2} {}^i f_{j_1 j_2}({}^{j_1}K_1)({}^{j_2}K_2)e_i$$

$$= \begin{bmatrix} {}^1f_{11}({}^1K_1)({}^1K_2) + {}^1f_{12}({}^1K_1)({}^2K_2) + {}^1f_{21}({}^2K_1)({}^1K_2) + {}^1f_{22}({}^2K_1)({}^2K_2) \\ {}^2f_{11}({}^1K_1)({}^1K_2) + {}^2f_{12}({}^1K_1)({}^2K_2) + {}^2f_{21}({}^2K_1)({}^1K_2) + {}^2f_{22}({}^2K_1)({}^2K_2) \end{bmatrix}.$$

Replace z by y and put $K_1 = K_2 = f$ to get

$$f^{(2)}(y)(f(y), f(y)) = \begin{bmatrix} {}^1f_{11}({}^1f)^2 + 2({}^1f_{12})({}^1f)({}^2f) + {}^1f_{22}({}^2f)^2 \\ {}^2f_{11}({}^1f)^2 + 2({}^2f_{12})({}^1f)({}^2f) + {}^2f_{22}({}^2f)^2 \end{bmatrix}. \tag{5.31}$$

where the notation on the left side indicates that the partial derivatives are with respect to components of y, not z. On comparing with (5.25), we see that the right side of (5.31) represents some, but not all, of the terms on the right side of (5.25), and that we obtain $y^{(3)}$ by adding to the right side of (5.31) the vector

$$\begin{bmatrix} {}^1f_1[{}^1f_1({}^1f) + {}^1f_2({}^2f)] + {}^1f_2[{}^2f_1({}^1f) + {}^2f_2({}^2f)] \\ {}^2f_1[{}^1f_1({}^1f) + {}^1f_2({}^2f)] + {}^2f_2[{}^2f_1({}^1f) + {}^2f_2({}^2f)] \end{bmatrix}. \tag{5.32}$$

Now if, in (5.29), we replace the operand $f(y)$ by the operand $f^{(1)}(y)(f(y))$ (given by (5.29) itself), the result is precisely (5.32). Hence, shortening the notation as in (5.30), (5.25) can be written as

$$y^{(3)} = f^{(2)}(f, f) + f^{(1)}(f^{(1)}(f)). \tag{5.33}$$

Comparing this with the last of (5.23) (the corresponding result for the scalar case) we see that we have achieved a generalization which appears natural; it is straightforward to show that when $m = 1$, $f^{(2)}(f, f)$ does reduce to $f_{yy}f^2$ and $f^{(1)}(f^{(1)}(f))$ to $f_y^2 f$.

We have, of course, proved the results (5.30) and (5.33) only for the case $m = 2$, but it is not hard to see that they will hold for all m. Thus, we have seen that $y^{(2)}$ is a single Frechet derivative of order 1 and that $y^{(3)}$ is a linear combination of Frechet derivatives of orders 1 and 2. In general, $y^{(p)}$ turns out to be a linear combination of Frechet derivatives of orders up to $p - 1$. The components in such linear combinations are called *elementary differentials*; they are the counterparts for systems of terms like F and G (see (5.15)) for scalar problems, and are the natural building blocks for Runge–Kutta methods. They are defined recursively as follows:

Definition The **elementary differentials** $F_s: \mathbb{R}^m \to \mathbb{R}^m$ *of* f *and their order are defined recursively by*

(i) f *is the only elementary differential of order 1, and*
(ii) *if* F_s, $s = 1, 2, \ldots, M$ *are elementary differentials of orders* r_s *respectively, then the*

Frechet derivative

$$f^{(M)}(F_1, F_2, \ldots, F_M) \tag{5.34}$$

is an elementary differential of order

$$1 + \sum_{s=1}^{M} r_s. \tag{5.35}$$

Notes

(1) The elementary differentials F_1, $F_2, \ldots, F_M$ appearing as operands in (5.34) need not be distinct; likewise, the orders r_s in (5.35) need not be distinct.

(2) Let us slim down the notation for elementary differentials as far as we sensibly can. By now we are familiar with the fact that an Mth Frechet derivative has M operands, and we do not need the notation to tell us *twice* what the order of the Frechet derivative is. In (5.34), we can see that there are M operands, so we do not need the superscript (M); nor do we need to be reminded at every stage that we are dealing with Frechet derivatives of f. All we need is a simple notation, such as the brackets $\{\cdots\}$, to indicate that a Frechet derivative of the order indicated by the number of operands within the brackets has been taken. Strictly, we do not even need the commas to separate the operands, so our shortened notation for (5.34) is

$$\{F_1 F_2, \ldots, F_M\} := f^{(M)}(F_1, F_2, \ldots, F_M). \tag{5.36}$$

(3) The order of the elementary differential (5.36) is, by (5.35), the sum of the orders of the elementary differentials F_s, $s = 1, 2, \ldots, M$ plus 1; thus the rule is, sum the orders and add 1 'for the brackets'.

We now identify all elementary differentials of orders up to 4 (and simplify the notation even further).

Order 1 There exists only one elementary differential, f.

Order 2 The only possibility is to take $M = 1$, and the single operand to have order 1; this identifies the operand as f, and there thus exists just one elementary differential $f^{(1)}(f) = \{f\}$

Order 3 There are now two options. We could take $M = 2$, in which case both of the operands must have order 1, and thus must be f, giving the elementary differential $f^{(2)}(f, f) = \{ff\}$. The other possibility is to take $M = 1$, in which case the single operand must have order 2, and can only be $f^{(1)}(f) = \{f\}$, giving the elementary differential $f^{(1)}(f^{(1)}(f)) = \{\{f\}\}$. We can shorten the notation further by writing

$$f^k \text{ for } \underbrace{fff\cdots f}_{k\text{-times}},\quad \{_k \text{ for } \underbrace{\{\{\{\cdots\{}_{k\text{-times}} \text{ and } \}_k \text{ for } \underbrace{\}\}\}\cdots\}}_{k\text{-times}}$$

Note that, with this notation, the order of an elementary differential will be the sum of the exponents of f plus the total number of left or right brackets. There are thus two elementary differentials of order 3, namely $\{f^2\}$ and $\{_2 f\}_2$.

Order 4 There are four elementary differentials:

$M = 3 \Rightarrow$ operands $f, f, f \Rightarrow$ elementary differential $\{f^3\}$

$M = 2 \Rightarrow$ operands f, $\{f\} \Rightarrow$ elementary differential $\{f\{f\}_2\ (\equiv \{_2f\}f\})$

$$M = 1 \Rightarrow \begin{cases} \text{operand } \{f^2\} \Rightarrow \text{elementary differential } \{_2f^2\}_2 \\ \quad \text{or} \\ \text{operand } \{_2f\}_2 \Rightarrow \text{elementary differential } \{_3f\}_3. \end{cases}$$

There is clearly a combinatorial problem in trying to determine all the elementary differentials of given order. This, and other questions, can be answered by investigating an analogy between elementary differentials and rooted trees, which we shall develop in the next section.

We conclude by noting that we now have even more concise expressions for the total derivatives $y^{(2)}$ and $y^{(3)}$; we can now rewrite (5.30) and (5.33) in the form

$$y^{(2)} = \{f\}, \qquad y^{(3)} = \{f^2\} + \{_2f\}_2. \tag{5.37}$$

The reader is invited to compare the latter of these with the horrendous equations (5.25) (which covered only the case $m = 2$) to appreciate how much progress has been made in taming these higher derivatives!

Exercise

5.5.1. (i) Given the differential system $u' = uv$, $v' = u + v$, calculate, by direct differentiation, $u^{(3)}$ and $v^{(3)}$ in terms of u and v,

(ii) Let $y = [u, v]^T$, $f = [uv, u + v]^T$. Calculate $f^{(1)}(f^{(1)}(f))$ and $f^{(2)}(f, f)$ and check that $y^{(3)} = f^{(1)}(f^{(1)}(f)) + f^{(2)}(f, f)$ gives the result obtained in (i).

(iii) Repeat (i) and (ii) for the system $u' = uvw$, $v' = u(v + w)$, $w' = v(v + w)$.

5.6 ROOTED TREES

Rooted trees are algebraic creatures, and a rigorous treatment would demand an approach via graph theory, such as can be found in Butcher (1987). However, we can get by with a very naive approach by simply not distinguishing between a rooted tree and its graph, and 'defining' the latter by pictures. Thus we say that a rooted tree (henceforth, just 'tree') of order n is a set of n points (or *nodes*) joined by lines to give a picture such as

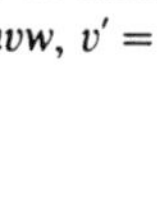

The only rules (which are shared by most trees of the horicultural variety) are that there must be just one root, and branches are not allowed to grow together again. Thus the following are **not** allowed:

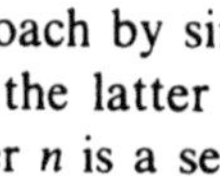

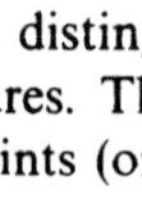

We shall use the notion of 'grafting' two or more trees on to a new root to produce a new tree, thus

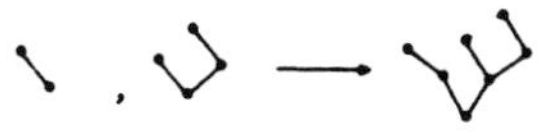

Let us draw all the tree of orders up to 4.

Order	*Trees*	*Number of trees*
1	•	1
2		1
3		2
4		4

In the preceding section, we saw that there were also precisely 1, 1, 2 and 4 elementary differentials of orders 1, 2, 3 and 4 respectively. The analogy between the proliferation of elementary differentials and that of trees stems (no pun intended) from the fact that if we graft trees t_1, $t_2, \dots, t_M$ of orders $r_1, r_2, \dots, r_M$ respectively on to a new root, then the result is a new tree of order $1 + \sum_{s=1}^{M} r_s$, that is, the sum of the orders of the constituent trees plus one for the new root; this is precisely the same rule as (5.35) for elementary differentials. We can see as follows that there is a one-to-one correspondence between elementary dfferentials and trees:

(i) Let f, the unique elementary differential of order 1 correspond to the unique tree of order 1, which consists of a single node.
(ii) If the elementary differentials F_s of order r_s, $s = 1, 2, \dots, M$ correspond to trees t_s of orders r_s, $s = 1, 2, \dots, M$, then let the elementary differential $\{F_1 F_2, \dots, F_M\}$ of order $1 + \sum_{s=1}^{M} r_s$ correspond to the tree of order $1 + \sum_{s=1}^{M} r_s$ obtained by grafting the M trees F_s, $s = 1, 2, \dots, M$ on to a new root.

Example

$F_1 \sim t_1 =$  $F_2 \sim t_2 =$  $F_3 \sim t_3 =$ 

then $\{F_1 F_2 F_3\} \sim$ 

In the next section we shall make this correspondence a little more formal, but for the moment we note that the number of elementary differentials of order r must be the same as the number of trees of order r. A result from combinatorics answers the question of how many trees there are of given order:

Let a_n be the number of trees of order n. Then $a_1, a_2, \dots,$ satisfy termwise the identity

$$a_1 + a_2 u + a_3 u^2 + \cdots \equiv (1-u)^{-a_1}(1-u^2)^{-a_2}(1-u^3)^{-a_3} \dots .$$

From this we can obtain the following table:

n	1	2	3	4	5	6	7	8
a_n	1	1	2	4	9	20	48	115

(We note in passing that it is rather easier to write down, say, all nine trees of order 5 than it is to build up all nine elementary differentials of order 5 in the way we did in the preceding section.)

Clearly, we need a notation for trees. In the preceding section, we saw that all elementary differentials could be labelled by combinations of the two symbols f (the unique elementary differential of order 1) and $\{\cdots\}$ meaning 'we have taken a Frechet derivative (of the order indicated) of the operands appearing between the brackets'. Likewise, all trees can be labelled with combinations of the symbol τ for the unique tree of order 1 (consisting of a single node) and the symbols $[\cdots]$ meaning 'we have grafted the trees appearing between the brackets onto a new root'. We shall denote n copies of the tree t_1 by t_1^n,

$$\underbrace{[[[\cdots[}_{k\text{-times}} \text{ by } [_k \text{ and } \underbrace{]\cdots]]]}_{k\text{-times}} \text{ by }]_k$$

For example,

$$t_1 = [\tau] \qquad =$$

$$t_2 = [\tau[\tau]](=[[\tau]\tau]) = [\tau[\tau]_2 =$$

$$t_3 = [t_1 t_2^2] = [[\tau][\tau[\tau]][\tau[\tau]]]$$
$$= [_2\tau][\tau[\tau]_2[\tau[\tau]_3 =$$

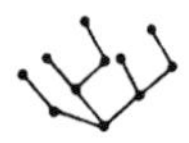

In each case, the order of the tree is the sum of the number of appearances of τ and of either] or [. Such labellings are clearly unique, provided we do not distinguish between, say, $[\tau[\tau]]$ and $[[\tau]\tau]$ (just as we do not distinguish between $\{f\{f\}\}$ and $\{\{f\}f\}$. In addition to the order of a tree, we shall also need the *symmetry* and the *density* of a tree, defined recursively as follows:

Definition *The* **order** $r(t)$, **symmetry** $\sigma(t)$ *and* **density** $\gamma(t)$ *of a tree t are defined by*

$$r(\tau) = \sigma(\tau) = \gamma(\tau) = 1,$$

and

$$r([t_1^{n_1} t_2^{n_2} \cdots]) = 1 + n_1 r(t_1) + n_2 r(t_2) + \cdots \quad (= \textit{number of nodes})$$

$$\sigma([t_1^{n_1} t_2^{n_2} \cdots]) = n_1! n_2! \cdots (\sigma(t_1))^{n_1} (\sigma(t_2))^{n_2} \cdots$$

$$\gamma([t_1^{n_1} t_2^{n_1} \cdots]) = r([t_1^{n_1} t_2^{n_2} \cdots])(\gamma(t_1))^{n_1} (\gamma(t_2))^{n_2} \cdots.$$

Finally, let $\alpha(t)$ be the number of essentially different ways of labelling the nodes of the tree t with the integers $1, 2, \ldots, r(t)$ such that along each outward arc the labels increase.

What is meant by 'essentially different' is illustrated in the following examples:

$$t_1 = [\tau^3] =$$ 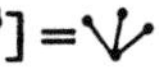. Then  and 

etc. are not regarded as essentially different labellings, and $\alpha(t_1) = 1$.

$$t_2 = [\tau[\tau]] =$$ 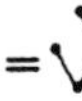. Then 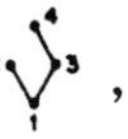, 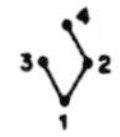 and 

are regarded as essentially different labellings, and $\alpha(t_2) = 3$. In any event, we have from combinatorics an easy way of computing $\alpha(t)$, namely

$$\alpha(t) = \frac{r(t)!}{\sigma(t)\gamma(t)}. \tag{5.38}$$

Table 5.1 displays $r(t)$, $\sigma(t)$, $\gamma(t)$ and $\alpha(t)$ for all trees t of order up to 4.

Table 5.1

Tree	Name	$r(t)$	$\sigma(t)$	$\gamma(t)$	$\alpha(t)$
•	τ	1	1	1	1
	$[\tau]$	2	1	2	1
	$[\tau^2]$	3	2	3	1
	$[[\tau]]$	3	1	6	1
	$[\tau^3]$	4	6	4	1
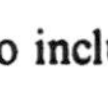	$[\tau[\tau]]$	4	1	8	3
	$[[\tau^2]]$	4	2	12	1
	$[[[\tau]]]$	4	1	24	1

Exercise

5.6.1. Extend Table 5.1 to include all trees of order 5.

5.7 ORDER CONDITIONS

Let us formalize the correspondence between elementary differentials and trees considered in the preceding section.

Definition *The function F is defined on the set T of all trees by*

$$\left.\begin{aligned} F(\tau) &= f \\ F([t_1 t_2, \ldots, t_M]) &= \{F(t_1)F(t_2), \ldots, F(t_M)\}. \end{aligned}\right\} \tag{5.39}$$

Table 5.2 shows $F(t)$ for all trees t of order up to 4.

Table 5.2

Order	Tree	t	$F(t)$
1		τ	f
2		$[\tau]$	$\{f\}$
3		$[\tau^2]$	$\{f^2\}$
		$[_2\tau]_2$	$\{_2f\}_2$
4		$[\tau^3]$	$\{f^3\}$
		$[\tau[\tau]_2$	$\{f\{f\}_2$
		$[_2\tau^2]_2$	$\{_2f^2\}_2$
		$[_3\tau]_3$	$\{_3f\}_3$

Recall that our first objective in seeking to derive order conditions for the general Runge–Kutta method was to find a means of expressing the total derivatives of y in terms of the partial derivatives of f. This is achieved in a remarkable theorem of Butcher, which states that, for general q, $y^{(q)}$ is a linear combination of all elementary differentials of order q and, moreoever, tells us what the coefficients in the linear combination are. A proof of the theorem can be found in Butcher (1987).

Theorem 5.1 *Let* $y' = f(y)$, $f: \mathbb{R}^m \to \mathbb{R}^m$. *Then*

$$y^{(q)} = \sum_{r(t)=q} \alpha(t)F(t), \tag{5.40}$$

where $F(t)$ *is defined by* (5.39) *and* $\alpha(t)$ *by* (5.38).

By way of illustration, let us apply this theorem for $p \leqslant 4$; we see at once that from Tables 5.1 and 5.2 that

$$y^{(2)} = \{f\}$$
$$y^{(3)} = \{f^2\} + \{_2f\}_2$$
$$y^{(4)} = \{f^3\} + 3\{f\{f\}_2 + \{_2f^2\}_2 + \{_3f\}_3.$$

Note that the first two of these were already given by (5.37).

Recall from (5.2) and (5.3) of §5.1 that the general s-stage Runge–Kutta method for the autonomous problem

$$y' = f(y), y(a) = \eta, \quad f: \mathbb{R}^m \to \mathbb{R}^m \tag{5.41}$$

is

$$y_{n+1} = y_n + h \sum_{i=1}^{s} b_i k_i \tag{5.42(i)}$$

where

$$k_i = f(y_n + h \sum_{j=1}^{s} a_{ij} k_j), \qquad i = 1, 2, \ldots, s \tag{5.42(ii)}$$

and

$$c_i = \sum_{j=1}^{s} a_{ij}, \qquad i = 1, 2, \ldots, s. \tag{5.42(iii)}$$

Let us define the right side of (5.42(i)) to be $y_n(h)$, which we then expand as a Taylor series about $h = 0$ to get

$$y_{n+1} = y_n(0) + h \frac{\mathrm{d}}{\mathrm{d}h} y_n(h)\bigg|_{h=0} + \tfrac{1}{2} h^2 \frac{d^2}{\mathrm{d}h^2} y_n(h)\bigg|_{h=0} + \cdots. \tag{5.43}$$

The corresponding expansion for the exact solution at x_{n+1} is

$$y(x_{n+1}) = y(x_n) + h y^{(1)}(x_n) + \tfrac{1}{2} h^2 y^{(2)}(x_n) + \cdots. \tag{5.44}$$

Making the localizing assumption that $y(x_n) = y_n\,(= y_n(0))$, we see that the method (5.42) will have order p if the expansions (5.43) and (5.44) match up to (and including) the terms in h^p. Theorem 5.1 enables us to express the derivatives $y^{(q)}$ appearing in (5.44) in terms of elementary differentials of f; we need a similar result for the derivatives

$$\frac{\mathrm{d}q}{\mathrm{d}h^q} y_n(h)\bigg|_{h=0}, \quad q = 1, 2, \ldots$$

appearing in (5.43). First, we modify the notation in (5.42) by defining $a_{s+1,i} := b_i$, $i = 1, 2, \ldots, s$, and then define on the set T of all trees, a new function $\psi(t)$, which depends on the elements of the Butcher array of (5.42).

Definition
(1) For $i = 1, 2, \ldots, s, s+1$ define on the set T of all trees the functions ψ_i by

$$\left.\begin{aligned} \psi_i(\tau) &= \sum_{j=1}^{s} a_{ij} \\ \psi_i([t_1 t_2, \ldots, t_M]) &= \sum_{j=1}^{s} a_{ij} \psi_j(t_1) \psi_j(t_2) \cdots \psi_j(t_M). \end{aligned}\right\} \tag{5.45}$$

(2) Define $\psi(t) := \psi_{s+1}(t)$.

In Table 5.3 we develop the functions $\psi_i(t)$ and $\psi(t)$ for all trees of order up to 4; all summations are from 1 to s.

Comments on Table 5.3
(1) The entries in Table 5.3 are obtained by repeatedly using (5.45) in the following way.

For $i = 1, 2, \ldots, s$, $\psi_i(\tau) = \sum_j a_{ij} = c_i$, by (5.42(iii));

$\psi(\tau) = \sum_j a_{s+1,j} = \sum_j b_j = \sum_i b_i$, on changing the (dummy) summation index;

For $i = 1, 2, \ldots, s$, $\psi_i([\tau]) = \sum_j a_{ij} \psi_j(\tau) = \sum_j a_{ij} c_j$;

Table 5.3

Tree	t	$\psi_i(t),\ i=1,2,\ldots,s$	$\psi(t)$
•	τ	$\sum_j a_{ij}\ (=c_i)$	$\sum_i b_i$
	$[\tau]$	$\sum_j a_{ij}c_j$	$\sum_i b_ic_i$
	$[\tau^2]$	$\sum_j a_{ij}c_j^2$	$\sum_i b_ic_i^2$
	$[[\tau]]$	$\sum_j a_{ij}\sum_k a_{jk}c_k$	$\sum_{ij} b_ia_{ij}c_j$
	$[\tau^3]$	$\sum_j a_{ij}c_j^3$	$\sum_i b_ic_i^3$
	$[\tau[\tau]]$	$\sum_j a_{ij}c_j\sum_k a_{jk}c_k$	$\sum_{ij} b_ic_ia_{ij}c_j$
	$[[\tau^2]]$	$\sum_j a_{ij}\sum_k a_{jk}c_k^2$	$\sum_{ij} b_ia_{ij}c_j^2$
	$[[[\tau]]]$	$\sum_j a_{ij}\sum_k a_{jk}\sum_n a_{kn}c_n$	$\sum_{ijk} b_ia_{ij}a_{jk}c_k$

$$\psi([\tau])=\sum_j a_{s+1,j}c_j=\sum_j b_jc_j=\sum_i b_ic_i;$$

For $i=1,2,\ldots,s$, $\psi_i([[\tau]])=\sum_j a_{ij}\psi_j([\tau])=\sum_j a_{ij}\sum_k a_{jk}c_k;$

$$\psi([[\tau]])=\sum_j a_{s+1,j}\sum_k a_{jk}c_k=\sum_{jk} b_ja_{jk}c_k=\sum_{ij} b_ia_{ij}c_j,\ \text{and so on.}$$

(2) When a tree t has an alternative (but necessarily equivalent) label, then the functions $\psi_i(t)$ are independent of which label we choose. Thus the fourth-order tree $[\tau[\tau]]$ could equally well be labelled $[[\tau]\tau]$, giving $\psi_i(t)=\sum_j a_{ij}\sum_k a_{jk}c_kc_j=\sum_{j,k} a_{ij}a_{jk}c_kc_j=\sum_j a_{ij}c_j\sum_k a_{jk}c_k$, as in Table 5.3.

(3) Eventually, all we will be interested in are the functions $\psi(t)$; but clearly we need the functions $\psi_i(t)$, $i=1,2,\ldots,s$ for trees of a given order to enable us to calculate $\psi(t)$ for trees of higher order.

(4) There is a useful rule-of-thumb which, if used *with care*, shortens the process of finding $\psi(t)$; it works for all trees except $t=\tau$. It goes as follows: Read the name of the tree from left to right, ignoring all second brackets]; interpret the first bracket [encountered as b, all subsequent brackets [as a, and τ as c. Finally, link the suffices in the natural way, noting that b and c have single suffices, whilst a has a double suffix. For example,

$$[\tau[\tau]]\rightarrow bcac\rightarrow\psi=\sum_{ij} b_ic_ia_{ij}c_j.$$

It is, however, possible to misinterpret this rule-of-thumb. Consider for example the fifth-order tree $t = [[\tau][\tau]]$. Applying (5.45), we have

$$\psi_i(t) = \sum_j a_{ij} \sum_k a_{jk} c_k \sum_n a_{jn} c_n = \sum_j a_{ij} \left(\sum_k a_{jk} c_k \right)^2, \qquad i = 1, 2, \ldots, s$$

and

$$\psi(t) = \sum_i b_i \left(\sum_j a_{ij} c_j \right)^2. \tag{5.46}$$

The rule-of-thumb gives $[[\tau][\tau]] \to bacac$, which could be interpreted to mean that $\psi(t) = \sum_{ijk} b_i a_{ij} c_j a_{jk} c_k$ (which is not, of course, the same as (5.46)). However, the structure $[[\tau][\tau]]$ should warn us that the correct interpretation of the rule-of-thumb is

$$[[\tau][\tau]] \to b(ac)^2 \to \psi(t) = \sum_i b_i \left(\sum_j a_{ij} c_j \right)^2.$$

A second theorem of Butcher enables us to calculate the derivatives appearing in (5.43) in terms of the functions F defined by (5.39) and the functions $\psi(t)$. A proof of the theorem can be found in Butcher (1987).

Theorem 5.2 Let the Runge–Kutta method (5.42) define the expansion (5.43). Then the derivatives on the right side of (5.43) are given by

$$\left. \frac{d^q}{dh^q} y_n(h) \right|_{h=0} = \sum_{r(t)=q} \alpha(t) \gamma(t) \psi(t) F(t)$$

where $F(t)$ is evaluated at y_n, $\psi(t)$ is defined by (5.45) and $\alpha(t)$ and $\gamma(t)$ are defined as in §5.6.

The main result follows immediately from (5.43) and (5.44) and Theorems 5.1 and 5.2:

Theorem 5.3 The Runge-Kutta method has order p if $\psi(t) = 1/\gamma(t)$ holds for all trees of order $r(t) \leqslant p$ and does not hold for some tree of order $p + 1$.

Hence to establish the conditions for any Runge–Kutta method (explicit or implicit) to have order p, all we need do is write down all the trees of order up to p, compute $\psi(t)$ and $\gamma(t)$ and set $\psi(t) = 1/\gamma(t)$ for each tree in the list. Using the data provided by Table 5.3 and Table 5.1 of §5.6, we obtain the order conditions given in Table 5.4, for orders up to 4.

Example

3-stage explicit: $a_{1j} = 0, j = 1, 2, 3, \Rightarrow c_1 = 0$, $a_{2j} = 0, j = 2, 3$, $a_{33} = 0$. From Table 5.4,

$$\begin{aligned} \text{Order } p = 1&: \quad b_1 + b_2 + b_3 = 1 \\ \text{Order } p = 2&: \quad b_2 c_2 + b_3 c_3 = \tfrac{1}{2} \\ \text{Order } p = 3&: \quad b_2 c_2^2 + b_3 c_3^2 = \tfrac{1}{3} \\ & \qquad\qquad b_3 a_{32} c_2 = \tfrac{1}{6}. \end{aligned}$$

Table 5.4

Tree	t	$r(t)$	Order conditions $\psi(t)$	$= 1/\gamma(t)$
	τ	1	$\sum_i b_i$	$= 1$
	$[\tau]$	2	$\sum_i b_i c_i$	$= \frac{1}{2}$
	$[\tau^2]$	3	$\sum_i b_i c_i^2$	$= \frac{1}{3}$
	$[[\tau]]$		$\sum_{ij} b_i a_{ij} c_j$	$= \frac{1}{6}$
	$[\tau^3]$	4	$\sum_i b_i c_i^3$	$= \frac{1}{4}$
	$[\tau[\tau]]$		$\sum_{ij} b_i c_i a_{ij} c_j$	$= \frac{1}{8}$
	$[[\tau^2]]$		$\sum_{ij} b_i a_{ij} c_j^2$	$= \frac{1}{12}$
	$[[[\tau]]]$		$\sum_{ijk} b_i a_{ij} a_{jk} c_k$	$= \frac{1}{24}$

These are the same conditions as we obtained in §5.3. Note that the left side of the last of the order 4 conditions in Table 5.4 becomes 0, and the condition cannot be satisfied, thus showing that order 4 cannot be attained by an explicit 3-stage method.

2-stage implicit: From Table 5.4, order 4 requires that eight conditions be satisfied; moreover, from (5.42(iii)) we must also have that $c_1 = a_{11} + a_{12}$, $c_2 = a_{21} + a_{22}$, giving ten conditions in all. There are just eight coefficients in the Butcher array, but it turns out that there does exist a unique solution given by the array

$$\begin{array}{c|cc} \frac{1}{2}+\frac{\sqrt{3}}{6} & \frac{1}{4} & \frac{1}{4}+\frac{\sqrt{3}}{6} \\ \frac{1}{2}-\frac{\sqrt{3}}{6} & \frac{1}{4}-\frac{\sqrt{3}}{6} & \frac{1}{4} \\ \hline & \frac{1}{2} & \frac{1}{2} \end{array}$$

(Strictly speaking, the Butcher array is not unique since replacing each plus sign in the first row by a minus and each minus sign in the second row by a plus also represents a solution of the order conditions; but since $b_1 = b_2$, both solutions give the same (unique) method.)

Finally, it follows at once from (5.43) and (5.44) that if the method (5.42) has order p then the local truncation error is given by

$$\text{LTE} = \frac{h^{p+1}}{(p+1)!} \sum_{r(t)=p+1} \alpha(t)[1 - \gamma(t)\psi(t)]F(t) + O(h^{p+2}) \tag{5.47}$$

where the functions $F(t)$ are evaluated at the value $y(x_n)$ of the argument.

Exercises

5.7.1. Extend Tables 5.3 and 5.4 to include all trees of order 5.

5.7.2. Show that each of the following Runge–Kutta methods has order 4.
(i) The popular explicit method (5.21) of §5.3,
(ii) The implicit methods

$$\begin{array}{c|cc} \frac{3-\sqrt{3}}{6} & \frac{1}{4} & \frac{3-2\sqrt{3}}{12} \\ \frac{3+\sqrt{3}}{6} & \frac{3+2\sqrt{3}}{12} & \frac{1}{4} \\ \hline & \frac{1}{2} & \frac{1}{2} \end{array} \qquad \begin{array}{c|ccc} 0 & 0 & 0 & 0 \\ \frac{1}{2} & \frac{5}{24} & \frac{1}{3} & \frac{-1}{24} \\ 1 & \frac{1}{6} & \frac{2}{3} & \frac{1}{6} \\ \hline & \frac{1}{6} & \frac{2}{3} & \frac{1}{6} \end{array}$$

5.7.3. Show that the 3-stage method in 5.7.2 (ii) can be written in the form

$$y_{n+1/2} - \tfrac{1}{2}y_{n+1} = \tfrac{1}{2}y_n + \frac{h}{8}(f_n - f_{n+1})$$

$$y_{n+1} = y_n + \frac{h}{6}(f_n + 4f_{n+1/2} + f_{n+1}).$$

Suggest a use to which this form could be put.

5.7.4. Write the following method as a Runge–Kutta method, and find its order:

$$y_{n+2/3} = y_n + \frac{h}{3}[f(y_{n+2/3}) + f(y_n)]$$

$$y_{n+1} = y_n + \frac{h}{4}[3f(y_{n+2/3}) + f(y_n)].$$

5.7.5. Show that the following method for the problem $y' = f(y)$, $y(x_0) = y_0$, $f: \mathbb{R}^m \to \mathbb{R}^m$, is equivalent to a 4-stage semi-implicit Runge–Kutta and hence show that it has order three:

$$y^{[1]}_{n+1} = y_n + \frac{h}{2}[f(y^{[1]}_{n+1}) + f(y_n)]$$

$$y^{[1]}_{n+2} = y^{[1]}_{n+1} + \frac{h}{2}[f(y^{[1]}_{n+2}) + f(y^{[1]}_{n+1})]$$

$$y_{n+1} - y_n = \frac{h}{3}[2f(y_{n+1}) + f(y_n)] - \tfrac{1}{6}[y^{[1]}_{n+2} - 2y^{[1]}_{n+1} + y_n].$$

(*Hint*: Use the alternative form (5.6) of §5.1 for a Runge–Kutta method.)

5.7.6. Define a *bush tree* to be a tree all of whose branches stem from the root of the tree, and a *trunk tree* to be a tree with only one branch stemming from the root. Using the tables found in Exercise 5.7.1, *demonstrate* that if $\sum_j a_{ij}c_j = \frac{1}{2}c_i^2$ for all i, then the order conditions for all bush trees of orders 3, 4, 5 can be ignored, and that if $\sum_i b_i a_{ij} = b_j(1 - c_j)$ for all j then the order conditions for all trunk trees of order 3, 4, 5 can likewise be ignored. Using the formal definitions of $\gamma(t)$ and $\psi(t)$, *prove* that these results hold for general order greater than 2.

5.7.7*. Consider the following semi-implicit Runge–Kutta method:

$$\left.\begin{array}{l} k_1 = f(y_n + \beta h k_1), \quad k_2 = f(y_n + hk_1 + \beta h k_2) \\ y_{n+1} - y_n = h[(\tfrac{1}{2} + \beta)k_1 + (\tfrac{1}{2} - \beta)k_2]. \end{array}\right\} \qquad (1)$$

(i) Find p, the order of the method, and show that it is independent of β.

(ii) Express the PLTE in the form of a linear combination of elementary differentials of order $p+1$, expressing the coefficients in the combination in terms of β only. To what does the PLTE reduce in the case when $f = Ay$, A a constant matrix?

(iii) Apply (1) to the problem $y' = Ay$ to obtain an expression of the form $y_{n+1} = R(hA, \beta)y_n$, where $R(hA, \beta)$ is a rational function of the matrix hA. By comparing y_{n+1} with the exact solution $y(x_{n+1})$, which satisfies $y(x_{n+1}) = \exp(hA)y(x_n)$, find the PLTE directly, and check that it is identical with the expression found in (ii).

(*Note*: $\exp(hA) = I + hA + (hA)^2/2! + (hA)^3/3! + \cdots$)

5.7.8*. The Trapezoidal Rule is being used to solve numerically the problem $y' = f(y)$, $y(x_0) = y_0$, $f: \mathbb{R}^m \to \mathbb{R}^m$, using a constant steplength h where $x_n = x_0 + nh$.

(i) Show that applying the Trapezoidal Rule on two successive steps to advance the solution from x_0 to x_2 is equivalent to applying a 3-stage semi-implicit Runge–Kutta method with steplength $H = 2h$, and write down the Butcher array for this equivalent method. Verify the order directly by applying the Runge–Kutta order conditions and, by considering the trees $[\tau^2]$ and $[[\tau]]$, show that the truncation error after two steps is $-\frac{1}{6}h^3 y^{(3)}(x_0) + 0(h^4)$.

(ii) Generalize the result of (i) to the case of N successive steps; that is, find the equivalent $(N+1)$-stage semi-implicit Runge–Kutta method with steplength $H = Nh$, verify the order and show that the truncation error at x_N is $-\frac{1}{12}(x_N - x_0)h^2 y^{(3)}(x_0) + 0(h^3)$.

5.7.9*. An explicit method for solving $y' = f(x, y)$, $y(x_0) = y_0$, $f: \mathbb{R} \times \mathbb{R}^m \to R^m$, consists of the following:

Step 1
$$\left\{\begin{array}{l} k_1 = f(x_n, y_n) \\ k_2 = f(x_n + h, y_n + hk_1) \\ y_{n+1} = y_n + \tfrac{1}{2}h(k_1 + k_2) \end{array}\right.$$

Step j, $j \geqslant 2$
$$\left\{\begin{array}{l} k_{j+1} = f(x_{n+j-1} + \tfrac{1}{2}h, y_{n+j-1} + (\tfrac{1}{2} - \alpha)hk_{j-1} + \alpha h k_j) \\ y_{n+j} = y_{n+j-1} + hk_{j+1}. \end{array}\right.$$

Applying the first two steps is equivalent to applying a 3-stage explicit Runge–Kutta method with steplength $H = 2h$. Write down the Butcher array for this equivalent method and show that it has order two. Continue the process by writing down the Butcher array for the 4-stage method with $H = 3h$ which is equivalent to the the first three steps of the given method etc., until the pattern is clear enough to enable you to write down the Butcher array for the $(j+1)$-stage method with steplength $H = jh$ which is equivalent to the first j steps of the given method. Show that this equivalent method has order two.

Calculate (for j sufficiently large) E_j, the PLTE of the equivalent $(j+1)$-stage method in terms of elementary differentials. Since E_j is the principal *accumulated* truncation error of the given method, we can define the principal local error of that method to be L_j, given by $L_j := E_j - E_{j-1}$. Show that (for sufficiently large j) L_j $(= L)$ is independent of j, and show that there exists a unique value of the free parameter α for which L takes the form $Kh^3 y^{(3)}(x)$, where K is constant.

5.8 SCALAR PROBLEMS AND SYSTEMS

A remarkable and (at the time) totally unexpected result that emerged from the Butcher theory is that a Runge–Kutta method which has order p for a scalar initial value problem may have order less than p when applied to a problem involving a system of differential equations. To see how this can come about, let us consider what the Frechet derivatives and the elementary differentials reduce to when $m = 1$. From the definition of §5.5 we see that when $m = 1$ the Mth Frechet derivative reduces to

$$f^{(M)}(K_1, K_2, \ldots, K_M) = f_{\underbrace{yy,\ldots,y}_{M\text{-times}}} K_1 K_2 \cdots K_M, \quad K_t \in \mathbb{R}^1, \qquad t = 1, 2, \ldots, M.$$

Table 5.5 shows the corresponding reduction of the elementary differentials of orders up to 4.

Recall the expansions (5.44) and (5.43) for the exact and the numerical solutions; by Theorems 5.1 and 5.2 these become

$$\left.\begin{aligned} y(x_{n+1}) - y(x_n) &= \sum_{q=1}^{\infty} \frac{h^q}{q!} \sum_{r(t)=q} \alpha(t) F(t) \\ y_{n+1} - y_n &= \sum_{q=1}^{\infty} \frac{h^q}{q!} \sum_{r(t)=q} \alpha(t)\gamma(t)\psi(t) F(t) \end{aligned}\right\} \tag{5.48}$$

where the $F(t)$ are evaluated at the value $y(x_n) = y_n = y_n(0)$ of the argument.

If all the $F(t)$ are distinct (which is certainly the case when $m > 1$) then the order is 4 iff $\psi(t) = 1/\gamma(t)$ for all trees of order up to 4. However, we see from Table 5.5 that when $m = 1$ two of the fourth-order elementary differentials reduce to the *same* scalar expression. Specifically, let $t_1 = [\tau[\tau]]_2$, $t_2 = [_2\tau^2]_2$; then when $m = 1$, $F(t_1) = F(t_2) = f_{yy} f_y f^2$. Hence, in the case of a scalar problem, the *two* conditions

$$\psi(t_1) = 1/\gamma(t_1), \quad \psi(t_2) = 1/\gamma(t_2) \tag{5.49}$$

Table 5.5

$r(t)$	t	$F(t)$	Scalar form
1	τ	f	f
2	$[\tau]$	$\{f\} = f^{(1)}(f)$	$f_y f$
3	$[\tau^2]$	$\{f^2\} = f^{(2)}(f, f)$	$f_{yy} f^2$
	$[_2\tau]_2$	$\{_2 f\}_2 = f^{(1)}(f^{(1)}(f))$	$(f_y)^2 f$
4	$[\tau^3]$	$\{f^3\} = f^{(3)}(f, f, f)$	$f_{yyy} f^3$
	$[\tau[\tau]]_2$	$\{f\{f\}\}_2 = f^{(2)}(f, \{f\})$	$f_{yy} f f_y f$
	$[_2\tau^2]_2$	$\{_2 f^2\}_2 = f^{(1)}(\{f^2\})$	$f_y f_{yy} f^2$
	$[_3\tau]_3$	$\{_3 f\}_3 = f^{(1)}(\{_2 f\}_2)$	$(f_y)^3 f$

could be replaced by the *single* condition

$$\alpha(t_1)\gamma(t_1)\psi(t_1)+\alpha(t_2)\gamma(t_2)\psi(t_2)=\alpha(t_1)+\alpha(t_2) \tag{5.50}$$

and the coefficients of $F(t_1)$ $(=F(t_2))$ in the expansions (5.48) would still match, so that order 4 would be achieved. It can of course happen that a solution which satisfies (5.50) does not satisfy (5.49). An example of this phenomenon is afforded by the 4-stage explicit method with Butcher array

$$\begin{array}{c|cccc} 0 & & & & \\ \frac{1}{2} & \frac{1}{2} & & & \\ -1 & \frac{1}{2} & -\frac{3}{2} & & \\ 1 & 0 & \frac{4}{3} & -\frac{1}{3} & \\ \hline & \frac{1}{6} & \frac{2}{3} & 0 & \frac{1}{6} \end{array} \tag{5.51}$$

This method has order 4 if applied to $y'=f(y), f:\mathbb{R}^1\to\mathbb{R}^1$, but order only 3 if applied to $y'=f(y), f:\mathbb{R}^m\to\mathbb{R}^m, m>1$. Of course, as we have seen in §1.4, the *general* scalar problem is not $y'=f(y)$, $y(a)=\eta$, $f:\mathbb{R}^1\to\mathbb{R}^1$, but

$$y'=f(x,y),\quad y(a)=\eta,\qquad f:\mathbb{R}^1\times\mathbb{R}^1\to\mathbb{R}^1 \tag{5.52}$$

which we can write (see §1.4 again) in the form

$$z'=\varphi(z),\quad z(a)=\zeta,\qquad \varphi:\mathbb{R}^2\to\mathbb{R}^2,$$

where $z=[y,x]^{\mathrm{T}}$, $\varphi=[f,1]^{\mathrm{T}}$, $\zeta=[\eta,a]^{\mathrm{T}}$. We then find that

$$F(t_1)=\{\varphi\{\varphi\}_2=[(f_{xx}+ff_{yy})(f_x+ff_y),0]^{\mathrm{T}},$$
$$F(t_2)=\{_2\varphi^2\}_2=[f_y(f_{xx}+2ff_{xy}+f^2f_{yy}),0]^{\mathrm{T}},$$

which are not identical. It follows that *all Runge–Kutta methods of order up to 4 have the same order for a system and for the general scalar problem* (5.52).

However, if we proceed to fifth-order conditions we find that there are two trees $[\tau[_2\tau]_3$ and $[_2\tau[\tau]_3$ for which the corresponding elementary differentials coincide for the *general* scalar problem (5.52). Thus there exist methods which have order p $(\geqslant 5)$ for the general scalar problem (5.52), but have order less than p for a system. As one may imagine, there was considerable consternation when this result first appeared, since there were in existence several Runge–Kutta methods of high order, invariably derived as in §5.3 for a scalar problem, but applied in practice to systems. (According to rumour, this was the case in the computation of the trajectories of some of the early space shots!) However, it turned out that none of these methods fell into the class of those whose order for a scalar problem differed from that for a system.

We summarize the above results as follows:

Statement A: The method has order p for $y'=f(y), f:\mathbb{R}^m\to\mathbb{R}^m$, $m>1$

Statement B: The method has order p for $y'=f(x,y), f:\mathbb{R}^1\times\mathbb{R}^1\to\mathbb{R}^1$

Statement C: The method has order p for $y'=f(y), f:\mathbb{R}^1\to\mathbb{R}^1$.

Table 5.6

	Problem (I)		Problem (II)	
h	Error	Ratio	Error	Ratio
0.6	1.52×10^{-3}		1.46×10^{-2}	
		22.8		10.2
0.3	6.66×10^{-5}		1.43×10^{-3}	
		18		9.1
0.15	3.70×10^{-6}		1.58×10^{-4}	
		16.5		8.6
0.075	2.24×10^{-7}		1.84×10^{-5}	

Then, for $1 \leqslant p \leqslant 3$, $\quad A \Leftrightarrow B \Leftrightarrow C$

for $p = 4$, $\quad A \Leftrightarrow B \Rightarrow C$, but $C \not\Rightarrow B$

for $p \geqslant 5$, $\quad A \Rightarrow B \Rightarrow C$, but $C \not\Rightarrow B$, $B \not\Rightarrow A$.

We conclude with a simple numerical illustration. Consider the following two scalar initial value problems:

$$\text{(I)}\ y' = \sqrt{y}, \quad y(0) = 1, \qquad \text{and} \qquad \text{(II)}\ y' = y/(1 + x/2), \quad y(0) = 1.$$

Both have the same exact solution $y(x) = (1 + x/2)^2$. We solve these problems using the Runge–Kutta method (5.51) which has order 4 when applied to the autonomous system (I), but order only 3 when applied to the non-autonomous problem (II). The global errors at $x = 3.0$ for a range of steplengths are given in Table 5.6.

The solution for problem (I) does appear to be more accurate than that for problem (II), but this is not very compelling evidence; the problems are different, even though they have the same exact solution. What is more persuasive is the column of entries headed 'Ratio', where we have calculated, for each h, the ratio of the error when the steplength is h to the error when the steplength is $h/2$. For a pth-order method, the local error is $0(h^{p+1})$, but the global error is $0(h^p)$. Thus, for a four-order method we expect this ratio to tend to $2^4 = 16$ as $h \to 0$, whereas for a third-order method we would expect it to tend to $2^3 = 8$. The results in Table 5.6 thus indicate that we are indeed achieving fourth-order behaviour for problem (I), but only third order for problem (II).

Exercises

5.8.1. Apply (5.51) to the two scalar initial value problems $y' = y$, $y(0) = 1$ and $y' = xy$, $y(0) = 1$ and compute $y(h) - y_1$ for each, where $y(x)$ is the exact solution. Deduce that (5.51) is exhibiting fourth-order behaviour for the first problem but third-order for the second.

5.8.2. Construct a table showing for each tree of order $r \leqslant 5$ the expressions to which the corresponding elementary differentials of f reduce when $f \in \mathbb{R}^1$.

5.8.3. (i) Deduce from the table found in the preceding exercise that when a fourth-order Runge–

Kutta method is applied to a scalar equation $y' = f(y)$ then four of the elementary differentials which occur in the expression for the PLTE reduce to the same expression $\hat{\varphi}(t)$.
(ii) Consider the explicit method

$$
\begin{array}{c|cccc}
0 & & & & \\
\alpha & \alpha & & & \\
\frac{1}{2} & \frac{1}{2} - \frac{1}{8\alpha} & \frac{1}{8\alpha} & & \\
1 & \frac{1}{2\alpha} - 1 & \frac{-1}{2\alpha} & 2 & \\
\hline
 & \frac{1}{6} & 0 & \frac{2}{3} & \frac{1}{6}
\end{array}
$$

Show that for all values of α the order is *precisely* four when applied to the general system $y' = f(y)$, $f \in \mathbb{R}^m$. In the case when $m = 1$, find the value of α for which the coefficient of $\hat{\varphi}(t)$, defined in (i), is zero.

5.9 EXPLICIT METHODS; ATTAINABLE ORDER

In §5.7 we saw how to establish the conditions for a Runge–Kutta method to have given order. We now turn to the problem of finding solutions of these order conditions, and in this section we consider the question of what order can be achieved by an s–stage explicit method. We start with two further results of Butcher, the first of which is a technical lemma which we shall need later in this section.

Lemma 5.1 *Let U and V be two 3×3 matrices such that*

$$
UV = [w_{ij}] = \begin{bmatrix} w_{11} & w_{12} & 0 \\ w_{21} & w_{22} & 0 \\ 0 & 0 & 0 \end{bmatrix}, \quad \text{where } w_{11}w_{22} \neq w_{21}w_{12}.
$$

Then **either** *the last row of U is the zero row vector* **or** *the last column of V is the zero column vector.*

Proof Clearly UV is singular and therefore either U is singular or V is singular. If U is singular, then there exists a non-zero row vector $p = [p_1, p_2, p_3]$ such that $pU = 0$, whence $pUV = 0$ and it follows that

$$
[p_1, p_2]\begin{bmatrix} w_{11} & w_{12} \\ w_{21} & w_{22} \end{bmatrix} = 0.
$$

Since $w_{11}w_{22} - w_{21}w_{12} \neq 0$, it follows that $p_1 = p_2 = 0$, and hence that $[0, 0, p_3]U = 0$; since $p_3 \neq 0$, it follows that the last row of U consists of zero elements. In the case when

V is singular, there must exist a non-zero column vector q such that $Vq = 0$, and a similar argument establishes that the last column of V contains only zero elements □

The second result establishes an upper bound to the order that can be attained by an explicit s-stage method. Its proof is a neat example of the power of the Butcher approach.

Theorem 5.4 *An s-stage explicit Runge–Kutta method cannot have order greater than s.*

Proof Let the s-stage method have order p, and consider the pth order tree $t = [_{p-1}\tau]_{p-1}$. It follows from §5.6 that $\gamma(t) = p!$ and from §5.7 that $\psi(t) = \sum_{i,j_1,j_2\cdots j_{p-2}} b_i a_{ij_1} a_{j_1 j_2} \cdots a_{j_{p-3} j_{p-2}} c_{j_{p-2}}$. Since the method is explicit, $a_{ij} = 0$ for $j \geqslant i$, and it follows that $\psi(t) = 0$ unless there exists a sequence $i, j_1, j_2, \ldots, j_{p-2}$ of the integers $1, 2, \ldots, s$ such that

$$i > j_1 > j_2 > \cdots > j_{p-2} > 1.$$

(Note that $j_{p-2} = 1$ would not do, since then $c_{j_{p-2}} = c_1 = 0$.) Hence $\psi(t) = 0$ (and the order condition $\psi(t) = 1/\gamma(t)$ contradicted) unless $i \geqslant p$, whence $p \leqslant i \leqslant s$. □

The obvious question now is whether order $p = s$ can be attained for all s. We consider in turn the cases $s = 2, 3, 4, 5$, and attempt to find the general solution of the order conditions.

Two-stage methods

Butcher array

$$\begin{array}{c|cc} 0 & & \\ c_2 & a_{21} & \\ \hline & b_1 & b_2 \end{array} \qquad a_{21} = c_2.$$

Order 2 conditions: $b_1 + b_2 = 1$, $b_2 c_2 = \frac{1}{2}$.

The general solution is

$$c_2 = \lambda \neq 0, \quad b_1 = 1 - 1/2\lambda, \quad b_2 = 1/2\lambda,$$

a single one-parameter family.

Three-stage methods

Butcher array

$$\begin{array}{c|ccc} 0 & & & \\ c_2 & a_{21} & & \\ c_3 & a_{31} & a_{32} & \\ \hline & b_1 & b_2 & b_3 \end{array} \qquad \begin{array}{l} \\ a_{21} = c_2 \\ a_{31} + a_{32} = c_3. \\ \\ \end{array}$$

Order 3 conditions

$$b_1 + b_2 + b_3 = 1$$
$$b_2c_2 + b_3c_3 = \tfrac{1}{2}$$
$$b_2c_2^2 + b_3c_3^2 = \tfrac{1}{3}$$
$$b_3a_{32}c_2 = \tfrac{1}{6}.$$

On solving these equations, we find that the solutions fall into three cases:

Case 1

$$c_2 = a_{21} = \lambda, \quad c_3 = \mu$$

$$b_1 = \frac{6\lambda\mu - 3(\lambda + \mu) + 2}{6\lambda\mu}, \quad b_2 = \frac{2 - 3\mu}{6\lambda(\lambda - \mu)}, \quad b_3 = \frac{3\lambda - 2}{6\mu(\lambda - \mu)}$$

$$a_{31} = \frac{\mu[3\lambda(\lambda - 1) + \mu]}{\lambda(3\lambda - 2)}, \quad a_{32} = \frac{\mu(\lambda - \mu)}{\lambda(3\lambda - 2)}$$

where $\lambda \neq 0, \frac{2}{3}, \mu$, and $\mu \neq 0$.

Case 2

$$c_2 = \frac{2}{3}, \quad c_3 = 0, b_1 = \frac{1}{4} - v, \quad b_2 = \frac{3}{4}, b_3 = v, \quad a_{31} = -a_{32} = \frac{-1}{4v}, \qquad v \neq 0.$$

Case 3

$$c_2 = c_3 = \frac{2}{3}, \quad b_1 = \frac{1}{4}, \quad b_2 = \frac{3}{4} - \omega, \quad b_3 = \omega, \quad a_{31} = \frac{2}{3} - \frac{1}{4\omega}, \quad a_{32} = \frac{1}{4\omega}, \qquad \omega \neq 0.$$

Thus there is one two-parameter family and two one-parameter families of solutions. We might conclude from this that the pattern is clear, and that for 4-stage methods there will be a three-parameter family of fourth-order methods, plus some families with fewer parameters; we would be wrong!

Four-stage methods

Butcher array

$$\begin{array}{c|cccc} 0 & & & & \\ c_2 & a_{21} & & & \\ c_3 & a_{31} & a_{32} & & \\ c_4 & a_{41} & a_{42} & a_{43} & \\ \hline & b_1 & b_2 & b_3 & b_4 \end{array} \qquad \begin{array}{l} \\ a_{21} = c_2 \\ a_{31} + a_{32} = c_3 \\ a_{41} + a_{42} + a_{43} = c_4. \\ \\ \end{array}$$

The order conditions are now too cumbersome to be written out in full, and we use a summation notation; for the remainder of this section, all summations run from 1 to s, the stage-number; where no suffices appear under the summation sign, the summation is taken over *all* subscripts appearing in the terms to be summed.

Order 4 conditions

$$\left.\begin{array}{ll} (1) & \sum b_i = 1 \\ (2) & \sum b_i c_i = \frac{1}{2} \\ (3) & \sum b_i c_i^2 = \frac{1}{3} \\ (4) & \sum b_i a_{ij} c_j = \frac{1}{6} \\ (5) & \sum b_i c_i^3 = \frac{1}{4} \\ (6) & \sum b_i c_i a_{ij} c_j = \frac{1}{8} \\ (7) & \sum b_i a_{ij} c_j^2 = \frac{1}{12} \\ (8) & \sum b_i a_{ij} a_{jk} c_k = \frac{1}{24}. \end{array}\right\} \tag{5.53}$$

Finding all solutions of this set of nonlinear equations is a formidable task. However, results due once again to Butcher lighten the task considerably. We apply Lemma 5.1 quoted at the start of this section with

$$U = \begin{bmatrix} c_2 & c_3 & c_4 \\ c_2^2 & c_3^2 & c_4^2 \\ \lambda_2 & \lambda_3 & \lambda_4 \end{bmatrix}, \quad V = \begin{bmatrix} b_2 & b_2 c_2 & \mu_2 - \beta_2 \\ b_3 & b_3 c_3 & \mu_3 - \beta_3 \\ b_4 & b_4 c_4 & \mu_4 - \beta_4 \end{bmatrix},$$

where

$$\left.\begin{array}{l} \lambda_i = \sum\limits_j a_{ij} c_j - \frac{1}{2} c_i^2 \\ \mu_j = b_j (1 - c_j) \\ \beta_j = \sum\limits_i b_i a_{ij}, \end{array}\right\} i,j = 2,3,4. \tag{5.54}$$

Let $UV = [w_{ij}]$; then, using conditions (2), (3) and (5), we get $w_{11} = \frac{1}{2}$, $w_{12} = \frac{1}{3} = w_{21}$, $w_{22} = \frac{1}{4}$. Also,

$$w_{13} = \sum_j c_j \left[b_j (1 - c_j) - \sum_i b_i a_{ij} \right] = \frac{1}{2} - \frac{1}{3} - \frac{1}{6} = 0, \text{ by (2), (3) and (4),}$$

$$w_{23} = \sum_j c_j^2 \left[b_j (1 - c_j) - \sum_i b_i a_{ij} \right] = \frac{1}{3} - \frac{1}{4} - \frac{1}{12} = 0, \text{ by (3), (5) and (7),}$$

$$w_{31} = \sum_i b_i \left[\sum_j a_{ij} c_j - \frac{1}{2} c_i^2 \right] = \frac{1}{6} - \frac{1}{2}\frac{1}{3} = 0, \text{ by (4) and (3),}$$

$$w_{32} = \sum_i b_i c_i \left[\sum_j a_{ij} c_j - \frac{1}{2} c_i^2 \right] = \frac{1}{8} - \frac{1}{2}\frac{1}{4} = 0, \text{ by (6) and (5),}$$

$$w_{33} = \sum_i \left[\sum_j a_{ij} c_j - \frac{1}{2} c_i^2 \right] \left[b_i (1 - c_i) - \sum_j b_j a_{ji} \right] = w_{31} - w_{32} - \frac{1}{24} + \frac{1}{2}\frac{1}{12} = 0,$$

by (7) and (8).

Thus

$$UV = [w_{ij}] = \begin{bmatrix} \frac{1}{2} & \frac{1}{3} & 0 \\ \frac{1}{3} & \frac{1}{4} & 0 \\ 0 & 0 & 0 \end{bmatrix}.$$

Since $w_{11}w_{22} - w_{21}w_{12} = \frac{1}{8} - \frac{1}{9} \neq 0$, the hypotheses of the lemma are satisfied, and we must have that either $\lambda_i = 0, i = 2, 3, 4$, or $\mu_j = \beta_j, j = 2, 3, 4$. We show that the first of these alternatives is impossible. Since $\lambda_2 = \sum_j a_{2j}c_j - \frac{1}{2}c_2^2, c_1 = 0$ and $a_{2j} = 0, j = 2, 3, 4$, $\lambda_2 = 0$ implies that $c_2 = 0$. Now, for an explicit method, the order condition (8) of (5.53) becomes

$$b_4 a_{43} a_{32} c_2 = \tfrac{1}{24}, \tag{5.55}$$

whence $c_2 \neq 0$. The second alternative, $\mu_j = \beta_i$, $j = 2, 3, 4$, must therefore hold, and this has two important consequences. By (5.54) this means that

$$\sum_i b_i a_{ij} = b_j(1 - c_j), \qquad j = 2, 3, 4. \tag{5.56}$$

It follows that

$$\begin{aligned}
\sum b_i a_{ij} c_j &= \sum b_j(1 - c_j)c_j = \tfrac{1}{2} - \tfrac{1}{3} = \tfrac{1}{6}, \text{ by (2) and (3)}, \\
\sum b_i a_{ij} c_j^2 &= \sum b_j(1 - c_j)c_j^2 = \tfrac{1}{3} - \tfrac{1}{4} = \tfrac{1}{12}, \text{ by (3) and (5)}, \\
\sum b_i a_{ij} a_{jk} c_k &= \sum b_j(1 - c_j)a_{jk}c_k = \tfrac{1}{6} - \tfrac{1}{8} = \tfrac{1}{24}, \text{ by (4) and (6)}.
\end{aligned}$$

In other words, conditions (4), (7) and (8) are automatically satisfied if the remaining order conditions are satisfied, and can be ignored.

The second consequence comes from setting $j = 4$ in (5.56) and noting that $\sum b_i a_{i4} = 0$, since $a_{ij} = 0$ if $j \geqslant i$. Thus $b_4(1 - c_4) = 0$, and since by (5.55) b_4 cannot be zero it follows that $c_4 = 1$. Thus we have the perhaps surprising result that *for all four-stage explicit Runge–Kutta methods of order 4*, $c_4 = 1$.

The fact that c_4 is fixed results in the general solution of the order conditions constituting a two-parameter family, not, as might have been anticipated, a three-parameter family. The full solution of the order conditions is still rather too cumbersome to reproduce here, and the reader is referred to Butcher (1987). The solution consists of one two-parameter family of solutions and four one-parameter families.

Five-stage methods

Butcher array

$$\begin{array}{c|ccccc}
0 & & & & & \\
c_2 & a_{21} & & & & \\
c_3 & a_{31} & a_{32} & & & \\
c_4 & a_{41} & a_{42} & a_{43} & & \\
c_5 & a_{51} & a_{52} & a_{53} & a_{54} & \\
\hline
 & b_1 & b_2 & b_3 & b_4 & b_5
\end{array}
\qquad
\begin{aligned}
& \\
a_{21} &= c_2 \\
a_{31} + a_{32} &= c_3 \\
a_{41} + a_{42} + a_{43} &= c_4 \\
a_{51} + a_{52} + a_{53} + a_{54} &= c_5.
\end{aligned}$$

The order 5 conditions are the eight conditions (5.53) (with the summations now running from 1 to 5) together with the following nine additional conditions

$$\left.\begin{array}{ll}
(9) & \sum b_i c_i^4 = \frac{1}{5} \\
(10) & \sum b_i c_i^2 a_{ij} c_j = \frac{1}{10} \\
(11) & \sum b_i c_i a_{ij} c_j^2 = \frac{1}{15} \\
(12) & \sum b_i c_i a_{ij} a_{jk} c_k = \frac{1}{30} \\
(13) & \sum b_i (\sum a_{ij} c_j)^2 = \frac{1}{20} \\
(14) & \sum b_i a_{ij} c_j^3 = \frac{1}{20} \\
(15) & \sum b_i a_{ij} c_j a_{jk} c_k = \frac{1}{40} \\
(16) & \sum b_i a_{ij} a_{jk} c_k^2 = \frac{1}{60} \\
(17) & \sum b_i a_{ij} a_{jk} a_{km} c_m = \frac{1}{120}.
\end{array}\right\} \tag{5.57}$$

Guessing from our previous results, we might expect that there would exist a one-parameter family of solutions; but, as we should have gathered by now, this is not an area in which it is wise to guess! For decades before the Butcher theory was established, many attempts were made to find a five-stage explicit method of order 5, and none were successful. The following theorem of Butcher put an end to the search.

Theorem 5.5 There exist no five-stage explicit Runge–Kutta methods of order 5.

Proof The proof closely follows the argument earlier in this section which led to the result that $c_4 = 1$ for all four-stage explicit methods of order 4, and we shall omit some of the detail. First apply Lemma 5.1 with

$$U = \begin{bmatrix} c_2 & c_3 & c_4 \\ c_2^2 & c_3^2 & c_4^2 \\ \lambda_2 & \lambda_3 & \lambda_4 \end{bmatrix}, \qquad V = \begin{bmatrix} \beta_2 & \beta_2 c_2 & v_2 \\ \beta_3 & \beta_3 c_3 & v_3 \\ \beta_4 & \beta_4 c_4 & v_4 \end{bmatrix},$$

where λ_i and β_i are given by (5.54) (but now for $i = 2, 3, 4, 5$) and

$$v_j = \tfrac{1}{2}\beta_j(1 - c_j) - \sum_i \beta_i a_{ij}, \qquad j = 2, 3, 4, 5. \tag{5.58}$$

Proceeding as before to use the order conditions (5.53) and (5.57) together with the fact that $\beta_5 = 0 = v_5$, we find that

$$UV = [w_{ij}] = \begin{bmatrix} \frac{1}{6} & \frac{1}{12} & 0 \\ \frac{1}{12} & \frac{1}{20} & 0 \\ 0 & 0 & 0 \end{bmatrix}. \tag{5.59}$$

Since $w_{11}w_{22} - w_{21}w_{12} = \frac{1}{120} - \frac{1}{144} \neq 0$, the hypotheses of the lemma are satisfied, and we must have that either $\lambda_i = 0, i = 2, 3, 4$, or $v_j = 0, j = 2, 3, 4$. The first is impossible; by exactly the same argument used earlier, $\lambda_2 = 0$ implies $c_2 = 0$, and since condition (17) reads

$$b_5 a_{54} a_{43} a_{32} c_2 = \tfrac{1}{120}, \tag{5.60}$$

it follows that $c_2 \neq 0$. Therefore $v_j = 0, j = 2, 3, 4$; but $v_4 = 0$ implies that $\beta_4(1 - c_4) = 2\sum \beta_i a_{i4} = 2\beta_5 a_{54} = 0$, since $\beta_5 = 0$ from (5.54). Now, also from (5.54), $\beta_4 = b_5 a_{54} \neq 0$, by (5.60). Hence we have that $c_4 = 1$.

We now apply Lemma 5.1 again with

$$U = \begin{bmatrix} c_2 & c_3 & c_5 \\ c_2^2 & c_3^2 & c_5^2 \\ \lambda_2 & \lambda_3 & \lambda_5 \end{bmatrix}, \qquad V = \begin{bmatrix} \mu_2 & \mu_2 c_2 & (\mu_2 - \beta_2)(1 - c_2) \\ \mu_3 & \mu_3 c_3 & (\mu_3 - \beta_3)(1 - c_3) \\ \mu_5 & \mu_5 c_5 & (\mu_5 - \beta_5)(1 - c_5) \end{bmatrix},$$

where λ_i, μ_j and β_j are given by (5.54) (for $i, j = 2, 3, 4, 5$). Using the order conditions (5.53) and (5.57) together with the fact that $c_4 = 1$ implies that $\mu_4 = 0 = (\mu_4 - \beta_4)(1 - c_4)$, we find that UV is once again given by (5.59). The argument used above shows that $c_2 \neq 0$, and it follows that $(\mu_5 - \beta_5)(1 - c_5) = 0$; since $\beta_5 = 0$, we have from (5.54) that $b_5(1 - c_5)^2 = 0$, and since $b_5 \neq 0$ by (5.60), it follows that $c_5 = 1$.

We have thus established that $c_4 = c_5 = 1$. Now consider

$$\sum b_i(1 - c_i) a_{ij} a_{jk} c_k = b_5(1 - c_5) \sum a_{5j} a_{jk} c_k + b_4(1 - c_4) a_{43} a_{32} c_2 = 0.$$

But, by order conditions (8) and (12), we also have that

$$\sum b_i(1 - c_i) a_{ij} a_{jk} c_k = \sum b_i a_{ij} a_{jk} c_k - \sum b_i c_i a_{ij} a_{jk} c_k = \tfrac{1}{24} - \tfrac{1}{30} = \tfrac{1}{120}.$$

We thus have a contradiction, and the theorem is proved. □

Theorem 5.5 can be extended to show that there exist no p-stage explicit methods of order p for $p \geqslant 5$; see Butcher (1987). The question of what order can be achieved by an explicit s-stage method is still an open one; the following is known (Butcher, 1987):

Order	1	2	3	4	5	6	7	8	9	10
Minimum stage number	1	2	3	4	6	7	9	11	$12 \leqslant s \leqslant 17$	$13 \leqslant s \leqslant 17$

The reason for the popularity of fourth-order methods is now clear. (By a somewhat illogical process, this may also explain the popularity of fourth-order predictor–corrector methods in the days before VSVO algorithms were developed!) The construction of explicit methods with order greater than four is quite involved; the best reference is once again Butcher (1987).

5.10 EXPLICIT METHODS; LOCAL ERROR ESTIMATION

As we have already remarked in §5.1, there exist no estimates for the local truncation error of explicit Runge–Kutta methods which are comparable in computational cheapness with the Milne estimate for predictor–corrector methods; all of the estimates discussed in this section require more function evaluations per step then are needed simply to advance the solution.

The first technique we discuss is *Richardson extrapolation*, also called *the deferred approach to the limit*; it is an old technique, and one which is applicable to any numerical method. Suppose that we have used a Runge–Kutta method of order p to obtain the numerical solution y_{n+1} at x_{n+1}. Under the usual localizing assumption that $y_n = y(x_n)$, it follows from (5.47) that the local truncation error T_{n+1} can be written in the form

$$T_{n+1} = y(x_{n+1}) - \tilde{y}_{n+1} = \Psi(y(x_n))h^{p+1} + 0(h^{p+2}), \tag{5.61}$$

where $\Psi(y(x_n))$ is a function of the elementary differentials of order $p+1$ evaluated at $y(x_n)$. (As in §3.5, the notation $\tilde{y}_{n+1}$ indicates the value for y at x_{n+1} given by the method under the localizing assumption.) Let us now compute a second numerical solution at x_{n+1} by applying the same method with steplength $2h$, but starting from x_{n-1}; denote the solution so obtained by $\tilde{z}_{n+1}$, the tilde indicating that the localizing assumption is again in force (but now at x_{n-1}). Then we may write

$$\begin{aligned} y(x_{n+1}) - \tilde{z}_{n+1} &= \Psi(y(x_{n-1}))(2h)^{p+1} + 0(h^{p+2}) \\ &= \Psi(y(x_n))(2h)^{p+1} + 0(h^{p+2}) \end{aligned} \tag{5.62}$$

on expanding $y(x_{n-1})$ about x_n. On subtracting (5.61) from (5.62) we obtain

$$(2^{p+1} - 1)h^{p+1}\Psi[y(x_n)] = \tilde{y}_{n+1} - \tilde{z}_{n+1} + 0(h^{p+2}),$$

whence we have, from (5.61), the following estimate for the principal local truncation error:

$$\text{PLTE} = (\tilde{y}_{n+1} - \tilde{z}_{n+1})/(2^{p+1} - 1).$$

This estimate works well in practice, and can be successfully used to monitor steplength, but it is expensive to implement; if the explicit Runge–Kutta method has s stages, then in general an additional $s-1$ function evaluations are needed, k_1 at x_{n-1} having been already computed. (The author was once asked by a member of a seminar audience—in a quite different context—why he didn't just use the 'usual' method for estimating the error of *any* numerical method; it transpired that the 'usual' method consisted of repeating a step with double the steplength, subtracting and dividing by the magic number 31. The magic number 31 is, of course, $2^{p+1} - 1$ when $p = 4$; such is the popularity of fourth-order methods!)

There exist in the literature a number of error estimates for explicit Runge–Kutta methods which do not involve additional function evaluations, but these are based on computed values at a number of consecutive integration steps. This approach obviously raises difficulties when the steplength is changed, and effectively sacrifices the major advantage of using Runge–Kutta methods, namely the freedom to change steplength with no attendant complications.

An early example of a Runge–Kutta method specially constructed to allow an error estimate in terms of the computed values k_i was proposed by Merson (1957). *Merson's method* is defined by the Butcher array

$$\begin{array}{c|ccccc} 0 & & & & & \\ \frac{1}{3} & \frac{1}{3} & & & & \\ \frac{1}{3} & \frac{1}{6} & \frac{1}{6} & & & \\ \frac{1}{2} & \frac{1}{8} & 0 & \frac{3}{8} & & \\ 1 & \frac{1}{2} & 0 & -\frac{3}{2} & 2 & \\ \hline & \frac{1}{6} & 0 & 0 & \frac{2}{3} & \frac{1}{6} \end{array} \tag{5.63}$$

This is a 5-stage method and it is easily checked that it has order 4. Merson proposed that the principal local truncation error be estimated by

$$h(-2k_1 + 9k_3 - 8k_4 + k_5)/30. \tag{5.64}$$

If this were a valid estimate of the principal local truncation error, then adding (5.64) to the value for y_{n+1} given by (5.63) would yield a 5th-order method for which c and A would be as in (5.63) and b^{T} would be

$$[\tfrac{1}{6}, 0, 0, \tfrac{2}{3}, \tfrac{1}{6}] + [\tfrac{-1}{15}, 0, \tfrac{3}{10}, \tfrac{-4}{15}, \tfrac{1}{30}] = [\tfrac{1}{10}, 0, \tfrac{3}{10}, \tfrac{2}{5}, \tfrac{1}{5}].$$

Now, we know from §5.9 that it is impossible for a 5-stage method to have order five, and we must conclude that the estimate (5.64) is not valid. Indeed one finds that the 5-stage method consisting of (5.63) with b^{T} modified as above has order only three; however, it has order five in the special case when the differential system is linear with constant coefficients. Although Merson's method played an important role in pointing the way to future developments, it is necessary to warn against using it for general problems, a warning that would appear to be necessary since its use appears still to be widespread. In practice, Merson's method usually overestimates the error, often grossly so at small steplength, and this has led to the belief that its use is always safe, albeit inefficient. However, some time ago England (1969) gave examples where Merson's method *under*estimates the error.

The essence of the Merson idea is to derive Runge–Kutta methods of orders p and $p+1$, which share the same set of vectors $\{k_i\}$; this process is known as *embedding*. In order to present embedded methods, we shall modify the Butcher array to the following form:

$$\begin{array}{c|c} c & A \\ \hline & b^{\mathrm{T}} \\ & \hat{b}^{\mathrm{T}} \\ \hline & E^{\mathrm{T}} \end{array} \tag{5.65}$$

This notation is to be interpreted to mean that the method defined by c, A and b^{T} has order p and that defined by c, A and $\hat{b}^{\mathrm{T}}$ has order $p+1$. The difference between the values for y_{n+1} generated by these two methods is then an estimate of the local truncation error. The vector E^{T} is $\hat{b}^{\mathrm{T}} - b^{\mathrm{T}}$, so that the error estimate is given by $h\sum_{i=1}^{s} E_i k_i$, where

$E^{\mathrm{T}} = [E_1, E_2, \ldots, E_s]$. It is convenient to attach to such an embedded method the label $(p, p+1)$. Note that the solution for y_{n+1} given by the pth order method is used as the initial value for the next step, so that the method has order p. One could use the $(p+1)$th-order value for y_{n+1} as the initial value for the next step, in which case the method has order $p+1$; it is appropriate in such cases to alter the label to $(p+1, p)$. This process is, of course, local extrapolation, discussed in §4.3 in the context of predictor–corrector methods; the caveats given there still apply.

It follows from §5.9 that for a fourth-order embedded method a minimum of six stages will be necessary. An example of such a method is the (4, 5) *England's method* (England, 1969), given by the modified Butcher array

$$
\begin{array}{c|cccccc}
0 & & & & & & \\
\frac{1}{2} & \frac{1}{2} & & & & & \\
\frac{1}{2} & \frac{1}{4} & \frac{1}{4} & & & & \\
1 & 0 & -1 & 2 & & & \\
\frac{2}{3} & \frac{7}{27} & \frac{10}{27} & 0 & \frac{1}{27} & & \\
\frac{1}{5} & \frac{28}{625} & -\frac{1}{5} & \frac{546}{625} & \frac{54}{625} & -\frac{378}{625} & \\
\hline
 & \frac{1}{6} & 0 & \frac{2}{3} & \frac{1}{6} & 0 & 0 \\
 & \frac{1}{24} & 0 & 0 & \frac{5}{48} & \frac{27}{56} & \frac{125}{336} \\
\hline
 & -\frac{1}{8} & 0 & -\frac{2}{3} & -\frac{1}{16} & \frac{27}{56} & \frac{125}{336}
\end{array}
$$

A feature of this method is that the last two elements of b^{T} are zero, implying that if the error estimate is not required then only four stages (the minimum possible for fourth order) need be computed. The method is thus economical if only occasional estimation of the error is intended.

Perhaps the most popular (4, 5) method is *RKF45*, one of a class of methods developed by Fehlberg (1968, 1969). In this class, the coefficients of the method are chosen so that the moduli of the coefficients of the functions $F(t)$ appearing in the principal part of the local truncation error (5.47) are small. We shall say that methods derived in this way are *error-tuned*. The modified Butcher array for RKF45 is

$$
\begin{array}{c|cccccc}
0 & & & & & & \\
\frac{1}{4} & \frac{1}{4} & & & & & \\
\frac{3}{8} & \frac{3}{32} & \frac{9}{32} & & & & \\
\frac{12}{13} & \frac{1932}{2197} & -\frac{7200}{2197} & \frac{7296}{2197} & & & \\
1 & \frac{439}{216} & -8 & \frac{3680}{513} & -\frac{845}{4104} & & \\
\frac{1}{2} & -\frac{8}{27} & 2 & -\frac{3544}{2565} & \frac{1859}{4104} & -\frac{11}{40} & \\
\hline
 & \frac{25}{216} & 0 & \frac{1408}{2565} & \frac{2197}{4104} & -\frac{1}{5} & 0 \\
 & \frac{16}{135} & 0 & \frac{6656}{12825} & \frac{28561}{56430} & -\frac{9}{50} & \frac{2}{55} \\
\hline
 & \frac{1}{360} & 0 & -\frac{128}{4275} & -\frac{2197}{75240} & \frac{1}{50} & \frac{2}{55}
\end{array}
$$

Note that if the error estimate is not required, then five stages are required to obtain the solution.

In most modern automatic codes based on embedded Runge–Kutta methods local extrapolation is used. Indeed, RKF45 is sometimes run as a (5, 4) method, even though it is not designed for such use, since error-tuning has been carried out on the fourth-order and not the fifth-order formula.

Embedded methods specifically designed for use with local extrapolation have been developed by Dormand and Prince (1980), Prince and Dormand (1981); see also Sharp (1989). In these methods it is the higher-order formula which is error-tuned and which carries the solution; the difference between the values given by the higher and lower order methods, though no longer a true estimate of the local truncation error, is used as a basis for monitoring steplength. Perhaps the most popular of these is a (5, 4) method, sometimes known as DOPRI (5, 4), defined by the modified Butcher array

$$
\begin{array}{c|ccccccc}
0 & & & & & & & \\
\frac{1}{5} & \frac{1}{5} & & & & & & \\
\frac{3}{10} & \frac{3}{40} & \frac{9}{40} & & & & & \\
\frac{4}{5} & \frac{44}{45} & -\frac{56}{15} & \frac{32}{9} & & & & \\
\frac{8}{9} & \frac{19372}{6561} & -\frac{25360}{2187} & \frac{64448}{6561} & -\frac{212}{729} & & & \\
1 & \frac{9017}{3168} & -\frac{355}{33} & \frac{46732}{5247} & \frac{49}{176} & -\frac{5103}{18656} & & \\
1 & \frac{35}{384} & 0 & \frac{500}{1113} & \frac{125}{192} & -\frac{2187}{6784} & \frac{11}{84} & \\
\hline
 & \frac{5179}{57600} & 0 & \frac{7571}{16695} & \frac{393}{640} & -\frac{92097}{339200} & \frac{187}{2100} & \frac{1}{40} \\
 & \frac{35}{384} & 0 & \frac{500}{1113} & \frac{125}{192} & -\frac{2187}{6784} & \frac{11}{84} & 0 \\
\hline
 & \frac{71}{57600} & 0 & -\frac{71}{16695} & \frac{71}{1920} & -\frac{17253}{339200} & \frac{22}{525} & -\frac{1}{40}
\end{array}
$$

(Note that we are sticking to the notation defined by (5.65), so that the vector $\hat{b}^{\mathrm{T}}$ for the method which carries the solution is the one starting 35/384, ...)

The above method has seven stages, as opposed to the six stages of England's method and RKF45; however, the last row of A is identical with the vector $\hat{b}^{\mathrm{T}}$, and we see, as follows, that this means that the method has effectively only six stages. Let the vectors k_i evaluated during the step from x_n to x_{n+1} be denoted by k_i^n. Then we have

$$k_7^n = f\left(x_n + h, y_n + h \sum_{j=1}^{6} a_{7j} k_j^n\right)$$

whence

$$k_1^{n+1} = f\left(x_n + h, y_n + h \sum_{j=1}^{6} \hat{b}_j k_j^n\right) = k_7^n,$$

and there is no need to compute k_1^{n+1}. Methods with this property are sometimes known as FSAL methods (First Same As Last).

Let us now compare the four embedded methods given above by applying each to the second-order initial value problem

$$\left.\begin{aligned} u'' &= -u(u^2+v^2)^{-3/2}, v'' = -v(u^2+v^2)^{3/2} \\ u(0) &= 1, u'(0) = 0, v(0) = 0, v'(0) = 1 \end{aligned}\right\} \quad (5.66)$$

which has exact solution $u(x) = \cos x$, $v(x) = \sin x$. Of course, we first rewrite (5.66) as an equivalent first-order system, as in §1.5. Since the exact solution is known, it is possible to implement, at each step, the localizing assumption $y_n = y(x_n)$, use the method to compute $\tilde{y}_{n+1}$ and thus compute the local truncation error from (5.12); the L_2-norms of the resulting exact local truncation errors are given in the columns headed LTE in Table 5.7. The L_2-norms of the estimates for the local truncation error provided by the embedded method are given in the columns headed EST. (In the case of RKF45, the bracketed numbers indicate LTE when the method is used as a (5,4) pair; EST is, of

Table 5.7

	Merson		England		RKF45		DOPRI (5,4)	
x	LTE	EST	LTE	EST	LTE	EST	LTE	EST
				$h = 0.8$ Errors $\times 10^3$				
0.8	2	3	11	10	4(4)	1	4	1
1.6	4	6	46	46	19(21)	3	17	4
2.4	7	6	36	45	20(19)	5	9	3
3.2	4	4	8	7	3(3)	1	1	1
4.0	2	3	14	13	5(4)	1	4	1
4.8	5	7	47	45	20(21)	3	17	4
5.6	7	6	33	38	18(16)	5	9	3
6.4	4	4	9	4	3(3)	0.4	1	1
				$h = 0.4$ Errors $\times 10^5$				
0.8	3	20	61	51	9(3)	7	3	5
1.6	10	28	182	175	19(11)	21	5	13
2.4	12	25	94	111	11(10)	14	1	7
3.2	7	23	27	19	5(3)	2	1	1
4.0	3	20	73	62	10(3)	8	4	6
4.8	11	29	183	176	19(12)	21	5	13
5.6	12	24	82	97	10(10)	13	1	7
6.4	6	23	27	25	5(3)	2	1	1
				$h = 0.2$ Errors $\times 10^7$				
0.8	12	129	228	206	37(8)	30	3	20
1.6	25	147	577	571	74(18)	73	4	44
2.4	26	140	259	285	30(14)	39	1	22
3.2	15	135	82	78	8(3)	5	2	3
4.0	12	130	266	243	41(9)	35	3	23
4.8	26	148	577	573	73(18)	73	4	44
5.6	26	139	222	247	26(13)	34	1	19
6.4	15	135	79	80	7(3)	6	2	3

course, independent of whether one uses the formulae as a (4, 5) or a (5, 4) pair.) The integration range is [0, 6.4], which covers one cycle of the periodic exact solution, and the computations are performed for three values of the steplength h.

We can draw several conclusions from this numerical experiment. Merson's method is remarkably accurate when one remembers that it uses only five function evaluations per step, whereas all the others use six. The error estimate is good for large steplength, but for small steplength the error is badly over-estimated. One can see why Merson's method remains popular, despite its shortcomings. England's method is the least accurate but it gives, for all steplengths, remarkably good estimates of the error. RKF45 has a tendency to underestimate the error, a tendency which is most noticeable at large steplength. This is the penalty incurred in error-tuning; error-tuning consists of trying to minimize the coefficients in the *principal* local truncation error, and results in the principal error being less than normally representative of the whole local truncation error, an effect obviously magnified when the steplength is large. In DOPRI (5,4), the error is a little underestimated at large steplength and clearly overestimated at small steplength; the smallness of the error reflects the use of local extrapolation. As one would expect, using RKF 45 as a (5, 4) pair (the bracketed numbers) results in a poorer error estimate but a more accurate solution—though not as accurate as that given by DOPRI (5, 4). It is of course dangerous to draw too many conclusions from a single example, but the author has conducted the above experiment on a number of problems, and the above conclusions always appeared valid.

In a simple automatic code based on embedded methods, the user supplies a tolerance TOL, and the algorithm successively halves the steplength until the error estimate is less than TOL; if the estimate is less than TOL/2^{p+1}, where p is the order, the steplength is doubled. (More sophisticated strategies are of course usually employed.) In such a context, it is not obvious that error-tuning—and for that matter the use of local extrapolation—is necessarily advantageous. Whether one does better with a less accurate method which has a very sharp error estimate (such as England's method) or with a more accurate error-tuned method for which the estimate is less sharp, resulting in the need for heuristic safeguards in the code, is ultimately a problem-dependent question.

There exist Fehlberg methods of orders up to eight. Unfortunately, all the Fehlberg methods of order greater than four suffer a peculiar deficiency, exemplified by the 8-stage (5, 6) Fehlberg method, for which the vectors c^{T} and E^{T} are

$$\left.\begin{aligned} c^{\mathrm{T}} &= [0, \ \tfrac{1}{6}, \ \tfrac{4}{15}, \ \tfrac{2}{3}, \ \tfrac{4}{5}, \ 1, \ 0, \ 1] \\ E^{\mathrm{T}} &= [\tfrac{-5}{66}, \ 0, \ 0, \ 0, \ 0, \ \tfrac{-5}{66}, \ \tfrac{5}{66}, \ \tfrac{5}{66}]. \end{aligned}\right\} \qquad (5.67)$$

Now suppose that such a method is applied to a system in which f depends only on x. Then the fifth- and sixth-order methods reduce to

$$y_{n+1} = y_n + h \sum_{j=1}^{8} b_j f(x_n + c_j h), \qquad \hat{y}_{n+1} = y_n + h \sum_{j=1}^{8} \hat{b}_j f(x_n + c_j h)$$

respectively, whence

$$\hat{y}_{n+1} - y_{n+1} = h \sum_{j=1}^{8} E_j f(x_n + c_j h) = 0$$

by (5.67). Thus, when $f(x, y) = f(x)$ the two methods give identically the same result, and the error estimate is in all cases zero, no matter what the size of the actual local truncation error. We can anticipate that such methods will give misleading results when applied to a system $y' = f(x, y)$, in which f depends much more strongly on x than it does on y. Alternative embedded methods of orders 5 to 8, which do not encounter this difficulty, are given by Verner (1978). Higher-order embedded methods using local extrapolation are derived by Dormand and Prince (1980); in that reference can be found an 8-stage (6, 5) method and a 13-stage (8, 7) method.

Computational experience shows that Runge–Kutta codes can be competitive with ABM codes for problems where function evaluations are not too expensive. Examples of such codes are DVERK (Hull, Enright and Jackson, 1976), which uses an 8-stage (5, 6) pair, RKF7 (Enright and Hull, 1976), based on a Fehlberg (7, 8) pair and XRK (Shampine and Baca, 1986) which uses the Dormand–Prince (8, 7) pair. The effectiveness of a Runge–Kutta code is much improved if the order of the pair is appropriate to the particular problem in hand; thus there have been developed variable order Runge–Kutta codes such a RKSW (Shampine and Wisniewski, 1978) which can switch between a (3, 4) and a (7, 8) pair.

Exercises

5.10.1. The scalar problem $y' = x^2 + y, y(0) = 0$ has exact solution $y(x) = 2\exp(x) - x^2 - 2x - 2$. Express $y(h)$ as a power series in h. Compare $y(h)$ with y_1 given by applying Merson's method (5.63) once. Deduce that the PLTE is $0(h^5)$, but that the estimate (5.64) does not correctly estimate the PLTE.

5.10.2*. Find the exact solution of the scalar problem $y' = ax + by + c$, $y(0) = 0$, where a, b and c are constants. Apply Merson's method, (5.63) to this problem and compute the solution y_1 at $x = h$. Calculate $y(h) - y_1$ and thus corroborate that the method is of order four. Show further that the PLTE is indeed given by (5.64). (The equation $y' = ay + bx + c$ is the most general scalar equation for which the Merson estimate is valid.)

5.11 IMPLICIT AND SEMI-IMPLICIT METHODS

As we noted in §5.1, implicit Runge–Kutta methods, even semi-implicit ones, are very expensive to implement and cannot rival predictor–corrector or explicit Runge–Kutta methods in efficiency when the problem to be solved is not stiff. Their use is almost exclusively restricted to stiff systems, in which context their superior stability properties justify the high cost of implementation. Consequently, much of our discussion of implicit methods and their implementation will be left to Chapters 6 and 7, where the problem of stiffness is addressed. In this section we merely list various categories of implicit methods and give examples of the more common methods. The reader who wishes to see a fuller discussion of the derivation of these methods from the order conditions derived in §5.7 is referred to the books by Butcher (1987) and Dekker and Verwer (1984).

If the general Runge–Kutta method (5.2) is applied to the scalar problem $y' = f(x)$, then the result is a quadrature formula

$$\int_{x_n}^{x_{n+1}} f(x)\,\mathrm{d}x \approx y_{n+1} - y_n = h \sum_{j=1}^{s} b_j f(x_n + c_j h)$$

which the reader may recognize (and it does not matter if he does not) as a Gaussian quadrature formula with ordinates (or abscissae) $x_n + c_j h$ and weights b_j, $j = 1, 2, \ldots, s$. The word 'Gaussian' is used somewhat loosely in this context, and there are several families, other than the original Gauss family, of quadrature formulae with unevenly spaced ordinates. Fully implicit Runge–Kutta methods are categorized by the class of quadrature formulae to which they revert when we put $f(x, y) = f(x)$. In the following we shall list low-order methods of various classes, quoting the stage-number s and the order p.

The first class of fully implicit methods consists of *Gauss* or *Gauss–Legendre methods* (Butcher, 1964). These squeeze out the highest possible order, and the s-stage Gauss method has order $2s$.

Gauss methods

$s = 1, p = 2$

$$\begin{array}{c|c} \frac{1}{2} & \frac{1}{2} \\ \hline & 1 \end{array}$$

$s = 2, p = 4$

$$\begin{array}{c|cc} \frac{3-\sqrt{3}}{6} & \frac{1}{4} & \frac{3-2\sqrt{3}}{12} \\ \frac{3+\sqrt{3}}{6} & \frac{3+2\sqrt{3}}{12} & \frac{1}{4} \\ \hline & \frac{1}{2} & \frac{1}{2} \end{array}$$

$s = 3, p = 6$

$$\begin{array}{c|ccc} \frac{5-\sqrt{15}}{10} & \frac{5}{36} & \frac{10-3\sqrt{15}}{45} & \frac{25-6\sqrt{15}}{180} \\ \frac{1}{2} & \frac{10+3\sqrt{15}}{72} & \frac{2}{9} & \frac{10-3\sqrt{15}}{72} \\ \frac{5+\sqrt{15}}{10} & \frac{25+6\sqrt{15}}{180} & \frac{10+3\sqrt{15}}{45} & \frac{5}{36} \\ \hline & \frac{5}{18} & \frac{4}{9} & \frac{5}{18} \end{array}$$

Note that the 1-stage Gauss method may be written

$$\begin{aligned} y_{n+1} &= y_n + hk_1 \\ &= y_n + hf(x_n + \tfrac{1}{2}h, y_n + \tfrac{1}{2}hk_1) \\ &= y_n + hf(x_n + \tfrac{1}{2}h, y_n + \tfrac{1}{2}(y_{n+1} - y_n)). \end{aligned}$$

The method can thus be written as

$$y_{n+1} = y_n + hf(x_n + \tfrac{1}{2}h, \tfrac{1}{2}(y_n + y_{n+1})), \tag{5.68}$$

in which form it is known as the *Implicit Mid-point Rule*. Note also that the 2-stage Gauss method was derived as an example in §5.7

The second category of methods reverts to the *Radau quadrature formulae*, characterized by the requirement that the ordinates include one or other of the ends of the interval of integration. This means that the corresponding implicit Runge–Kutta methods have either $c_1 = 0$ (*Radau I*) or $c_s = 1$ (*Radau II*). The maximum attainable order of an s-stage method is now $2s - 1$, and it turns out that this order can be achieved by a number of different choices of coefficients. We quote only the classes that turn out to be of most interest, namely the *Radau IA* and *Radau IIA* methods (Ehle, 1969; Chipman, 1971).

Radau IA

$s = 1, p = 1$

$$\begin{array}{c|c} 0 & 1 \\ \hline & 1 \end{array}$$

$s = 2, p = 3$

$$\begin{array}{c|cc} 0 & \frac{1}{4} & -\frac{1}{4} \\ \frac{2}{3} & \frac{1}{4} & \frac{5}{12} \\ \hline & \frac{1}{4} & \frac{3}{4} \end{array}$$

$s = 3, p = 5$

$$\begin{array}{c|ccc} 0 & \frac{1}{9} & \frac{-1-\sqrt{6}}{18} & \frac{-1+\sqrt{6}}{18} \\ \frac{6-\sqrt{6}}{10} & \frac{1}{9} & \frac{88+7\sqrt{6}}{360} & \frac{88-43\sqrt{6}}{360} \\ \frac{6+\sqrt{6}}{10} & \frac{1}{9} & \frac{88+43\sqrt{6}}{360} & \frac{88-7\sqrt{6}}{360} \\ \hline & \frac{1}{9} & \frac{16+\sqrt{6}}{36} & \frac{16-\sqrt{6}}{36} \end{array}$$

Note that the 1-step Radau IA method does not satisfy the row sum condition (5.3). The row sum condition was imposed because it greatly simplifies the derivation of the order conditions and there is nothing to be gained in terms of extra order by not imposing it—well, almost! For very low order only, it is possible to use this extra freedom to improve order, a curiosity first observed by Oliver (1975). The 1-step Radau IA method is an example of this phenomenon.

Radau IIA

$s = 1, p = 1$

$$\begin{array}{c|c} 1 & 1 \\ \hline & 1 \end{array}$$

$s = 2, p = 3$

$$\begin{array}{c|cc} \frac{1}{3} & \frac{5}{12} & -\frac{1}{12} \\ 1 & \frac{3}{4} & \frac{1}{4} \\ \hline & \frac{3}{4} & \frac{1}{4} \end{array}$$

$s = 3, p = 5$

$\frac{4-\sqrt{6}}{10}$	$\frac{88-7\sqrt{6}}{360}$	$\frac{296-169\sqrt{6}}{1800}$	$\frac{-2+3\sqrt{6}}{225}$
$\frac{4+\sqrt{6}}{10}$	$\frac{296+169\sqrt{9}}{1800}$	$\frac{88+7\sqrt{6}}{360}$	$\frac{-2-3\sqrt{6}}{225}$
1	$\frac{16-\sqrt{6}}{36}$	$\frac{16+\sqrt{6}}{36}$	$\frac{1}{9}$
	$\frac{16-\sqrt{6}}{36}$	$\frac{16+\sqrt{6}}{36}$	$\frac{1}{9}$

Note that, by an argument similar to that which showed that the 1-stage Gauss method could be rewritten as the Implicit Mid-point Rule, we see that the 1-stage Radau IIA method can be written as the Backward Euler method

$$y_{n+1} = y_n + hf(x_n + h, y_{n+1}),$$

which we met in §3.9. (This humble method is thus simultaneously the first-order Adams–Moulton method, the 1-step BDF method and the 1-stage Radau IIA method; to confuse things further, it is also called the *Implicit Euler Rule*.) Note also that in the Radau IIA methods the last row of the matrix A is identical with b^T. When this happened in an explicit method, as we saw in the preceding section it did for DOPRI(5, 4), a function evaluation was saved, and the stage-number effectively reduced by 1. Alas, there is no such benefit here, since for an implicit method it is no longer true that k_1 evaluated at the step starting from x_{n+1} is identical with $f(x_{n+1}, y_{n+1})$.

The last category of methods is associated with the *Lobatto quadrature formulae*, for which the ordinates include both ends of the interval of integration. The corresponding implicit Runge–Kutta methods (which obviously must have stage-number at least 2) have $c_1 = 0$ and $c_s = 1$. The maximum attainable order is now $2s - 2$. According to a classification of Butcher (1964a), these methods are of the third type, which is why they are called *Lobato III methods*. Again, various possibilities arise, the most useful of which are the Lobatto IIIA and IIIB methods of Ehle (1969) and the Lobatto IIIC methods of Chipman (1971); all attain order $2s - 2$.

Lobatto IIIA

$s = 2, p = 2$

0	0	0
1	$\frac{1}{2}$	$\frac{1}{2}$
	$\frac{1}{2}$	$\frac{1}{2}$

$s = 3, p = 4$

0	0	0	0
$\frac{1}{2}$	$\frac{5}{24}$	$\frac{1}{3}$	$-\frac{1}{24}$
1	$\frac{1}{6}$	$\frac{2}{3}$	$\frac{1}{6}$
	$\frac{1}{6}$	$\frac{2}{3}$	$\frac{1}{6}$

$s=4, p=6$

$$
\begin{array}{c|cccc}
0 & 0 & 0 & 0 & 0 \\
\frac{5-\sqrt{5}}{10} & \frac{11+\sqrt{5}}{120} & \frac{25-\sqrt{5}}{120} & \frac{25-13\sqrt{5}}{120} & \frac{-1+\sqrt{5}}{120} \\
\frac{5+\sqrt{5}}{10} & \frac{11-\sqrt{5}}{120} & \frac{25+13\sqrt{5}}{120} & \frac{25+\sqrt{5}}{120} & \frac{-1-\sqrt{5}}{120} \\
1 & \frac{1}{12} & \frac{5}{12} & \frac{5}{12} & \frac{1}{12} \\
\hline
 & \frac{1}{12} & \frac{5}{12} & \frac{5}{12} & \frac{1}{12}
\end{array}
$$

Note that the 2-stage Lobotto IIIA method is just the Trapezoidal Rule. Note also that the Lobatto IIIA methods, unlike the Radau IIA, are genuinely FSAL methods (see §5.10).

Lobatto IIIB

$s=2, p=2$

$$
\begin{array}{c|cc}
0 & \frac{1}{2} & 0 \\
1 & \frac{1}{2} & 0 \\
\hline
 & \frac{1}{2} & \frac{1}{2}
\end{array}
$$

$s=3, p=4$

$$
\begin{array}{c|ccc}
0 & \frac{1}{6} & -\frac{1}{6} & 0 \\
\frac{1}{2} & \frac{1}{6} & \frac{1}{3} & 0 \\
1 & \frac{1}{6} & \frac{5}{6} & 0 \\
\hline
 & \frac{1}{6} & \frac{2}{3} & \frac{1}{6}
\end{array}
$$

$s=4, p=6$

$$
\begin{array}{c|cccc}
0 & \frac{1}{12} & \frac{-1-\sqrt{5}}{24} & \frac{-1+\sqrt{5}}{24} & 0 \\
\frac{5-\sqrt{5}}{10} & \frac{1}{12} & \frac{25+\sqrt{5}}{120} & \frac{25-13\sqrt{5}}{120} & 0 \\
\frac{5+\sqrt{5}}{10} & \frac{1}{12} & \frac{25+13\sqrt{5}}{120} & \frac{25-\sqrt{5}}{120} & 0 \\
1 & \frac{1}{12} & \frac{11-\sqrt{5}}{24} & \frac{11+\sqrt{5}}{24} & 0 \\
\hline
 & \frac{1}{12} & \frac{5}{12} & \frac{5}{12} & \frac{1}{12}
\end{array}
$$

Note that the 2-stage Lobatto IIIB method is another example of the row-sum condition not being met; note also that this method is semi-implicit.

Lobatto IIIC

$s=2, p=2$

$$
\begin{array}{c|cc}
0 & \frac{1}{2} & -\frac{1}{2} \\
1 & \frac{1}{2} & \frac{1}{2} \\
\hline
 & \frac{1}{2} & \frac{1}{2}
\end{array}
$$

$s = 3, p = 4$

$$
\begin{array}{c|ccc}
0 & \frac{1}{6} & -\frac{1}{3} & \frac{1}{6} \\
\frac{1}{2} & \frac{1}{6} & \frac{5}{12} & -\frac{1}{12} \\
1 & \frac{1}{6} & \frac{2}{3} & \frac{1}{6} \\
\hline
 & \frac{1}{6} & \frac{2}{3} & \frac{1}{6}
\end{array}
$$

$s = 4, p = 6$

$$
\begin{array}{c|cccc}
0 & \frac{1}{12} & \frac{-\sqrt{5}}{12} & \frac{\sqrt{5}}{12} & -\frac{1}{12} \\
\frac{5-\sqrt{5}}{10} & \frac{1}{12} & \frac{1}{4} & \frac{10-7\sqrt{5}}{60} & \frac{\sqrt{5}}{60} \\
\frac{5+\sqrt{5}}{10} & \frac{1}{12} & \frac{10+7\sqrt{5}}{60} & \frac{1}{4} & -\frac{\sqrt{5}}{60} \\
1 & \frac{1}{12} & \frac{5}{12} & \frac{5}{12} & \frac{1}{12} \\
\hline
 & \frac{1}{12} & \frac{5}{12} & \frac{5}{12} & \frac{1}{12}
\end{array}
$$

Implicit Runge–Kutta methods can also be categorized according to whether or not they are *collocation methods.* Collocation is an old idea, widely applicable in numerical analysis, and consists of choosing a function (usually a polynomial) and a set of *collocation points*, and then demanding that the function does, at the collocation points, whatever is necessary to make it mimic the behaviour of the unknown function we are trying to approximate numerically. In the context of solving the initial value problem $y' = f(x, y), y(a) = \eta$, we can advance the numerical solution from x_n to x_{n+1} by choosing a polynomial P of degree s, with coefficients in $\mathbb{R}^m$, and a set of distinct collocation points $\{x_n + c_i h, i = 1, 2, \ldots, s\}$ and demanding that

$$
\begin{aligned}
&P(x_n) = y_n, \\
&P'(x_n + c_i h) = f(x_n + c_i h, P(x_n + c_i h)), \qquad i = 1, 2, \ldots, s.
\end{aligned}
$$

Note that this defines $P(x)$ uniquely. We then complete the step by taking $y_{n+1} = P(x_n + h)$. It was originally shown by Wright (1970) that this process is identical with an s-stage implicit Runge–Kutta method. To see this, observe that $P'(x)$ is a polynomial of degree $s - 1$ which interpolates the s data points $(x_n + c_i h, P'(x_n + c_i h)), i = 1, 2, \ldots, s$. We can therefore write it in the form of a Lagrange interpolation polynomial (see §1.10). Define $k_i := P'(x_n + c_i h), i = 1, 2, \ldots, s$; then, on writing $x = x_n + th$, we have

$$P'(x_n + th) = \sum_{j=1}^{s} L_j(t) k_j \tag{5.69}$$

where

$$L_j(t) = \prod_{\substack{i=1 \\ i \neq j}}^{s} \frac{t - c_i}{c_j - c_i}. \tag{5.70}$$

Now integrate (5.69) with respect to x from $x = x_n$ to $x = x_n + c_i h, i = 1, 2, \ldots, s$ and from $x = x_n$ to $x = x_{n+1}$ to get

$$P(x_n + c_i h) - P(x_n) = h \sum_{j=1}^{s} \left(\int_0^{c_i} L_j(t)\,\mathrm{d}t \right) k_j, \qquad i = 1, 2, \ldots, s \tag{5.71}$$

and

$$P(x_n + h) - P(x_n) = h \sum_{j=1}^{s} \left(\int_0^1 L_j(t)\mathrm{d}t \right) k_j. \qquad (5.72)$$

Now, for $j = 1, 2, \ldots, s$, define

$$a_{ij} := \int_0^{c_i} L_j(t)\mathrm{d}t, \qquad i = 1, 2, \ldots, s, \qquad b_j := \int_0^1 L_j(t)\mathrm{d}t \qquad (5.73)$$

and (5.71) and (5.72) give

$$\begin{aligned} k_i = P'(x_n + c_i h) &= f(x_n + c_i h, P(x_n + c_i h)) \\ &= f\left(x_n + c_i h, y_n + h \sum_{j=1}^{s} a_{ij} k_j \right), \qquad i = 1, 2, \ldots, s, \end{aligned}$$

and

$$y_{n+1} = y_n + h \sum_{j=1}^{s} b_j k_j$$

and we have an implicit Runge–Kutta method with the elements of c being the collocation points and b and A given by (5.73). The class of collocation methods consists of those implicit Runge–Kutta methods which can be derived in this fashion.

There is another interpretation we can put on collocation methods. Consider what happens to the alternative form (5.6) of the general Runge–Kutta method if we put $f(x, y) = f(x)$. The second of (5.6) can be interpreted as a quadrature formula for $\int_{x_n}^{x_n + c_i h} f(x)\mathrm{d}x, i = 1, 2, \ldots, s$. It follows from the above (see (5.71)) that if the Runge–Kutta method is a collocation method, then each of these quadratures will be exact if f is a polynomial in x of degree $\leqslant s - 1$, a property sometimes used to define the class of collocation methods. We can take this argument further to produce a useful characterization of collocation methods. Since a polynomial is linear in its coefficients, to say that

$$\int_{x_n}^{x_n + c_i h} f(x)\mathrm{d}x = h \sum_{j=1}^{s} a_{ij} f(x_n + c_j h)$$

is an exact quadrature formula when $f(x)$ is a polynomial of degree $\leqslant s - 1$ is the same as to say that it is exact when $f(x) = x^r, r = 0, 1, \ldots, s - 1$, leading to the identity

$$[(x_n + c_i h)^{r+1} - x_n^{r+1}]/(r + 1) \equiv h \sum_{j=1}^{s} a_{ij}(x_n + c_j h)^r, \qquad r = 0, 1, \ldots, s - 1.$$

On equation powers of h, we easily obtain the condition

$$\sum_{j=1}^{s} a_{ij} c_j^{\sigma - 1} = c_i^{\sigma}/\sigma, \qquad \sigma = 1, 2, \ldots, s, \qquad i = 1, 2, \ldots, s. \qquad (5.74)$$

(Note that for $\sigma = 1$, (5.74) is just the row-sum condition (5.3).) Indeed, for an s-stage Runge–Kutta method of order at least s and with distinct c_i, (5.74) is a necessary and sufficient condition for the method to be a collocation method (see, for example, Hairer, Nørsett and Wanner (1980)).

The families of implicit Runge–Kutta methods quoted earlier in this section split into two groups; the Gauss, Radau IIA and Lobatto IIIA methods are collocation methods, while the Radau IA, Lobatto IIIB and Lobatto IIIC are not. If the reader feels in need of mathematical exercise, he can verify that (5.74) holds for each of the quoted methods in the first group and is contradicted for each method in the second group. If he is really desperate for exercise, he may also verify that (5.73) holds for each method in the first group.

We turn now to semi-implicit methods. As we have already remarked in §5.1, the computational effort in implementing these methods is substantially less than for a fully implicit method, but still sufficiently onerous for the methods to be of interest only for stiff systems. We shall discuss their implementation in that context in Chapter 6, where it will emerge that considerable gains in efficiency occur in the case when all of the elements on the main diagonal of the coefficient matrix A of a semi-implicit method are identical. This defines the class of *diagonally implicit Runge–Kutta methods or DIRK methods*, developed by Nørsett (1974), Crouziex (1976) and Alexander (1977). (There is some confusion over nomenclature in this area; some authors use the term 'diagonally implicit' to describe *any* semi-implicit method, and then refer to the DIRK methods we have just defined as *singly diagonally implicit*.)

It is readily established that the following 2-stage semi-implicit method has order 3 for all values of the parameter μ other then $\mu = 0$:

$$\begin{array}{c|cc} \dfrac{3\mu-1}{6\mu} & \dfrac{3\mu-1}{6\mu} & 0 \\[2ex] \dfrac{1+\mu}{2} & \mu & \dfrac{1-\mu}{2} \\[2ex] \hline & \dfrac{3\mu^2}{3\mu^2+1} & \dfrac{1}{3\mu^2+1} \end{array} \tag{5.75}$$

There exists no value of μ for which the method has order greater than 3, but taking $\mu = \mp\sqrt{3}/3$ gives the following pair of DIRK methods:

$$\begin{array}{c|cc} \dfrac{3\pm\sqrt{3}}{6} & \dfrac{3\pm\sqrt{3}}{6} & 0 \\[2ex] \dfrac{3\mp\sqrt{3}}{6} & \dfrac{\mp\sqrt{3}}{3} & \dfrac{3\pm\sqrt{3}}{6} \\[2ex] \hline & \dfrac{1}{2} & \dfrac{1}{2} \end{array} \tag{5.76}$$

There exist three 3-stage DIRK methods of order 4 given by

$$
\begin{array}{c|ccc}
\dfrac{1+\nu}{2} & \dfrac{1+\nu}{2} & 0 & 0 \\
\dfrac{1}{2} & -\dfrac{\nu}{2} & \dfrac{1+\nu}{2} & 0 \\
\dfrac{1-\nu}{2} & 1+\nu & -1-2\nu & \dfrac{1+\nu}{2} \\
\hline
 & \dfrac{1}{6\nu^2} & 1-\dfrac{1}{3\nu^2} & \dfrac{1}{6\nu^2}
\end{array}
\qquad (5.77)
$$

where ν takes one of the three values $(2/\sqrt{3})\cos(10^\circ)$, $-(2/\sqrt{3})\cos(50^\circ)$, $-(2/\sqrt{3})\cos(70^\circ)$, the roots of $3\nu^3 - 3\nu = 1$.

Finally, we briefly mention a further class of implicit methods, the *singly-implicit Runge–Kutta* or *SIRK methods*, developed by Nørsett (1976) and Burrage (1978a, 1978b, 1982). Although these are fully implicit methods, they can be regarded as generalizations of DIRK methods. The trouble with DIRK methods is that it is very difficult indeed to construct such methods with high stage-number (those appearing in the literature have order at most four), making them unsuitable as the basis of a variable order code. Now, it is clear that the spectrum of eigenvalues of the matrix A for a DIRK method consists of the single eigenvalue a_{ii} repeated s times. SIRK methods are defined by the requirement that the matrix A, though not lower triangular, should have a spectrum consisting of the single eigenvalue μ repeated s times, where s is the stage-number. As we shall see in §6.5, this has the consequence that the methods can be implemented at a cost not much greater than that for a DIRK method. SIRK methods of arbitrary order can be derived; see Dekker and Verwer (1984). An example of a SIRK method is the 2-stage method

$$
\begin{array}{c|cc}
(2-\sqrt{2})\mu & \dfrac{(4-\sqrt{2})\mu}{4} & \dfrac{(4-3\sqrt{2})\mu}{4} \\
(2+\sqrt{2})\mu & \dfrac{(4+3\sqrt{2})\mu}{4} & \dfrac{(4+\sqrt{2})\mu}{4} \\
\hline
 & \dfrac{4(1+\sqrt{2})\mu-\sqrt{2}}{8\mu} & \dfrac{4(1-\sqrt{2})\mu+\sqrt{2}}{8\mu}
\end{array}
\qquad (5.78)
$$

which has order 2 in general and order 3 if $\mu = (3 \pm \sqrt{3})/6$.

Exercise

5.11.1. Use (5.74) to show that the SIRK method (5.78) is a collocation method.

5.12 LINEAR STABILITY THEORY FOR RUNGE–KUTTA METHODS

We first met linear stability theory in the context of linear multistep methods (see §3.8), where we chose as test system the system $y' = Ay$, where A is an $m \times m$ matrix with distinct eigenvalues $\{\lambda_t, t = 1, 2, \dots, m\}$ lying strictly in the negative half-plane, a condition which ensures that all solutions of the test system tend to zero as x tends to infinity. Since the eigenvalues of A are distinct there exists a non-singular matrix Q such that $Q^{-1}AQ = \Lambda = \text{diag}[\lambda_1, \lambda_2, \dots, \lambda_m]$ and, by using a transformation $y = Qz$, we showed that it was enough to consider only the scalar test equation $y' = \lambda y$, where $\lambda \in \mathbb{C}$ and $\text{Re}(\lambda) < 0$. Linear stability was concerned with the question of whether or not the numerical solution of this scalar test equation tended to zero as n tended to infinity.

The application of the transformation $y = Qz$ uncoupled not only the original test system but also the difference system arising from the linear multistep method, and it was obvious that this was also true for predictor–corrector methods. It is perhaps less obvious that it holds also for Runge–Kutta methods, and we start by showing that this is indeed the case. We had better change the notation, and write the test system as

$$y' = \mathscr{A} y \tag{5.79}$$

to avoid confusion with the coefficient matrix A appearing in the Butcher array of a Runge–Kutta method. The matrix $\mathscr{A}$ is assumed to have distinct eigenvalues λ_t where $\text{Re}(\lambda_t) < 0$, $t = 1, 2, \dots, m$, and the non-singular matrix Q is defined by $Q^{-1}\mathscr{A}Q = \Lambda = \text{diag}[\lambda_1, \lambda_2, \dots, \lambda_m]$. Applying the general Runge–Kutta method (5.2) to (5.79) gives

$$\left.\begin{aligned} &y_{n+1} = y_n + h \sum_{i=1}^{s} b_i k_i \\ &\text{where} \\ &k_i = \mathscr{A}\left[y_n + h \sum_{j=1}^{s} a_{ij} k_j \right], \qquad i = 1, 2, \dots, s. \end{aligned}\right\} \tag{5.80}$$

Now define z_n and l_i by

$$y_n = Qz_n, \qquad k_i = Ql_i, \quad i = 1, 2, \dots, s.$$

Substituting for y_n and k_i in (5.80) and premultiplying by Q^{-1} gives

$$\left.\begin{aligned} &z_{n+1} = z_n + h \sum_{i=1}^{s} b_i l_i \\ &\text{where} \\ &l_i = \Lambda\left[z_n + h \sum_{j=1}^{s} a_{ij} l_j \right], \quad i = 1, 2, \dots, s \end{aligned}\right\} \tag{5.81}$$

which is precisely the result we would get from applying the method (5.2) to the system

$$z' = \Lambda z. \tag{5.82}$$

It is clear from (5.82) and (5.81) that we have indeed uncoupled both the differential system and the difference system. We are thus justified in using as test equation the scalar problem

$$y' = \lambda y, \qquad \lambda \in \mathbb{C}, \quad \operatorname{Re}(\lambda) < 0. \tag{5.83}$$

If we apply the general Runge–Kutta method (5.2) to (5.83), we are clearly going to obtain a one-step difference equation of the form

$$y_{n+1} = R(\hat{h}) y_n,$$

where, as before, $\hat{h} = h\lambda$. We shall call $R(\hat{h})$ the *stability function* of the method. Clearly $y_n \to 0$ as $n \to \infty$ if and only if

$$|R(\hat{h})| < 1 \tag{5.84}$$

and the method is absolutely stable for those values of $\hat{h}$ for which (5.84) holds. The region $\mathscr{R}_A$ of the complex $\hat{h}$-plane for which (5.84) holds is then the region of absolute stability of the method. Let us now investigate the form that $R(\hat{h})$ takes. It is marginally easier to work with the alternative form (5.6) of the general s-stage Runge–Kutta method. Applying this to the test equation (5.83) (where, we recall, y_n is scalar) yields

$$\left.\begin{aligned} Y_i &= y_n + \hat{h} \sum_{j=1}^{s} a_{ij} Y_j, \qquad i = 1, 2, \ldots, s \\ y_{n+1} &= y_n + \hat{h} \sum_{i=1}^{s} b_i Y_i. \end{aligned}\right\} \tag{5.85}$$

Now define $Y, e \in \mathbb{R}^s$ by $Y := [Y_1, Y_2, \ldots, Y_s]^{\mathrm{T}}$ and $e := [1, 1, \ldots, 1]^{\mathrm{T}}$; we may then write (5.85) in the form

$$Y = y_n e + \hat{h} A Y, \quad y_{n+1} = y_n + \hat{h} b^{\mathrm{T}} Y.$$

Solving the first of these for Y and substituting in the second gives

$$y_{n+1} = y_n [1 + \hat{h} b^{\mathrm{T}} (I - \hat{h} A)^{-1} e],$$

where I is the $s \times s$ unit matrix. The stability function is therefore given by

$$R(\hat{h}) = 1 + \hat{h} b^{\mathrm{T}} (I - \hat{h} A)^{-1} e. \tag{5.86}$$

An approach due to Dekker and Verwer (1984) gives an alternative form for $R(\hat{h})$. To avoid filling the page with large determinants, we develop this alternative for the case $s = 2$, when (5.85) may be written as

$$\begin{bmatrix} 1 - \hat{h} a_{11} & -\hat{h} a_{12} & 0 \\ -\hat{h} a_{21} & 1 - \hat{h} a_{22} & 0 \\ -\hat{h} b_1 & -\hat{h} b_2 & 1 \end{bmatrix} \begin{bmatrix} Y_1 \\ Y_2 \\ y_{n+1} \end{bmatrix} = \begin{bmatrix} y_n \\ y_n \\ y_n \end{bmatrix}.$$

The solution for y_{n+1} by Cramer's rule is $y_{n+1} = N/D$ where

$$N = \det\begin{bmatrix} 1-\hat{h}a_{11} & -\hat{h}a_{12} & y_n \\ -\hat{h}a_{21} & 1-\hat{h}a_{22} & y_n \\ -\hat{h}b_1 & -\hat{h}b_2 & y_n \end{bmatrix}, \quad D = \det\begin{bmatrix} 1-\hat{h}a_{11} & -\hat{h}a_{12} & 0 \\ -\hat{h}\alpha_{21} & 1-\hat{h}a_{22} & 0 \\ -\hat{h}b_1 & -\hat{h}b_2 & 1 \end{bmatrix}.$$

Subtracting the last row of N from each of the first two rows leaves N unaltered, whence

$$N = \det\begin{bmatrix} 1-\hat{h}a_{11}+\hat{h}b_1 & -\hat{h}a_{12}+\hat{h}b_2 & 0 \\ -\hat{h}a_{21}+\hat{h}b_1 & 1-\hat{h}a_{22}+\hat{h}b_2 & 0 \\ -\hat{h}b_1 & -\hat{h}b_2 & y_n \end{bmatrix} = y_n \det[I - \hat{h}A + \hat{h}eb^{\mathrm{T}}].$$

Clearly, $D = \det[I - \hat{h}A]$, and we obtain $y_{n+1} = R(\hat{h})y_n$, where

$$R(\hat{h}) = \frac{\det[I - \hat{h}A + \hat{h}eb^{\mathrm{T}}]}{\det[I - \hat{h}A]}. \tag{5.87}$$

It is clear that the above derivation can be extended to the case of general s, and that (5.87) holds in general. The alternative forms (5.86) and (5.87) for $R(\hat{h})$ are complementary; sometimes one is more convenient, sometimes the other.

Let us consider what form $R(\hat{h})$ takes when the method is explicit, that is, when A is a strictly lower triangular matrix. The matrix $I - \hat{h}A$ is then lower triangular, with all the elements of its main diagonal being unity. It follows that $\det(I - \hat{h}A) = 1$, and by (5.87) we see that for all explicit Runge–Kutta methods the stability function is a polynomial in $\hat{h}$. For implicit and semi-implicit methods, however, $\det(I - \hat{h}A)$ is no longer 1, but is itself a polynomial in $\hat{h}$, so that the stability function becomes a rational function of $\hat{h}$ (which is why we called the stability function R).

If $R(\hat{h})$ is a polynomial in $\hat{h}$, then there is no way in which the condition (5.84) for absolute stability can be satisfied when $|\hat{h}| \to \infty$, and it follows that all explicit methods have finite regions of absolute stability. (We found this to be the case for explicit linear multistep methods and for predictor–corrector methods when the corrector was applied a finite number of times—essentially an explicit process.) When $R(\hat{h})$ is a rational function of $\hat{h}$, however, it is at least possible that (5.84) can be satisfied when $|\hat{h}| \to \infty$, holding out the possibility that implicit and semi-implicit Runge–Kutta methods can have infinite regions of absolute stability.

As we have already mentioned, the role of implicit and semi-implicit Runge–Kutta methods is in attempting to solve stiff systems, so we shall delay until Chapter 6 any further discussion of the linear stability properties of such methods; in the remainder of this section we consider only explicit methods. Recall from §5.9 that explicit methods of the maximum attainable order for a given stage-number contain a number of free parameters. We have not as yet been able to find a means of using this freedom to advantage, and at first sight linear stability theory would appear to be a happy hunting ground; why not choose the free parameters to optimize the region of absolute stability? Let us try to do this in the case of the family of explicit 3-stage methods of order three. The Butcher array and the order conditions are

$$\begin{array}{c|ccc} c_1 & 0 & 0 & 0 \\ c_2 & c_2 & 0 & 0 \\ c_3 & c_3 - a_{32} & a_{32} & 0 \\ \hline & b_1 & b_2 & b_3 \end{array}$$

$$\begin{aligned} b_1 + b_2 + b_3 &= 1 \\ b_2 c_2 + b_3 c_3 &= \tfrac{1}{2} \\ b_2 c_2^2 + b_3 c_3^2 &= \tfrac{1}{3} \\ b_3 a_{32} c_2 &= \tfrac{1}{6}. \end{aligned} \tag{5.88}$$

We recall from §5.9 that there exist one two-parameter family and two one-parameter families of solutions of (5.88). The easiest way to compute $R(\hat{h})$ is to use (5.86) and define $d := (I - \hat{h}A)^{-1}e$, whence

$$\begin{bmatrix} 1 & 0 & 0 \\ -c_2\hat{h} & 1 & 0 \\ (a_{32} - c_3)\hat{h} & -a_{32}\hat{h} & 1 \end{bmatrix} \begin{bmatrix} d_1 \\ d_2 \\ d_3 \end{bmatrix} = \begin{bmatrix} 1 \\ 1 \\ 1 \end{bmatrix}.$$

Solving this triangular system gives

$$d_1 = 1, \quad d_2 = 1 + c_2\hat{h}, \qquad d_3 = 1 + c_3\hat{h} + c_2 a_{32}\hat{h}^2.$$

From (5.86), $R(\hat{h}) = 1 + \hat{h}b^{\mathrm{T}}d$, whence

$$R(\hat{h}) = 1 + (b_1 + b_2 + b_3)\,\hat{h} + (b_2c_2 + b_3c_3)\hat{h}^2 + b_3a_{32}c_2\hat{h}^3. \tag{5.89}$$

On applying the order conditions (5.88) (note that only three of them are needed) we find that

$$R(\hat{h}) = 1 + \hat{h} + \hat{h}^2/2 + \hat{h}^3/6$$

for *all* 3-stage methods of order 3. So much for our hopes of choosing free parameters to improve the linear stability properties!

The above result can be generalized as follows. Let the s-stage explicit Runge–Kutta method have order p. It follows from §5.7 that, under the localizing assumption that $y_n = y(x_n)$, the value y_{n+1} given by the method applied to the test equation (5.83) differs from the Taylor expansion of the exact solution $y(x_{n+1})$ of (5.83) by terms of order h^{p+1}. Now, it readily follows from repeatedly differentiating (5.83) that the expansion for $y(x_{n+1})$ is

$$y(x_{n+1}) = y(x_n) + h\lambda y(x_n) + \frac{1}{2!}h^2\lambda^2 y(x_n) + \cdots + \frac{1}{p!}h^p\lambda^p y(x_n) + O(h^{p+1})$$

whence we must have that

$$y_{n+1} = \left[1 + h\lambda + \frac{1}{2!}h^2\lambda^2 + \cdots + \frac{1}{p!}h^p\lambda^p\right]y_n + O(h^{p+1})$$

or

$$y_{n+1}/y_n = 1 + \hat{h} + \frac{1}{2!}\hat{h}^2 + \cdots + \frac{1}{p!}\hat{h}^p + O(\hat{h}^{p+1}). \tag{5.90}$$

On the other hand, it is clear from (5.87) that for an s-stage explicit method $R(\hat{h})$ will be a polynomial in $\hat{h}$ of degree at most s. This fact, together with (5.90), implies that if $s = p$ (and we know from §5.9 that this can only happen for $s = 1, 2, 3, 4$) then

$$R(\hat{h}) = y_{n+1}/y_n = 1 + \hat{h} + \frac{1}{2!}\hat{h}^2 + \cdots + \frac{1}{s!}\hat{h}^s. \tag{5.91}$$

Thus, for $s = 1, 2, 3, 4$, all s-stage explicit Runge–Kutta methods of order s have the same stability function, and therefore the same stability regions. These stability functions are given by (5.91) and we could compute the boundaries of the corresponding regions of absolute stability by adapting the boundary locus technique described in §3.8. An alternative approach is to write a program along the following lines. Let $R(\hat{h}) = x + iy$, scan the line $x =$ constant and set the point (x, y) if and only if $|R(\hat{h})| < 1$, then increment x and repeat the process. Note that this approach, which we shall call the *scanning technique*, is practicable only for one-step methods. It is more expensive on computing time than the boundary locus technique, but is easier to program—an advantageous exchange if one has a desk-top microcomputer! The regions of absolute stability for

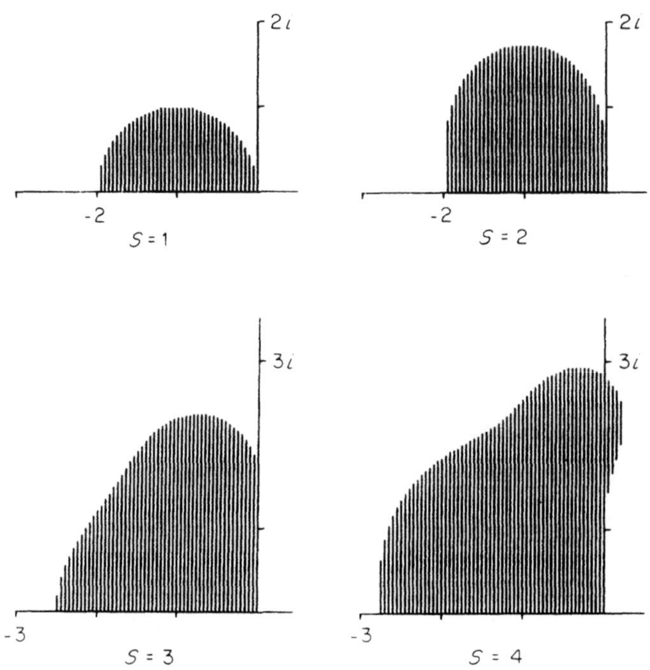

Figure 5.1 Regions of absolute stability for s-stage explicit Runge–Kutta methods of order s.

s-stage explicit Runge–Kutta methods of order s, $s = 1, 2, 3, 4$, obtained using the scanning technique, are displayed in Figure 5.1; the regions are symmetric about the real axis, and Figure 5.1 shows only the regions in the half-plane $\text{Im}(\hat{h}) > 0$. Note that for $s = 1$ (Euler's Rule), the boundary of the region is a circle. It is of interest to note that as the order increases, the stability regions become larger; the opposite happened for linear multistep methods.

If the s-stage method has order $p < s$ (and this will always be the case for $s > 4$) then, from (5.90) and (5.91), the stability function clearly takes the form

$$R(\hat{h}) = 1 + \hat{h} + \frac{1}{2!}\hat{h}^2 + \cdots + \frac{1}{p!}\hat{h}^p + \sum_{q=p+1}^{s} \gamma_q \hat{h}^q \tag{5.92}$$

where the coefficients γ_q are functions of the coefficients of the method. There is now scope for attempting to improve the stability region, but it has to be said that attempts to exploit this possibility do not produce any spectacular results. The stability function can be computed as a function of $\hat{h}$ from (5.86), but if we go at this task like a bull at a gate we land ourselves in a lot of needless work. First observe that we need not compute the inverse of $(I - \hat{h}A)$; all that is needed is the product $(I - \hat{h}A)^{-1}e$ which, as before, we define to be d. Then $(I - \hat{h}A)d = e$, a triangular system which can be readily solved for d. Moreover, if the method has order p, then we know that $R(\hat{h}) = 1 + \hat{h}b^{\mathrm{T}}d$ has the form given by (5.92), and we need only to find the terms in $\hat{h}^q$, $q = p, p+1, \dots, s-1$ in d. Let us illustrate by finding $R(\hat{h})$ for an s-stage (explicit) method of order $s - 1$. We need only find the term in $\hat{h}^{s-1}$ in d, which means that we need save only the highest power of $\hat{h}$ at each stage of the solution of the system $(I - \hat{h}A)d = e$. Using the row-sum condition and indicating terms involving lower powers of $\hat{h}$ by L.P., we obtain

$$\begin{aligned}
d_1 &= 1\\
d_2 &= 1 + \hat{h}a_{21}d_1 = c_2\hat{h} + \text{L.P.}\\
d_3 &= 1 + \hat{h}a_{31}d_1 + \hat{h}a_{32}d_2 = a_{32}c_2\hat{h}^2 + \text{L.P.}\\
&\vdots\\
d_s &= 1 + \hat{h}a_{s1}d_1 + \hat{h}a_{s2}d_2 + \cdots + \hat{h}a_{s,s-1}d_s\\
&= a_{s,s-1}a_{s-1,s-2}\cdots a_{32}c_2\hat{h}^{s-1} + \text{L.P.}
\end{aligned}$$

The term in $\hat{h}^s$ in $R(\hat{h}) = 1 + \hat{h}b^{\mathrm{T}}d$ is then $\gamma_s\hat{h}^s$ where

$$\gamma_s = b_s a_{s,s-1}a_{s-1,s-2}\cdots a_{32}c_2$$

which, in the notation of §5.7, is just the function $\psi([_{s-1}\tau]_{s-1})$. Note that γ_s is easily computed; it is just the product of the elements of the first sub-diagonal of A multiplied by b_s. We thus have

$$R(\hat{h}) = 1 + \hat{h} + \hat{h}^2/2! + \cdots + \hat{h}^{s-1}/(s-1)! + \hat{h}^s\psi([_{s-1}\tau]_{s-1}).$$

Note that if the method had order s, then the order conditions require that $\psi([_{s-1}\tau]_{s-1}) = 1/s!$, which merely corroborates (5.91). By a similar approach (now saving only terms in the two highest powers of $\hat{h}$) we can show that for an s-stage method of order $s-2$

$$R(\hat{h}) = 1 + \hat{h} + \hat{h}^2/2! + \cdots + \hat{h}^{s-2}/(s-2)! + \hat{h}^{s-1}\psi([_{s-2}\tau]_{s-2}) + \hat{h}^s\psi([_{s-1}\tau]_{s-1}).$$

The obvious extension to the general case $p < s$ holds.

Using this approach, we can easily investigate the linear stability properties of the embedded methods discussed in §5.10. In the following, the stage-number s and the order p refer to the method carrying the solution and not to the pair of embedded methods. The stability functions of the various methods are

Merson's method; $s = 5$, $p = 4$

$$R(\hat{h}) = 1 + \hat{h} + \hat{h}^2/2 + \hat{h}^3/6 + \hat{h}^4/24 + \hat{h}^5/144.$$

England's method; $s = 4$, $p = 4$

$$R(\hat{h}) = 1 + \hat{h} + \hat{h}^2/2 + \hat{h}^3/6 + \hat{h}^4/24.$$

RKF45; $s = 5$, $p = 4$

$$R(\hat{h}) = 1 + \hat{h} + \hat{h}^2/2 + \hat{h}^3/6 + \hat{h}^4/24 + \hat{h}^5/104.$$

DOPRI(5, 4); $s = 6$, $p = 5$

$$R(\hat{h}) = 1 + \hat{h} + \hat{h}^2/2 + \hat{h}^3/6 + \hat{h}^4/24 + \hat{h}^5/120 + \hat{h}^6/600.$$

Figure 5.2 shows the corresponding regions of absolute stability, computed by the scanning technique described earlier in this section. The region for England's method is, of course, identical with that given in Figure 5.1 for $s = 4$ and is included only for comparison.

The presence of a 'moon' in Figure 5.2(d) is a surprise! It is of no particular practical significance, but it does raise two points of interest. First, it demonstrates that there exists an explicit Runge–Kutta method whose region of absolute stability is a union of disjoint subsets. Secondly, it shows up an unexpected advantage that the scanning technique has over the boundary locus technique; there is no way that the latter would ever have discovered the 'moon'! The region for the fifth-order DOPRI(5, 4) method is perhaps a little smaller than we might have expected. An alternative (5, 4) pair with improved region of absolute stability is offered by Dormand and Prince (1980).

Exercises

5.12.1. Illustrate the effect of absolute stability by using the popular fourth-order explicit method (5.21) of §5.3 to compute numerical solutions of the problem $y' = Ay$, $y(0) = [1, 0, -1]^T$, where

$$A = \begin{bmatrix} -21 & 19 & -20 \\ 19 & -21 & 20 \\ 40 & -40 & -40 \end{bmatrix}$$

using two fixed steplengths, such that $\hat{h}$ is inside $\mathscr{R}_A$ for one of the values and outside it for the other.

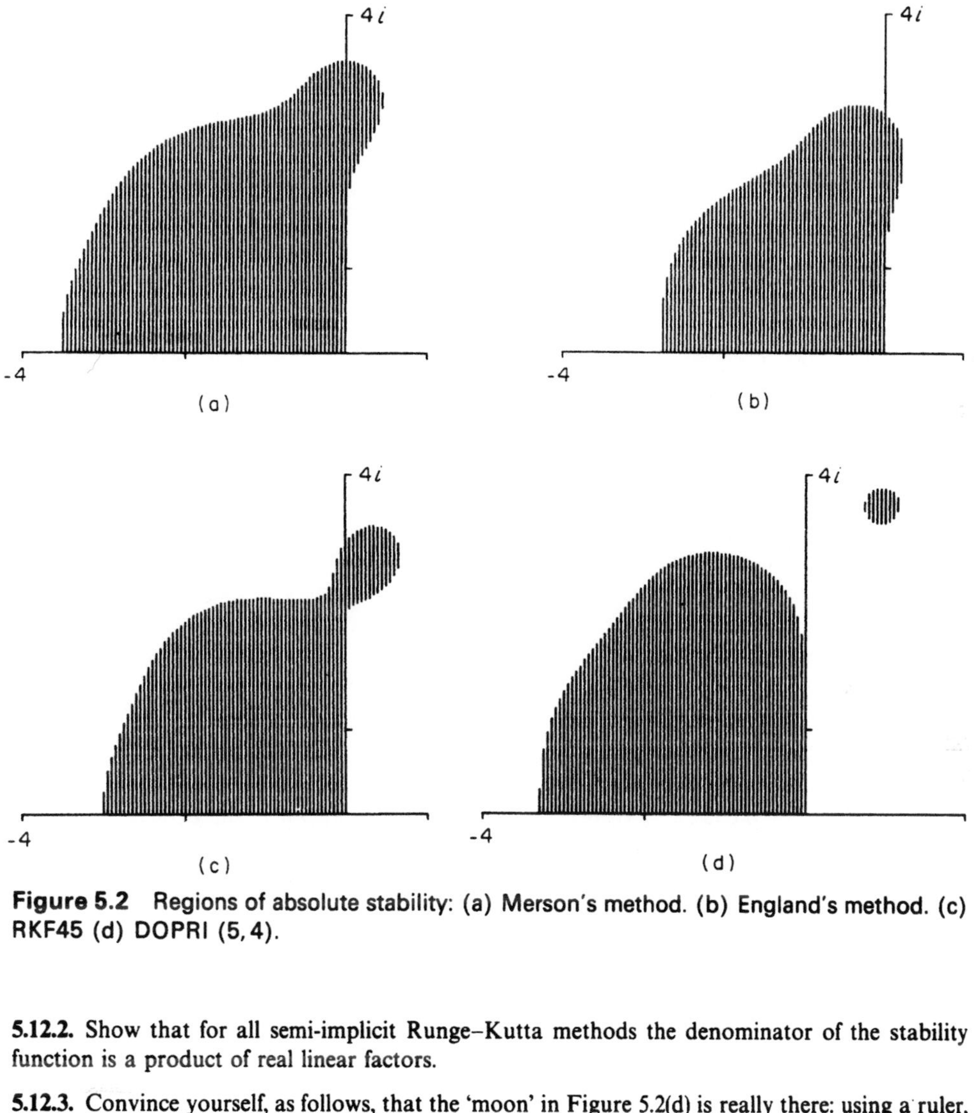

Figure 5.2 Regions of absolute stability: (a) Merson's method. (b) England's method. (c) RKF45 (d) DOPRI (5,4).

5.12.2. Show that for all semi-implicit Runge–Kutta methods the denominator of the stability function is a product of real linear factors.

5.12.3. Convince yourself, as follows, that the 'moon' in Figure 5.2(d) is really there: using a ruler, estimate from Figure 5.2(d) the coordinates of a point inside the 'moon', and show that $|R(\hat{h})| < 1$ at that point. In a similar way, convince yourself that the 'moon' is disjoint from the main region of absolute stability.

5.13 ORDER CONDITIONS; THE ALTERNATIVE APPROACH OF ALBRECHT

So far in this chapter, we have made much use of the well-established Butcher theory. A quite different approach to the problem of finding the conditions for a Runge–Kutta method to have given order has been proposed by Albrecht (1987). An interesting feature of this work is that it applies to Runge–Kutta methods the ideas we developed in

Chapter 3 for linear multistep methods. In this section we give only an outline of Albrecht's approach; full details can be found in Albrecht (1987).

Albrecht (1985) defined a general class of methods, the *A-methods*, by

$$Y_{n+1} = \mathscr{A} Y_n + h\Phi_f(x_n, Y_n, Y_{n+1}; h) \tag{5.93}$$

where Y_n, $\Phi_f \in \mathbb{R}^{\sigma m}$, $\mathscr{A}$ is a $\sigma m \times \sigma m$ matrix (where, as usual, m is the dimension of the differential system) and Φ_f satisfies a Lipschitz condition with respect to its second and third arguments; the subscript f indicates that the dependence of Φ on Y_n, Y_{n+1} is through the function $f(x, y)$ defining the differential system. (We have written $\mathscr{A}$ in place of the more natural A to avoid later confusion with the matrix A of coefficients of a Runge–Kutta method.) It is of interest to compare (5.93) with the general class of methods

$$\sum_{j=0}^{k} \alpha_j y_{n+j} = h\phi_f(y_{n+k}, y_{n+k-1}, \dots, y_n; h) \tag{5.94}$$

which we considered in Chapter 2 (see (2.4) of §2.2). At first sight, (5.93) might appear less general than (5.94), but this is not so; Y_n, which has dimension σm whilst y_n has dimension m, can itself be a function of $y_n, y_{n+1}, \dots$. The class of A-methods is indeed a very broad one and encompasses, in addition to the methods discussed in this book, other classes of methods such as cyclic methods; it also turns out to be an appropriate alternative framework for the study of the Nordsieck vector approach of §4.9; see Albrecht (1985).

Re-casting a Runge–Kutta method for an m-dimensional system as an A-method leads to a somewhat cumbersome notation which is considerably simplified in the case $m = 1$. One of the features of the Albrecht approach is that, unlike the Butcher theory, analysis in the case of the scalar initial value problem yields *all* of the order conditions for the systems case (cf. §5.8). Thus nothing essential is lost if, for the remainder of this section, we consider only the scalar initial value problem $y' = f(x, y)$, $y(a) = \eta$. The reader who wishes to see the full analysis for the m-dimensional case is referred to Albrecht (1987).

Consider the general s-stage Runge–Kutta method defined by the Butcher array

$$\begin{array}{c|c} c & A \\ \hline & b^{\mathrm{T}} \end{array} \quad = \quad \begin{array}{c|cc} c_1 & a_{11} & a_{12}\cdots a_{1s} \\ c_2 & a_{21} & a_{22}\cdots a_{2s} \\ \vdots & & \\ c_s & a_{s1} & a_{s2}\cdots a_{ss} \\ \hline & b_1 & b_2\cdots b_s \end{array}$$

Writing this method in the alternative form (5.6) (with a slight notational change) we have

$$y_{n+c_i} = y_n + h\sum_{j=1}^{s} a_{ij} f(x_n + c_j h, y_{n+c_j}) \qquad i = 1, 2, \dots, s \tag{5.95(i)}$$

$$y_{n+1} = y_n + h\sum_{i=1}^{s} b_i f(x_n + c_i h, y_{n+c_i}). \tag{5.95(ii)}$$

Define $Y_{n+1}, F(x_n, Y_{n+1}; h) \in \mathbb{R}^{s+1}$ by

$$
\begin{aligned}
Y_{n+1} &:= [y_{n+c_1}, y_{n+c_2}, \ldots, y_{n+c_s}, y_{n+1}]^{\mathrm{T}} \\
F(x_n, Y_{n+1}; h) &:= [f(x_n + c_1 h, y_{n+c_1}), f(x_n + c_2 h, y_{n+c_2}), \ldots, f(x_n + c_s h, y_{n+c_s}), f(x_{n+1}, y_{n+1})]^{\mathrm{T}}
\end{aligned}
\tag{5.96}
$$

and define $\mathscr{A}$ and $\mathscr{B}$ by

$$
\mathscr{A} = \begin{bmatrix} 0 & 0 & \cdots & 0 & 1 \\ 0 & 0 & \cdots & 0 & 1 \\ \vdots & & & & \\ 0 & 0 & \cdots & 0 & 1 \end{bmatrix} = \begin{bmatrix} 0_{ss} & e \\ 0_s & 1 \end{bmatrix} \tag{5.97(i)}
$$

$$
\mathscr{B} = \begin{bmatrix} a_{11} & a_{12} & \cdots & a_{1s} & 0 \\ a_{21} & a_{22} & \cdots & a_{2s} & 0 \\ \vdots & & & & \\ a_{s1} & a_{s2} & \cdots & a_{ss} & 0 \\ b_1 & b_2 & \cdots & b_s & 0 \end{bmatrix} = \begin{bmatrix} A & 0_s \\ b^{\mathrm{T}} & 0 \end{bmatrix} \tag{5.97(ii)}
$$

where 0_{ss} is the $s \times s$ null matrix, $0_s = [0, 0, \ldots, 0]^{\mathrm{T}} \in \mathbb{R}^s$ and $e = [1, 1, \ldots, 1]^{\mathrm{T}} \in \mathbb{R}^s$.

The method (5.95) can now be written in the form of an A-method,

$$
Y_{n+1} = \mathscr{A} Y_n + h \mathscr{B} F(x_n, Y_{n+1}; h). \tag{5.98}
$$

The essence of the Albrecht approach is to observe that each of the s internal stages (5.95(i)) and the final stage (5.95(ii)) of a Runge–Kutta method are linear, in the sense that a linear multistep method is linear. Nonlinearity arises only when we substitute from one stage into another. We can regard each of the $s + 1$ stages as being a generalized linear multistep method (on an unevenly distributed discrete point set) and associate with it a linear difference operator, in exactly the same way as we did for linear multistep methods in §3.2 (see (3.13)). Let $z(x)$ be a sufficiently differentiable arbitrary function and define $\mathscr{L}_i$, $i = 1, 2, \ldots, s$ and $\hat{\mathscr{L}}$ by

$$
\left.
\begin{aligned}
\mathscr{L}_i[z(x); h] &:= z(x + c_i h) - z(x) - h \sum_{j=1}^{s} a_{ij} z'(x + c_j h), \quad i = 1, 2, \ldots, s \\
\hat{\mathscr{L}}[z(x); h] &:= z(x + h) - z(x) - h \sum_{i=1}^{s} b_i z'(x + c_i h).
\end{aligned}
\right\} \tag{5.99}
$$

Proceeding as we did for linear multistep methods, we expand $z(x + c_i h)$, $z(x + h)$ and $z'(x + c_j h)$ about x and collect powers in h to obtain

$$
\left.
\begin{aligned}
\mathscr{L}_i[z(x); h] &= C_{i1} h z^{(1)}(x) + C_{i2} h^2 z^{(2)}(x) + \cdots, \qquad i = 1, 2, \ldots, s \\
\hat{\mathscr{L}}[z(x); h] &= \hat{C}_1 h z^{(1)}(x) + \hat{C}_2 h^2 z^{(2)}(x) + \cdots
\end{aligned}
\right\} \tag{5.100}
$$

where

$$
\left.
\begin{aligned}
C_{iq} &= \frac{c_i^q}{q!} - \frac{1}{(q-1)!} \sum_{j=1}^{s} a_{ij} c_j^{q-1}, \qquad i = 1, 2, \ldots, s \\
\hat{C}_q &= \frac{1}{q!} - \frac{1}{(q-1)!} \sum_{i=1}^{s} b_i c_i^{q-1}
\end{aligned}
\right\} q = 1, 2, \ldots \tag{5.101}
$$

which can be seen as a generalization of the corresponding result ((3.15) of §3.2) for linear multistep methods. We could define order in the same way as we did for a linear multistep method and say that the ith internal stage has order p_i if $C_{i1} = C_{i2} = \cdots = C_{ip_i} = 0$, $C_{i,p_i+1} \neq 0$, and the final stage has order p if $\hat{C}_1 = \hat{C}_2 = \cdots = \hat{C}_p = 0$, $\hat{C}_{p+1} \neq 0$. Note that the row-sum condition (5.3) implies that $C_{i1} = 0$, $i = 1, 2, \ldots, s$, so that each internal stage has order at least 1, that is, is consistent.

It will prove helpful to write (5.100) and (5.101) more compactly. To this end we introduce, for the purposes of this section only, the following notational convention. Let $u = [u_1, u_2, \ldots, u_s]^{\mathrm{T}}$ and $v = [v_1, v_2, \ldots, v_s]^{\mathrm{T}}$ be two vectors in $\mathbb{R}^s$. Then we denote by uv the vector in $\mathbb{R}^s$ obtained by componentwise multiplication; that is

$$uv := [u_1 v_1, u_2 v_2, \ldots, u_s v_s]^{\mathrm{T}}. \tag{5.102}$$

Note in particular that $u^\sigma = [u_1^\sigma, u_2^\sigma, \ldots, u_s^\sigma]^{\mathrm{T}}$. Define the vectors $\mathscr{L}[z(x); h]$, $C_q \in \mathbb{R}^s$ by

$$\left.\begin{aligned} \mathscr{L}[z(x); h] &:= [\mathscr{L}_1[z(x); h], \mathscr{L}_2[z(x); h], \ldots, \mathscr{L}_s[z(x); h]]^{\mathrm{T}} \\ C_q &:= [C_{1q}, C_{2q}, \ldots, C_{sq}]^{\mathrm{T}}. \end{aligned}\right\} \tag{5.103}$$

We can now write (5.100) and (5.101) in the form

$$\left.\begin{aligned} \mathscr{L}[z(x); h] &= C_1 h z^{(1)}(x) + C_2 h^2 z^{(2)}(x) + \cdots \\ \hat{\mathscr{L}}[z(x); h] &= \hat{C}_1 h z^{(1)}(x) + \hat{C}_2 h^2 z^{(2)}(x) + \cdots \end{aligned}\right\} \tag{5.104}$$

where, using the notation defined by (5.102),

$$C_q := \frac{1}{q!} c^q - \frac{1}{(q-1)!} A c^{q-1}, \qquad \hat{C}_q = \frac{1}{q!} - \frac{1}{(q-1)!} b^{\mathrm{T}} c^{q-1}, \qquad q = 1, 2, \ldots \tag{5.105}$$

and $C_1 = 0_s$.

If the Runge–Kutta method is to have order p, then clearly a necessary (but far from sufficient) condition is that the final stage should have order p. We thus obtain from the second of (5.105) the following necessary condition for the method (5.95) to have order p:

$$b^{\mathrm{T}} c^{q-1} = 1/q, \qquad q = 1, 2, \ldots, p. \tag{5.106}$$

(Note that (5.106) is equivalent to $\sum_i b_i c_i^{q-1} = 1/q$, the order condition which corresponds, for $q \geqslant 2$, to the tree $[\tau^{q-1}]$ in the Butcher theory.)

In order to obtain sufficient conditions for the method to have order p, we consider the global truncation error. This error being the difference between the exact and the numerical solution, it is natural to proceed by defining vectors that bear the same relation to the exact solution as the vectors Y_{n+1} and $F(x_n, Y_{n+1}; h)$ defined by (5.96) do to the numerical solution. Thus, we define $Y(x_{n+1})$, $F(x_n, Y(x_{n+1}); h) \in \mathbb{R}^{s+1}$ by

$$\begin{aligned} Y(x_{n+1}) &:= [y(x_n + c_1 h), y(x_n + c_2 h), \ldots, y(x_n + c_s h), y(x_n + h)]^{\mathrm{T}} \\ F(x_n, Y(x_{n+1}); h) &:= [f(x_n + c_1 h, y(x_n + c_1 h)), f(x_n + c_2 h, y(x_n + c_2 h)) \cdots, \\ &\quad \cdot f(x_n + c_s h, y(x_n + c_s h)), f(x_n + h, y(x_n + h)),]^{\mathrm{T}}. \end{aligned} \tag{5.107}$$

On putting $f(x, y(x)) = y'(x)$ and using (5.98), (5.99) and (5.103), we obtain

$$Y(x_{n+1}) - \mathscr{A}Y(x_n) - h\mathscr{B}F(x_n, Y(x_{n+1}); h) = [\mathscr{L}^{\mathrm{T}}[y(x_n); h], \hat{\mathscr{L}}[y(x_n); h]]^{\mathrm{T}} =: T_{n+1} \in \mathbb{R}^{s+1}. \tag{5.108}$$

The parallel with linear multistep theory continues. The vector T_{n+1}, the residual when the 'exact' vector $Y(x_{n+1})$ replaces the 'numerical' vector Y_{n+1} in (5.98), is the local truncation error of the A-method (5.98) equivalent to the Runge–Kutta method (5.95), a natural extension of the definition $T_{n+k} := \mathscr{L}[y(x_n; h]$ ((3.23) of §3.5). It is natural that T_{n+1} should be a vector of dimension $s+1$, since each of the s internal stages and the final stage have different local truncation errors and, in general, different orders.

Recall that for a method of order p the global truncation error is of order h^p. We could therefore define the A-method (5.98) to have order p if $\sup_n \| Y(x_{n+1}) - Y_{n+1} \| = 0(h^p)$ However, recalling the structure of Y_{n+1}, this would clearly be asking too much; all that is needed is that

$$\sup_n |y(x_{n+1}) - y_{n+1}| = 0(h^p). \tag{5.109}$$

Subtracting (5.98) from (5.108) gives

$$Y(x_{n+1}) - Y_{n+1} = \mathscr{A}[Y(x_n) - Y_n] + h\mathscr{B}[F(x_n, Y(x_{n+1}); h) - F(x_n, Y_{n+1}; h)] + T_{n+1}. \tag{5.110}$$

Let us simplify the notation by defining $Q_{n+1} := Y(x_{n+1}) - Y_{n+1}$ and $U_{n+1} := F(x_n, Y(x_{n+1}); h) - F(x_n, Y_{n+1}; h)$. We can partition these vectors and T_{n+1} as follows:

$$Q_{n+1} = \begin{bmatrix} q_{n+1} \\ \hat{q}_{n+1} \end{bmatrix}, \qquad U_{n+1} = \begin{bmatrix} u_{n+1} \\ \hat{u}_{n+1} \end{bmatrix}, \qquad T_{n+1} = \begin{bmatrix} t_{n+1} \\ \hat{t}_{n+1} \end{bmatrix}$$

where $q_{n+1}, u_{n+1}, t_{n+1} \in \mathbb{R}^s$ and $\hat{q}_{n+1}, \hat{u}_{n+1}, \hat{t}_{n+1} \in \mathbb{R}$. Note that $\hat{q}_{n+1} = y(x_{n+1}) - y_{n+1}$. Substituting in (5.110) and using the partitioned forms of the matrices $\mathscr{A}$ and $\mathscr{B}$ (see ((5.97)), we obtain

$$\left.\begin{aligned} q_{n+1} &= \hat{q}_n e + hAu_{n+1} + t_{n+1}, \\ \hat{q}_{n+1} &= \hat{q}_n + hb^{\mathrm{T}}u_{n+1} + \hat{t}_{n+1}. \end{aligned}\right\} \tag{5.111}$$

Now, if the conditions (5.106) are satisfied then $\hat{t}_{n+1} = 0(h^{p+1})$, by (5.105) and (5.104). If, in addition, $b^{\mathrm{T}}u_{n+1} = 0(h^p)$ then the second of (5.111) reads

$$\hat{q}_{n+1} = \hat{q}_n + 0(h^{p+1})$$

which is enough for (5.109) to be satisfied (just enough, when one recalls that $n \cdot 0(h^{p+1}) = 0(h^p)$, since $nh = x_n - a$). Thus we arrive at the following conditions for the Runge–Kutta method to have order p:

$$b^{\mathrm{T}}c^{q-1} = 1/q, \qquad q = 1, 2, \ldots, p, \qquad b^{\mathrm{T}}u_{n+1} = 0(h^p), \qquad n = 0, 1, \ldots, . \tag{5.112}$$

The analysis by which the second of these conditions is brought into an implementable form can be summarized as follows. From the definitions of U_{n+1} and Q_{n+1} it is possible to expand u_{n+1} in the form

$$u_{n+1} = G_1 q_{n+1} + G_2 q_{n+1}^2 + G_3 q_{n+1}^3 + \cdots \tag{5.113}$$

where G_i, $i = 1, 2, \ldots$ are diagonal matrices (and where the convention (5.102) pertains). Moreover we can assume that u_{n+1} and q_{n+1} can be expanded, in a neighbourhood of $h = 0$, as power series in h. It follows from the fact that $C_1 = 0_s$ that such series start with the term in h^2, giving

$$u_{n+1} = w_2(x_n)h^2 + w_3(x_n)h^3 + \cdots + w_{p-1}(x_n)h^{p-1} + 0(h^p)$$

$$q_{n+1} = r_2(x_n)h^2 + r_3(x_n)h^3 + \cdots + r_{p-1}(x_n)h^{p-1} + 0(h^p).$$

The s-dimensional vectors $w_i(x_n)$ and $r_i(x_n)$ have the form

$$w_i(x_n) = \sum_{j=1}^{M(i)} \alpha_{ij} e_{ij}(x_n), \quad r_i(x_n) = \sum_{j=1}^{N(i)} \beta_{ij} e_{ij}(x_n)$$

where α_{ij}, $\beta_{ij} \in \mathbb{R}^s$ and the $e_{ij}(x_n) \in \mathbb{R}$ are nothing other than the scalar forms of the elementary differentials of the Butcher theory, though this last fact is not made use of. It can be shown that the e_{ij} are all distinct (and it is this that makes it possible to get the full set of order conditions from an analysis restricted to the scalar problem) and it is enough to observe that the conditions (5.112) are equivalent to

$$\left.\begin{array}{lll} b^{\mathrm{T}} c^{q-1} = 1/q, & q = 1, 2, \ldots, p, & \\ b^{\mathrm{T}} \alpha_{ij} = 0, & j = 1, 2, \ldots, M(i), \quad i = 2, 3, \ldots, p-1 & \text{if } p \geqslant 3. \end{array}\right\} \tag{5.114}$$

Finally, (5.113) can be used to set up a recurrence relation between the $w_i(x_n)$ and $r_i(x_n)$, from which the following procedure for implementing the second of (5.114) can be derived.

Let w_i^* and r_i^*, $i = 1, 2, \ldots$ be defined by the following recurrence:

$$w_1^* = 0$$

$$r_i^* = C_i + A w_{i-1}^*, \qquad i \geqslant 2$$

$$w_i^* = \sum_{j=0}^{i-2} \frac{1}{j!} D^j \left[r_{i-j}^* + \sum_{\substack{\lambda,\mu \geqslant 2 \\ \lambda+\mu=i-j}} r_\lambda^* r_\mu^* + \sum_{\substack{\lambda,\mu,\nu \geqslant 2 \\ \lambda+\mu+\nu=i-j}} r_\lambda^* r_\mu^* r_\nu^* + \cdots \right], \qquad i \geqslant 2$$

where the notational convention (5.102) is assumed. The C_i are given by (5.105) and $D = \mathrm{diag}(c_1, c_2, \ldots, c_s)$, the coefficients c_i and the matrix A are defined by the Butcher array of the s-stage Runge–Kutta method. Each w_i^* is a sum of terms, and these individual terms are the α_{ij}, giving

$$w_i^* = \alpha_{i1} + \alpha_{i2} + \alpha_{i3} + \cdots.$$

This procedure enables us to identify the vectors α_{ij} and thus apply (5.114). Let us work out the first few w_i^*

$$
\begin{aligned}
w_1^* &= 0\\
r_2^* &= C_2\\
w_2^* &= r_2^* = C_2\\
r_3^* &= C_3 + AC_2\\
w_3^* &= r_3^* + Dr_2^* = C_3 + AC_2 + DC_2\\
r_4^* &= C_4 + AC_3 + A^2C_2 + ADC_2\\
w_4^* &= r_4^* + r_2^*r_2^* + Dr_3^* + D^2r_2^*/2\\
&= C_4 + AC_3 + A^2C_2 + ADC_2 + C_2^2 + DC_3 + DAC_2 + D^2C_2/2\\
&\vdots
\end{aligned}
$$

We can now apply the conditions (5.114), recalling from (5.105) that

$$C_q := \frac{1}{q!}c^q - \frac{1}{(q-1)!}Ac^{q-1}.$$

Order 1 $$b^{\mathsf{T}}e = 1 \Leftrightarrow \sum_i b_i = 1$$

Order 2 $$b^{\mathsf{T}}c = \tfrac{1}{2} \Leftrightarrow \sum_i b_i c_i = \tfrac{1}{2}$$

Order 3 $$b^{\mathsf{T}}c^2 = \tfrac{1}{3} \Leftrightarrow \sum_i b_i c_i^2 = \tfrac{1}{3}$$

$$0 = b^{\mathsf{T}}C_2 = b^{\mathsf{T}}(c^2/2 - Ac) \Leftrightarrow b^{\mathsf{T}}Ac = b^{\mathsf{T}}c^2/2 = \tfrac{1}{6}$$

$$\Leftrightarrow \sum_{ij} b_i a_{ij} c_j = \tfrac{1}{6}$$

Order 4 $$b^{\mathsf{T}}c^3 = \tfrac{1}{4} \Leftrightarrow \sum_i b_i c_i^3 = \tfrac{1}{4}$$

$$0 = b^{\mathsf{T}}C_3 = b^{\mathsf{T}}(c^3/6 - Ac^2/2) \Leftrightarrow b^{\mathsf{T}}Ac^2 = b^{\mathsf{T}}c^3/3 = \tfrac{1}{12}$$

$$\Leftrightarrow \sum_{ij} b_i a_{ij} c_j^2 = \tfrac{1}{12}$$

$$0 = b^{\mathsf{T}}AC_2 = b^{\mathsf{T}}A(c^2/2 - Ac) \Leftrightarrow b^{\mathsf{T}}A^2c = b^{\mathsf{T}}Ac^2/2 = \tfrac{1}{24}$$

$$\Leftrightarrow \sum_{ijk} b_i a_{ij} a_{jk} c_k = \tfrac{1}{24}$$

$$0 = b^{\mathsf{T}}DC_2 = b^{\mathsf{T}}D(c^2/2 - Ac) \Leftrightarrow b^{\mathsf{T}}DAc = b^{\mathsf{T}}Dc^2/2 = b^{\mathsf{T}}c^3/2 = \tfrac{1}{8}$$

$$\Leftrightarrow \sum_{ij} b_i c_i a_{ij} c_j = \tfrac{1}{8}.$$

These are precisely the conditions for order 4 derived earlier from the Butcher theory; see, for example, (5.53) of §5.9. Note that w_4^* is the sum of eight terms; these, together with the first of (5.114) with $q = 5$, give rise to the nine additional conditions needed to attain order 5, thus corroborating that the above approach, although based on the scalar problem, does indeed generate *all* of the order conditions for a system.

As we have seen above, the second of the conditions (5.114) gives rise to a series of orthogonality conditions. These can be employed to provide alternative proofs of the theoretical results we obtained in §5.9, specifically the non-existence of an explicit 5-stage method of order 5 and the fact that $c_4 = 1$ for all 4-stage explicit methods of order four; see Albrecht (1989).

Exercise

5.13.1. Use the approach described in the above section to derive the additional nine conditions for a Runge–Kutta method to have order 5.

6 Stiffness: Linear Stability Theory

6.1 A PRELIMINARY NUMERICAL EXPERIMENT

Let us consider two initial value problems each involving a linear constant coefficient inhomogeneous system of dimension 2.

Problem 1

$$\begin{bmatrix} {}^1y' \\ {}^2y' \end{bmatrix} = \begin{bmatrix} -2 & 1 \\ 1 & -2 \end{bmatrix} \begin{bmatrix} {}^1y \\ {}^2y \end{bmatrix} + \begin{bmatrix} 2\sin x \\ 2(\cos x - \sin x) \end{bmatrix}, \quad \begin{bmatrix} {}^1y(0) \\ {}^2y(0) \end{bmatrix} = \begin{bmatrix} 2 \\ 3 \end{bmatrix}. \tag{6.1}$$

Problem 2

$$\begin{bmatrix} {}^1y' \\ {}^2y' \end{bmatrix} = \begin{bmatrix} -2 & 1 \\ 998 & -999 \end{bmatrix} \begin{bmatrix} {}^1y \\ {}^2y \end{bmatrix} + \begin{bmatrix} 2\sin x \\ 999(\cos x - \sin x) \end{bmatrix}, \quad \begin{bmatrix} {}^1y(0) \\ {}^2y(0) \end{bmatrix} = \begin{bmatrix} 2 \\ 3 \end{bmatrix}. \tag{6.2}$$

Both problems have identically the same exact solution, given by

$$\begin{bmatrix} {}^1y(x) \\ {}^2y(x) \end{bmatrix} = 2\exp(-x)\begin{bmatrix} 1 \\ 1 \end{bmatrix} + \begin{bmatrix} \sin x \\ \cos x \end{bmatrix}. \tag{6.3}$$

The graph of this common solution in the interval $[0, 10]$ of x is shown in Figure 6.1(a). The object of this numerical experiment is to attempt to generate numerically graphical solutions of Problems 1 and 2 which are acceptable representations of Figure 6.1(a).

For our first attempt we use a simple code based on the fourth-order explicit embedded Runge–Kutta method RKF45, described in §5.10. The code controls steplength in the following way. Call the local error estimate produced by the method EST. The user provides a tolerance TOL (and an initial steplength h_0) and if EST > TOL the steplength is successively halved until EST $\leqslant$ TOL. If at any step EST < TOL/2^5 (recall that the local error is $O(h^5)$), the steplength is successively doubled until TOL/$2^5 \leqslant$ EST $\leqslant$ TOL. Since a graphical solution is limited by the resolution of the computer graphics, there is little point in asking for high accuracy and we set TOL at the modest value of 0.01; we choose $h_0 = 0.1$ (for no particularly good reason). The resulting graphical solution when the code is so applied to Problem 1 is shown in Figure 6.1(b); the numerically generated points are marked + and N is the number of steps taken to complete the integration from $x = 0$ to $x = 10$. The code accepts the initial steplength of 0.1 until $x = 2.0$, when it doubles the steplength to 0.2; it takes 60 steps to complete the solution, and clearly gives an adequate respresentation of Figure 6.1(a).

We now apply the same code, with the same values for h_0 and TOL to Problem 2. The results, shown in Figure 6.1(c), are very different from those for Problem 1. Before computing the first step, the code halves the steplength three times to 0.0 125 and before

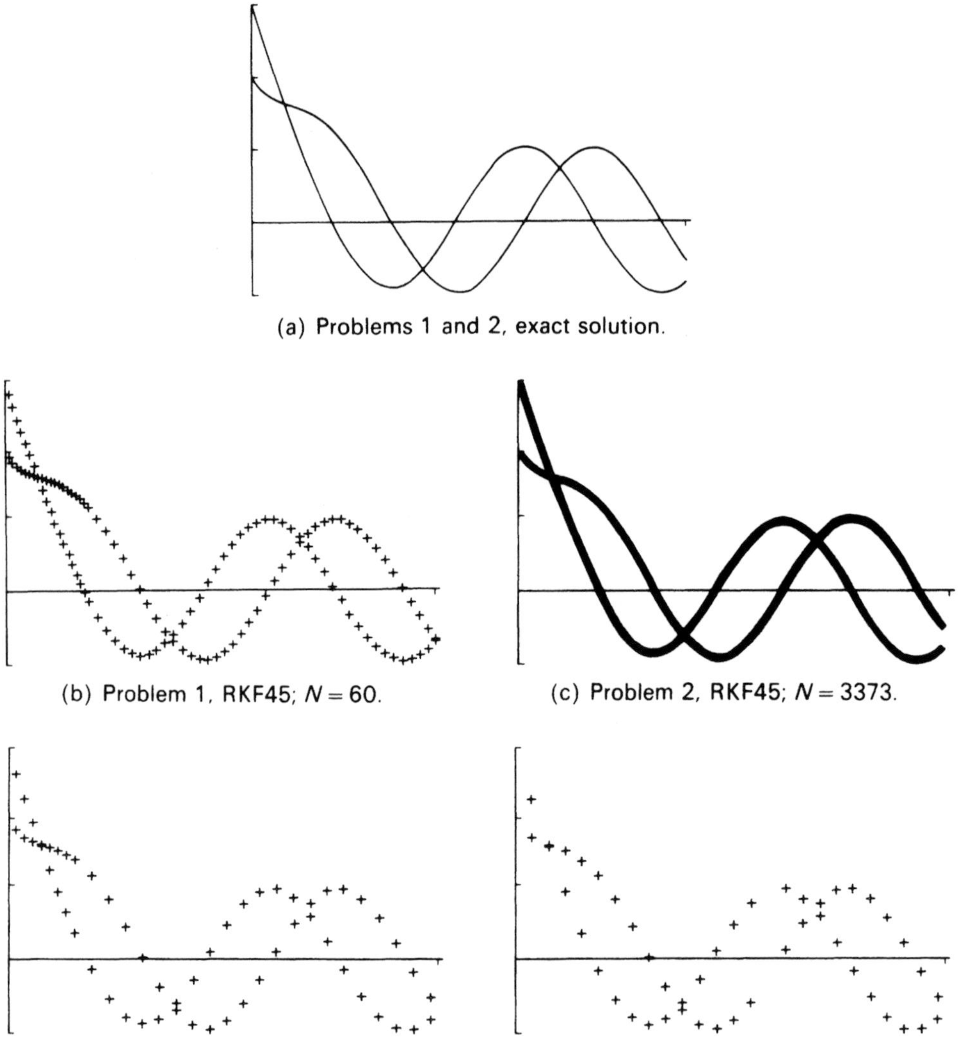

(a) Problems 1 and 2, exact solution.

(b) Problem 1, RKF45; $N = 60$.

(c) Problem 2, RKF45; $N = 3373$.

(d) Problem 1, 2-stage Gauss; $N = 29$.

(e) Problem 2, 2-stage Gauss; $N = 24$.

Figure 6.1

taking the second step it halves it twice more to 0.003 125; thereafter the steplength oscillates between that value and 0.001 5625. The code takes a total of 3373 steps to complete the solution, resulting in the saturation of crosses shown in Figure 6.1(c). The result is an accurate numerical solution—a much more accurate one than we had intended, and indeed a much more expensive one; recalling that RKF45 is a six-stage method, the code has made a total of over 20 000 function evaluations. It is impossible to get a cheap solution of this problem with an explicit Runge–Kutta method (or, indeed, with any explicit method). If we dispense with automatic step-control and attempt to use an explicit method with a fixed steplength of around 0.1, then the numbers produced are nonsense, and very soon overflow.

Let us now try a different code, based on the 2-step Gauss method described in §5.11. It has the same order (four) as RKF45, but is implicit. The step-control mechanism is the same as in the first code, the local error estimate now being provided by Richardson extrapolation (see §5.10). Applying this code to Problems 1 and 2 (with the same values for TOL and h_0 as we used previously) gives the results shown in Figures 6.1(d) and 6.1(e) respectively. For Problem 1, the code initially doubles the steplength to the value 0.2 which is retained until $x = 2.0$, when it is doubled again to 0.4; a total of 29 steps are needed to complete the solution. For Problem 2, the code initially doubles the steplength twice to the value 0.4, which is retained for the remainder of the computation, except for a solitary step of length 0.8 at $x = 6.4$; the code takes a total of 24 steps to complete the solution.

So, we have two similar problems which behave very differently when tackled numerically. The explicit RKF45 method solves Problem 1 easily, with a fairly large average steplength, but can solve Problem 2 only at the cost of cutting the steplength down to an unacceptably small level. In contrast, the implicit 2-stage Gauss method solves both problems with moderately large average steplength (and actually appears to find Problem 2 marginally the easier). The phenomenon being exhibited here is known as *stiffness*; Problem 2 is *stiff*, while Problem 1 is *non-stiff*. Clearly the phenomenon cannot be a function of the exact solution, since this is the same for both problems, and must be a property of the differential system itself. It is thus more appropriate to talk of *stiff systems* rather than of *stiff problems*. This thought suggests that we consider, not the particular solutions of Problems 1 and 2 satisfying the given initial conditions, but the general solutions of the systems, which in turn requires us (see §1.6) to look at the eigenvalues of the coefficient matrix of the systems. (Henceforth we shall refer to such eigenvalues simply as eigenvalues of the system, or of the problem.) For Problem 1, the eigenvalues are -1 and -3, and the general solution is

$$\begin{bmatrix} {}^1y \\ {}^2y \end{bmatrix} = \kappa_1 \exp(-x)\begin{bmatrix} 1 \\ 1 \end{bmatrix} + \kappa_2 \exp(-3x)\begin{bmatrix} 1 \\ -1 \end{bmatrix} + \begin{bmatrix} \sin x \\ \cos x \end{bmatrix} \tag{6.4}$$

where κ_1 and κ_2 are arbitrary constants. For Problem 2, the eigenvalues are -1 and -1000, and the general solution is

$$\begin{bmatrix} {}^1y \\ {}^2y \end{bmatrix} = \kappa_1 \exp(-x)\begin{bmatrix} 1 \\ 1 \end{bmatrix} + \kappa_2 \exp(-1000x)\begin{bmatrix} 1 \\ -998 \end{bmatrix} + \begin{bmatrix} \sin x \\ \cos x \end{bmatrix}. \tag{6.5}$$

An explanation of the results of our numerical experiment can be made in terms of linear stability theory. Since for both problems the eigenvalues are real, we need consider only intervals of absolute stability. From Figure 5.2 of §5.12 we see that the interval of absolute stability of RKF45 is approximately $(-3, 0)$ so that for Problem 1 absolute stability is achieved if $-3h\in(-3, 0)$, or $h < 1.0$; even the modest tolerance of 0.01 requires h to be less than that, so that it is the demands of accuracy and not of linear stability that constrain the steplength. For Problem 2, however, absolute stability is achieved only if $-1000h\in(-3, 0)$, or $h < 0.003$, and the demands of linear stability rather than those of accuracy constrain the steplength. (In the next section we shall modify this last remark somewhat.) Earlier in this book we remarked that automatic codes for the numerical solution of initial value problems do not normally test for absolute stability, but rely on their error-monitoring features to detect the increase in error that occurs if the steplength corresponds to operating outside a region of absolute stability, and consequently reduce the steplength until absolute stability is obtained. Our numerical experiment shows the simple-minded code based on RKF45 doing just that; it cuts down the steplength so that it oscillates between 0.003 125 and 0.0015 625, consistent with the linear stability requirement that $h < 0.003$. In contrast, the 2-stage implicit Gauss method has, as we shall see later, a region of absolute stability which includes the whole of the left half-plane, so that the linear stability requirement $h\lambda \in \mathscr{R}_A$ holds for all positive h when the eigenvalue λ has negative real part; thus for both problems linear stability makes no demands on the steplength. The reason that the 2-stage Gauss code solves Problem 1 in less steps than does the RKF45 code is simply that the Gauss method has a smaller local truncation error (for this particular problem).

6.2 THE NATURE OF STIFFNESS

In this section we consider (at some length!) various aspects of the phenomenon of stiffness. 'Phenomenon' is probably a more appropriate word than 'property', since the latter rather implies that stiffness can be defined in precise mathematical terms; it turns out not to be possible to do this in a satisfactory manner, even for the restricted class of linear constant coefficient systems. We shall also examine critically various qualitative statements that can be (and mostly have been) made in an attempt to encapsulate the notion of stiffness, and select the most satisfactory of these as a 'definition' of stiffness. We start by generalizing the linear stability analysis of the preceding section.

Consider the linear constant coefficient inhomogeneous system

$$y' = Ay + \varphi(x), \tag{6.6}$$

where $y, \varphi \in \mathbb{R}^m$ and A is a constant $m \times m$ matrix with eigenvalues $\lambda_t \in \mathbb{C}$, $t = 1, 2, \ldots, m$ (assumed distinct) and corresponding eigenvectors $c_t \in \mathbb{C}^m, t = 1, 2, \ldots, m$. The general solution of (6.6) (see §1.6) takes the form

$$y(x) = \sum_{t=1}^{m} \kappa_t \exp(\lambda_t x) c_t + \psi(x), \tag{6.7}$$

where the κ_t are arbitrary constants and $\psi(x)$ is a particular integral. Now let us suppose that

$$\operatorname{Re} \lambda_t < 0, \qquad t = 1, 2, \ldots, m, \tag{6.8}$$

which implies that each of the terms $\exp(\lambda_t x) c_t \to 0$ as $x \to \infty$, so that the solution $y(x)$ approaches $\psi(x)$ asymptotically as $x \to \infty$; the term $\exp(\lambda_t x) c_t$ will decay monotonically if λ_t is real and sinusoidally if λ_t is complex. Interpreting x to be time (as it often is in physical problems) it is appropriate to call $\sum_{t=1}^{m} \kappa_t \exp(\lambda_t x) c_t$ the *transient solution* and $\psi(x)$ the *steady-state solution*. If $|\operatorname{Re} \lambda_t|$ is large then the corresponding term $\kappa_t \exp(\lambda_t x) c_t$ will decay quickly as x increases and is thus called a *fast transient*; if $|\operatorname{Re} \lambda_t|$ is small the corresponding term $\kappa_t \exp(\lambda_t x) c_t$ decays slowly and is called a *slow transient*. Let $\bar{\lambda}, \underline{\lambda} \in \{\lambda_t, t = 1, 2, \ldots, m\}$ be defined by

$$|\operatorname{Re} \bar{\lambda}| \geqslant |\operatorname{Re} \lambda_t| \geqslant |\operatorname{Re} \underline{\lambda}|, \qquad t = 1, 2, \ldots, m$$

so that $\kappa_t \exp(\bar{\lambda}x)c_t$ is the fastest transient and $\kappa_t \exp(\underline{\lambda}x)c_t$ the slowest. If we solve numerically an initial value problem involving the system (6.6) and our aim is to reach the steady-state solution, then we must keep integrating until the slowest transient is negligible. The smaller $|\mathrm{Re}\,\underline{\lambda}|$ is, the longer we must keep integrating. If, however, the method we are using has a finite region of absolute stability (as did the RKF45 code of the preceding section), we must ensure that the steplength h is sufficiently small for $h\lambda_t \in \mathscr{R}_A, t = 1, 2, \ldots, m$ to hold. Clearly a large value of $|\mathrm{Re}\,\bar{\lambda}|$ implies a small steplength. We therefore get into a difficult situation if $|\mathrm{Re}\,\bar{\lambda}|$ is very large and $|\mathrm{Re}\,\underline{\lambda}|$ is very small; we are forced to integrate for a very long time with an excessively small steplength. This is precisely what happened when we attempted to integrate Problem 2 of the previous section by the RKF45 code; it was necessary to integrate to around $x = 10$ if we wanted to see the form of the steady-state solution, but the eigenvalue -1000 forced the steplength to be excessively small. Note the irony that the eigenvalue $\bar{\lambda}$ which causes the stability difficulties has a very short-term effect on the exact solution (6.7), and indeed none at all in the case of Problem 2, where the initial conditions happened to wipe out the fast transient altogether.

It would therefore appear that stiffness arises when $|\mathrm{Re}\,\bar{\lambda}|$ is very large and $|\mathrm{Re}\,\underline{\lambda}|$ very small; it seems natural to take the ratio $|\mathrm{Re}\,\bar{\lambda}|/|\mathrm{Re}\,\underline{\lambda}|$, the *stiffness ratio*, as a measure of the stiffness of the system. We are now in a position to make the first of the statements which are candidates for adoption as a definition of stiffness:

Statement 1 A linear constant coefficient system is stiff if all of its eigenvalues have negative real part and the stiffness ratio is large.

In Lambert (1973) (and elsewhere) this statement is adopted as a definition of stiffness. However, as we hope to show by the following examples, such a definition is not entirely satisfactory—nor indeed is the definition of stiffness ratio. Let us consider the following three systems, quoted together with their general solutions. The first is the system involved in Problem 2 of the preceding section; in order to avoid confusion, we shall call it System 2, and the remaining examples Systems 3 and 4.

System 2

$$\left.\begin{aligned}
\begin{bmatrix} {}^1y' \\ {}^2y' \end{bmatrix} &= \begin{bmatrix} -2 & 1 \\ 998 & -999 \end{bmatrix}\begin{bmatrix} {}^1y \\ {}^2y \end{bmatrix} + \begin{bmatrix} 2\sin x \\ 999(\cos x - \sin x) \end{bmatrix} \\
\begin{bmatrix} {}^1y \\ {}^2y \end{bmatrix} &= \kappa_1 \exp(-x)\begin{bmatrix} 1 \\ 1 \end{bmatrix} + \kappa_2 \exp(-1000x)\begin{bmatrix} 1 \\ -998 \end{bmatrix} + \begin{bmatrix} \sin x \\ \cos x \end{bmatrix}.
\end{aligned}\right\} \tag{6.9}$$

System 3

$$\left.\begin{aligned}
\begin{bmatrix} {}^1y' \\ {}^2y' \end{bmatrix} &= \begin{bmatrix} -2 & 1 \\ -1.999 & 0.999 \end{bmatrix}\begin{bmatrix} {}^1y \\ {}^2y \end{bmatrix} + \begin{bmatrix} 2\sin x \\ 0.999(\sin x - \cos x) \end{bmatrix} \\
\begin{bmatrix} {}^1y \\ {}^2y \end{bmatrix} &= \kappa_1 \exp(-x)\begin{bmatrix} 1 \\ 1 \end{bmatrix} + \kappa_2 \exp(-0.001x)\begin{bmatrix} 1 \\ 1.999 \end{bmatrix} + \begin{bmatrix} \sin x \\ \cos x \end{bmatrix}.
\end{aligned}\right\} \tag{6.10}$$

System 4

$$\begin{bmatrix} {}^1y' \\ {}^2y' \end{bmatrix} = \begin{bmatrix} -0.002 & 0.001 \\ 0.998 & -0.999 \end{bmatrix} \begin{bmatrix} {}^1y \\ {}^2y \end{bmatrix} + \begin{bmatrix} 0.002\sin(0.001x) \\ 0.999[\cos(0.001x) - \sin(0.001x)] \end{bmatrix} \tag{6.11}$$

$$\begin{bmatrix} {}^1y \\ {}^2y \end{bmatrix} = \kappa_1 \exp(-x) \begin{bmatrix} 1 \\ -998 \end{bmatrix} + \kappa_2 \exp(-0.001x) \begin{bmatrix} 1 \\ 1 \end{bmatrix} + \begin{bmatrix} \sin(0.001x) \\ \cos(0.001x) \end{bmatrix}.$$

For System 2, $\bar{\lambda} = -1000$ and $\underline{\lambda} = -1$, while for Systems 3 and 4, $\bar{\lambda} = -1$ and $\underline{\lambda} = -0.001$; all three systems thus have a stiffness ratio of 1000.

We have already seen in the preceding section that System 2 cannot be solved by the RKF45 code unless the steplength is excessively small, and that this happens even in the case when $\kappa_2 = 0$ and the fast transient is not present in the exact solution. The same happens even if we choose initial conditions such that $\kappa_1 = \kappa_2 = 0$; in that case the RKF45 code is unable to integrate even the very smooth solution $y(x) = [\sin x, \cos x]^T$ with a steplength greater than roughly 0.003. This system is exhibiting genuine stiffness, and the difficulties arise whatever the choice of initial values. Note that if the initial conditions are such that the fast transient is present in the exact solution, then we would expect to have to use a very small steplength in able to follow that fast transient; the effect of stiffness is that we have to continue using that small steplength long after the fast transient has died.

Now let us consider System 3. If we impose initial conditions such as $y(0) = [2, 3.999]^T$, which corresponds to $\kappa_1 = \kappa_2 = 1$, and apply the RKF45 code (with the previously used values for TOL and h_0), then the steplength settles down to 0.4. This is broadly what we would expect, since the modulus of neither of the eigenvalues is sufficiently large to impose a stability restriction on the steplength. Of course, if we wish to reach the steady-state solution, we must continue the integration until the term in $\exp(-0.001x)$ is negligible—say equal to TOL; this implies integrating from $x = 0$ to $x = 1010$ at a cost of around 2500 steps. The total computational effort is comparable with that for Problem 2, thus supporting the view that stiffness ratio is a consistent measure of stiffness. However, if we change the initial conditions to $y(0) = [2, 3]^T$, for which $\kappa_1 = 2, \kappa_2 = 0$, the slow transient in (6.10) is annihilated and there is no need to integrate a long way to reach the steady-state solution. There is no stability restriction on the steplength, so there arise none of the difficulties we associate with solving a stiff system by a method with a finite region of absolute stability. The RKF45 code with the previously used values of TOL and h_0 integrates from $x = 0$ to $x = 10$ (now well into the steady-state phase) at a cost of only 25 steps. The problem is effectively not stiff at all! Thus, if Statement 1 were adopted as a definition, the stiffness of a system would depend on the initial conditions imposed by a particular problem—a state of affairs that would not be acceptable for a linear constant coefficient system.

The inadequacy of the concept of stiffness ratio can perhaps best be seen by considering what happens in the limiting case when the eigenvalue with smallest modulus real part is in fact zero. The contribution of that eigenvalue to the exact solution is then a constant. If the moduli of the real parts of the remaining eigenvalues are not particularly large, the system exhibits no signs of stiffness, yet the stiffness ratio is now infinite!

The 'stiffness' exhibited by System 3, which is caused solely by the presence of a very slow transient, is not the same sort of phenomenon as the stiffness exhibited by System 2.

It is debatable whether such systems should be called stiff at all, and we shall call them *pseudo-stiff*. In contrast with System 2, for all choices of initial conditions such systems can be integrated by methods with finite regions of absolute stability using the sort of steplength that the exact solution would suggest was reasonable. Note, however, that the associated homogeneous system $y' = Ay$, where A is the coefficient matrix in System 3, *is* genuinely stiff, indeed precisely as stiff as are the homogeneous systems associated with Systems 2 and 4. It is the presence of a steady-state solution which varies at a rate comparable with that of the *fastest* transient that motivates us to regard System 3 as pseudo-stiff and not genuinely stiff. This remark forces us to abandon the notion that stiffness of a linear constant coefficient system can be described solely in terms of the spectrum of the matrix A; it is essential to consider the full system $y' = Ay + \varphi(x)$. Thus Statement 1 fails, on another count, to be acceptable as a definition of stiffness.

These conclusions might suggest that true stiffness requires that $|\mathrm{Re}(\bar{\lambda})|$ must be large in some absolute sense, say $|\mathrm{Re}(\bar{\lambda})| \gg 1$. Consideration of System 4, for which $|\mathrm{Re}\,\bar{\lambda}| = 1$, soon dispels that notion. Here, the steady-state solution varies extremely slowly with x and we might hope to be able to integrate in the steady-state region with a very large steplength, of the order of 100. The presence of the eigenvalue -1 precludes this possibility, even in the case when $\kappa_1 = \kappa_2$, where the RKF45 code chooses a steplength of 0.4; it takes a very long time even to see the form of the steady-state solution, which is produced with much greater accuracy than we want. System 4 is stiff in exactly the same sense as is System 2. (In fact if in System 2 we make the transformation $x = 0.001\xi$ and then write x for ξ, we obtain System 4.)

A statement which is frequently made in an attempt to tie down the concept of stiffness is

Statement 2 Stiffness occurs when stability requirements, rather than those of accuracy, constrain the steplength.

One can certainly observe this happening when Systems 2 and 4 are solved by the RKF45 code. It does not happen for System 3, and the statement properly separates out the genuinely stiff system from the pseudo-stiff. However, the statement is not entirely accurate. Stability is concerned with the *accumulation* of error, yet we recall from the preceding section that when Problem 2 was solved by the RKF45 the initial steplength of 0.1 was *immediately* cut down to 0.0125, whereas for the non-stiff Problem 1, the initial steplength was accepted. Thus the local error at the very first step was substantially higher for the stiff problem than for the non-stiff. It is not possible to separate stability from accuracy in quite as clear-cut a manner as the statement implies.

Another such statement that has been made is

Statement 3 Stiffness occurs when some components of the solution decay much more rapidly than others

The difficulty with this statement is that it does not differentiate between the genuinely stiff Systems 2 and 4 and the pseudo-stiff System 3. Statements based on comparing the rate of change of the fastest transient with that of the steady-state solution come up against another difficulty. Consider a homogeneous system $y' = Ay$; the steady-state solution is zero, and *any* transient varies infinitely rapidly compared with the rate of change of the steady-state solution, so that all homogeneous systems become stiff!

Perhaps the best statement we can come up with—and we shall adopt it as our 'definition'—is one which merely relates what we observe happening in practice:

Definition If a numerical method with a finite region of absolute stability, applied to a system with any initial conditions, is forced to use in a certain interval of integration a steplength which is excessively small in relation to the smoothness of the exact solution in that interval, then the system is said to be **stiff** *in that interval.*

This does differentiate between the genuinely stiff and the pseudo-stiff system, and moreover introduces the idea that stiffness may vary over the total interval of integration. What we mean by an 'excessively small' steplength depends on what stage in the integration we have reached. In the phase when the fastest transients are still alive, the exact solution is not at all smooth; a very small steplength is natural and should not be seen as 'excessively small'. When the fast transients are dead, the exact solution is smooth, and we can properly regard the same very small steplength as being 'excessively small'. Note that if we adopt this definition, then we can relax the requirement (6.8) that all the eigenvalues of A have negative real part. The definition still makes sense if A has some eigenvalues with positive real parts which are small relative to $|\mathrm{Re}\,\bar{\lambda}|$.

Let us now consider another aspect of the phenomenon of stiffness, first pointed out by Curtiss and Hirschfolder (1952). The non-stiff and stiff Problems 1 and 2 of the preceding section were chosen to have the same exact solution. However, if we look not just at the exact solution but at the neighbouring solution curves as well, we see very different pictures. For the non-stiff Problem 1, these neighbouring curves approach the exact solution curve at a moderately slow rate, whereas for the stiff Problem 2, they approach the solution curve so steeply that it is necessary to use a large magnifying glass to see what is happening. We blow up Figure 6.1(a) and look at a very small interval, $[4.684, 4.696]$ of x, near where ${}^2y(x)$, the second component of the common exact solution (6.3) of Problems 1 and 2, crosses the x-axis. The graph of ${}^2y(x)$ is shown as the heavy line labelled (0) in Figure 6.2. The line labelled (1) is a neighbouring solution curve for Problem 1 and that labelled (2) a neighbouring solution curve for Problem 2; both pass through the point A, where $x = 4.685$, $y = -0.004$. The point A is very close to the exact solution, the global error at $x = 4.685$ (the length of the line A_0A) being -0.005. The effect of the inevitable errors in any numerical method is that the numerical solution point lies not on the exact solution curve but at a neighbouring point such as A. When we evaluate the function $f(x, y) = Ay + \varphi(x)$ at the point A (or at points very close to A in the case of an explicit Runge–Kutta method such as RKF45) we are simply evaluating the slope of the neighbouring solution curve through A (or at points very close to A). In the case of the neighbouring solution curve (1) of the non-stiff Problem 1, this gradient information is a good approximation to the gradient information on the exact solution curve (0), since the two curves are virtually parallel. Not so for the neighbouring solution curve (2) for the stiff Problem 2, where the slope of (2) at A is wildly different from the slope of (0) at A_0. We would need to move much further up the curve (2)—at least as far as B—before we begin to get reasonable gradient information. In order to stay that close to the exact solution, we need to employ a very small steplength.

More can be gleaned from Figure 6.2. The argument we have given above assumes

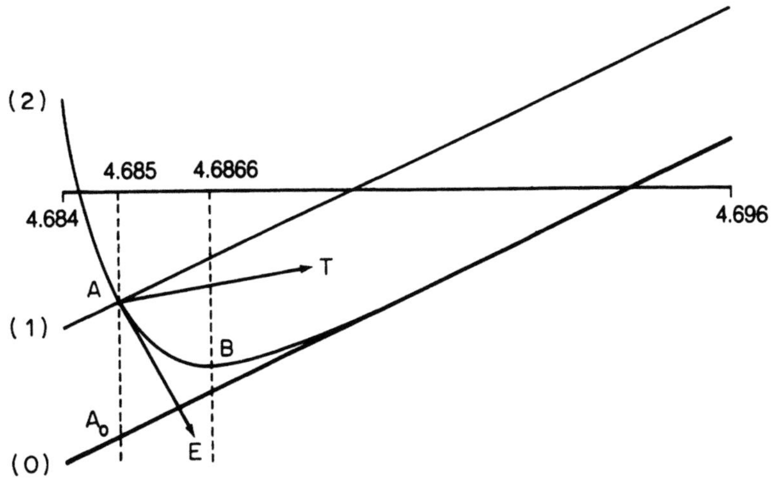

Figure 6.2

that the step ahead from A will depend only on gradient information evaluated at A (or points very close to it in the case of an explicit Runge–Kutta method). This is not the case if the method is implicit. To simplify ideas, consider what would happen if we employed the explicit Euler's Rule, starting from point A with steplength 0.01. Euler's Rule is equivalent to proceeding down the tangent to the solution curve through A, that is, in the direction labelled E in Figure 6.2, until x reaches the value 4.695. This takes us very far away on the other side of the curve (0), and it is not hard to envisage that a further Euler step from $x = 0.695$ will take us even further away, demonstrating instability. Recalling from §5.1 the interpretation of an explicit Runge–Kutta method as a sequence of Euler steps, it is clear that RKF45 will fare no better than does Euler's Rule. For the implicit Trapezoidal Rule however, we proceed along a direction which is the mean of the slope at A and the slope at the new solution point. A simple calculation shows that this is the direction labelled T in Figure 6.2, a direction which suggests that stability can be achieved without forcing the steplength to be excessively small. Thus, by considering the geometry of the neighbouring solution curves, we can anticipate that no explicit method will be able to cope efficiently with a stiff system, but that implicit methods may be able to do so.

The above arguments motivate yet another statement concerning stiffness:

Statement 4 *A system is said to be stiff in a given interval of x if in that interval the neighbouring solution curves approach the solution curve at a rate which is very large in comparison with the rate at which the solution varies in that interval.*

Like our definition, this statement properly separates out the genuinely stiff system from the pseudo-stiff and stresses that whether or not a system is stiff depends on the stage of integration we have reached. For a genuinely stiff system the neighbouring curves will indeed approach the solution curve at a rate which is very large (relative to the rate of change of the solution in its post-transient phase), but which is not particularly large relative to the current rate of change of the solution in the phase when the fastest transients are still alive. Moreover, like our definition, Statement 4 makes sense even if we allow the matrix A to have some eigenvalues with small positive real part. However, Statement 4 obviously would not make a very practicable definition in the sense that it would tell us whether or not a problem with unknown solution was exhibiting stiffness only *after* we had computed the solution (and at least one neighbouring solution). The same sort of criticism cannot be made of our definition of stiffness; if we attempt to solve a problem involving a stiff system using a code based on a method with a finite region of absolute stability, we will quickly become aware of the presence of stiffness and abort the computation.

An analytic interpretation that can be put on the phenomenon of neighbouring curves that we have just discussed is that the function $f(x, y) = Ay + \varphi(x)$ is *ill-conditioned*, that is, has a large Lipschitz constant. For such a system, small changes in the value of y lead to large changes in $f(x, y)(= y')$. For our linear constant coefficient system, the Lipschitz constant may be taken to be $L = \|\partial f/\partial y\| = \|A\| \geqslant \max_t |\lambda_t|$, for any norm. Thus, for a stiff system, L is large, and indeed some authors use the phrase 'systems with large Lipschitz constants' to describe stiff systems. However, this leaves open the question 'Large relative to what?'. Attempts can be made to answer this question, but such attempts do not, in the author's opinion, result in definitions which are superior to that we have adopted.

The reader who is familiar with singular perturbation theory will see some connection between that phenomenon and stiffness; indeed systems exhibiting singular perturbation can be seen as a sub-class of stiff systems. We do not pursue this connection further, other than to quote a very simple example. Consider the homogeneous linear constant coefficient system

$$\begin{bmatrix} {}^1y' \\ {}^2y' \end{bmatrix} = \begin{bmatrix} a_{11} & a_{12} \\ a_{21} & a_{22} \end{bmatrix} \begin{bmatrix} {}^1y \\ {}^2y \end{bmatrix},$$

where the matrix A has real eigenvalues λ_1, λ_2 such that $\lambda_1 \ll \lambda_2 < 0$. By eliminating ${}^2y$ and ${}^2y'$ we obtain the equivalent second-order scalar equation

$${}^1y'' - (a_{11} + a_{22}){}^1y' + (a_{11}a_{22} - a_{12}a_{21}){}^1y = 0.$$

Since λ_1, λ_2 are the zeros of the quadratic $\lambda^2 - (a_{11} + a_{22})\lambda + (a_{11}a_{22} - a_{12}a_{21})$ this scalar equation can be rewritten as

$$(1/\lambda_1){}^1y'' - (1 + \lambda_2/\lambda_1){}^1y' + \lambda_2{}^1y = 0.$$

As $\lambda_1 \to -\infty$, we have the classical singular perturbation situation.

All of our discussions on the nature of stiffness have so far centred around the linear constant coefficient system. Variable coefficient linear systems $y' = A(x)y + \varphi(x)$ and nonlinear systems $y' = f(x, y)$ can also exhibit stiffness, and it is such systems that present the real computational challenge. In our discussion in §3.8 of linear stability theory for linear multistep methods, we reproduced an argument which purported to extend linear stability theory to variable coefficient and nonlinear problems, and showed by example that such an argument could lead to invalid conclusions. The flaw in the argument was the assumption that the Jacobian $A(x)$ or $\partial f/\partial y$ could be assumed piecewise constant (or 'frozen') and that the behaviour of the resulting linear constant coefficient system gave an indication of the local behaviour of the variable coefficient or nonlinear system. The same 'frozen Jacobian' argument can be used to attempt to describe the stiffness of a variable coefficient or nonlinear system in terms of the distribution of the eigenvalues of the frozen Jacobian. In Chapter 7, on nonlinear stability theory for stiff systems, we shall give some further examples illustrating the invalidity of this approach. (It was the inability of linear stability theory to deal satisfactorily with variable coefficient or nonlinear systems that led to the development of an alternative stability theory.) It follows that Statement 1 does not hold for variable coefficient or nonlinear systems; the remaining Statements do hold for such systems (subject to the reservations expressed earlier). In particular, it is invariably the case that stiff variable coefficient and nonlinear systems cannot be integrated by a method with a finite region of absolute stability unless the steplength is excessively small; thus our definition of stiffness (despite its use of the language of linear stability theory) does hold for variable coefficient and nonlinear systems. Consider the following nonlinear example:

$$\left.\begin{aligned} {}^1y' &= 1/{}^1y - {}^2y\exp(x^2)/x^2 - x, & {}^1y(1) &= 1 \\ {}^2y' &= 1/{}^2y - \exp(x^2) - 2x\exp(-x^2), & {}^2y(1) &= \exp(-1). \end{aligned}\right\} \tag{6.12}$$

The exact solution of this problem is ${}^1y(x) = 1/x$, ${}^2y(x) = \exp(-x^2)$. We see that for $x \geqslant 1$ the solution decays monotonically and is smooth; the larger x is, the smoother the solution becomes. If we now attempt to solve this problem by the RKF45 code with $\mathrm{TOL} = 10^{-4}$, we find that the total number of steps $N(x)$ taken to reach the point x is as follows:

x	1.2	1.4	1.6	1.8	2.0	2.2
$N(x)$	20	40	60	92	192	701

Our definition of stiffness would indicate that the system is not particularly stiff for $1.0 \leqslant x \leqslant 1.6$; thereafter, the steplength becomes increasingly small relative to the smoothness of the exact (or of the computed) solution, and increasing stiffness is indicated. The numerical solution so produced is quite acceptable; at $x = 2.2$ the L_2-norm of the global error is 4×10^{-4}.

Despite the existence of counter-examples, it has to be said that analysing the stiffness of a variable coefficient or nonlinear system by freezing the Jacobian and applying Statement 1, can (and very often does) give valid qualitative (and sometimes quantitative) information about the stiffness of the system. Applying such an argument to (6.12), the numerical solution in some interval containing the point x will be deemed stable if the steplength there is such that $h\lambda_t(x) \in \mathscr{R}_A$, $t = 1, 2$, where $\mathscr{R}_A$ is the region of absolute stability of the RKF45 method, and λ_t, $t = 1, 2$, are the eigenvalues of the Jacobian of the system in (6.12), evaluated on the exact solution. These eigenvalues are readily found to be $-1/({}^1y)^2$ and $-1/({}^2y)^2$ so that on substituting from the exact solution we have

$$\bar{\lambda} = -\exp(2x^2), \qquad \underline{\lambda} = -x^2.$$

Table 6.1 compares $h(\mathrm{MAX})$, the maximum steplength that the above stability requirement allows at x, with $h(\mathrm{ACT})$, the actual steplength the code typically used in a neighbourhood of x.

Table 6.1

x	1.0	1.2	1.4	1.6	1.8	2.0	2.2
$\bar{\lambda}$	−7.4	−17.81	−50.4	−167	−652	−2,981	−15,994
h(MAX)	0.41	0.17	0.06	0.018	0.005	0.0010	0.000 19
h(ACT)	0.01	0.01	0.01	0.01	0.005	0.0013	0.000 16

The 'frozen Jacobian' approach thus affords a virtually perfect explanation of what happened in practice. Stiffness has no effect until $x > 1.6$; for $x \leqslant 1.6$ it is accuracy, not stability that dictates the steplength (see Statement 2). Thereafter, h(ACT) follows h(MAX) quite convincingly. Despite the excellent results that the 'frozen Jacobian' argument gives in this example, we will find, in §7.1, that there exist examples in which the above procedure yields results which are not only poor, but frankly ludicrous.

In this section, the impression might have been given that stiffness is an example of that sort of bizarre, pathological problem which so fascinates the mathematician but which seldom arises in real-life situations. Let the reader be assured that this is very far from the truth. The author is unaware of precisely when the word 'stiffness' first entered the literature, but it probably had its origins in control theory. (Any control mechanism which could be modelled by a stiff system would have a strong tendency to seek the equilibrium solution and would feel 'stiff' in the mechanical sense.) Stiffness arises in a wide array of real-life problems, and areas such as chemical kinetics, reactor kinetics, control theory, electronics and mathematical biology regularly throw up stiff systems, some of an awesome degree of stiffness. The author sometimes gets the impression that the degree of stiffness that real-life problems exhibit becomes greater year by year. Perhaps within the mathematical community there are groups of mathematical modellers striving to produce ever more accurate models which include the 'switching-on' phenomena that produce very fast transient solutions, as well as groups of numerical analysts striving equally hard to get rid of these transients!

6.3 LINEAR STABILITY DEFINITIONS PERTINENT TO STIFFNESS

It is clear from the considerations of the preceding section that, as far as the linear constant coefficient system $y' = Ay + \varphi(x)$ is concerned, if the method employed has a region of absolute stability which includes the whole of the left half-plane, then there will be no stability-imposed restriction on the steplength. Denoting $h\lambda$ by $\hat{h}$, (as in §3.8), we have the following definition (Dahlquist, 1963)

Definition *A method is said to be* **A-stable** *if* $\mathscr{R}_A \supseteq \{\hat{h} \mid \operatorname{Re} \hat{h} < 0\}$.

A-stability turns out to be a demanding requirement (particularly for linear multistep methods) and it is natural to restrict the class of problems in some way and seek alternative and less demanding requirements which will remove the restriction on the steplength for that restricted class of problems. Consider the case when the eigenvalues

$\lambda_t, t = 1, 2, \ldots, m$, of A all lie within an infinite wedge in the left half of the $\hat{h}$-plane, bounded by the rays $\arg \hat{h} = \pi - \alpha$, $\arg \hat{h} = \pi + \alpha$. (Recall that complex eigenvalues occur as complex conjugate pairs.) Multiplying $\lambda_t \in \mathbb{C}$ by $h \in \mathbb{R}$ is equivalent to moving along the ray from the origin to λ_t, and it follows that $h\lambda_t$ also lies within this wedge, for all positive values of h. This observation motivates the following definition (Widlund, 1967):

Definition *A method is said to be* ***A(α)-stable**, $\alpha \in (0, \pi/2)$ if $\mathcal{R}_A \supseteq \{\hat{h} \mid -\alpha < \pi - \arg \hat{h} < \alpha\}$; it is said to be* ***A(0)-stable*** *if it is $A(\alpha)$-stable for some $\alpha \in (0, \pi/2)$.*

Clearly, $A(0)$-stability has some relevance to the class of problems for which all of the eigenvalues are real and negative. In that case, however, we can do better by requiring only that $\mathcal{R}_A$ contains the negative real axis, thus motivating the following definition (Cryer, 1973):

Definition *A method is said to be* ***A_0-stable*** *if $\mathcal{R}_A \supseteq \{\hat{h} \mid \operatorname{Re} \hat{h} < 0, \operatorname{Im} \hat{h} = 0\}$.*

An alternative way of slackening the requirements of A-stability is to argue that, for many problems, the eigenvalues which produce the fastest transients all lie to the left of a line $\operatorname{Re} \hat{h} = -a$, where $a > 0$, the remaining eigenvalues (which are responsible for the slower transients) being clustered fairly close to the origin. We are thus assuming that there are no eigenvalues with small negative real part and large imaginary part. This motivates the following definition (Gear, 1969):

Definition *A method is said to be* ***stiffly stable*** *if $\mathcal{R}_A \supseteq \mathcal{R}_1 \cup \mathcal{R}_2$ where $\mathcal{R}_1 = \{\hat{h} \mid \operatorname{Re} \hat{h} < -a\}$ and $\mathcal{R}_2 = \{\hat{h} \mid -a \leqslant \operatorname{Re} \hat{h} < 0, -c \leqslant \operatorname{Im} \hat{h} \leqslant c\}$, and a and c are positive real numbers.*

The minimum region that is necessary to ensure A-stability, $A(\alpha)$-stability and stiff stability are shown in Figure 6.3. It is at once obvious that stiff stability implies $A(\alpha)$-stability with $\alpha = \arctan(c/a)$.

It is possible to argue that there is a sense in which A-stability, far from being over-restrictive, is not restrictive enough. Consider the Trapezoidal Rule, $y_{n+1} = y_n + \frac{1}{2}h(f_{n+1} + f_n)$, which is A-stable (see Figure 3.2 of §3.8), applied to the test equation $y' = Ay$, where A is an $m \times m$ matrix with distinct eigenvalues λ_t satisfying $\operatorname{Re} \lambda_t < 0$, $t = 1, 2, \ldots, m$; as before, we indicate by $\bar{\lambda}$ the eigenvalue with the maximum modulus real part. We obtain the system of difference equations

$$y_{n+1} = By_n, \quad B = (I - hA/2)^{-1}(I + hA/2). \tag{6.13}$$

It is straightforward to check that the general solution of (6.13) takes the form

$$y_n = \sum_{t=1}^{m} K_t(\mu_t)^n d_t, \tag{6.14}$$

where the K_t are arbitrary constants and μ_t, d_t are respectively the eigenvalues (assumed distinct) and the eigenvectors of B. The numerical solution $\{y_n\}$ is an approximation to the exact solution

$$y(x_n) = \sum_{t=1}^{m} \kappa_t \exp(\lambda x_n) c_t = \sum_{t=1}^{m} \kappa_t \exp(\lambda x_0)[\exp(\lambda h)]^n c_t, \tag{6.15}$$

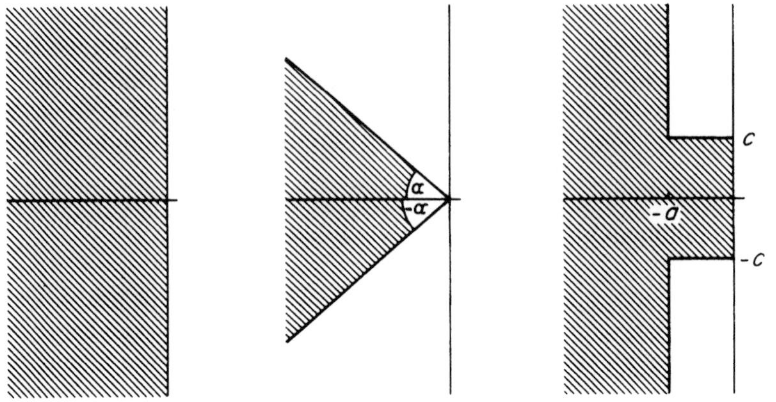

Figure 6.3 *A*-stability; *A*(α)-stability; Stiff stability

where the $c_t(\neq d_t)$ are the eigenvectors of A. Now, if a matrix A has eigenvalues λ_t, and $R(\cdot)$ is a rational function, then the eigenvalues of $R(A)$ are $R(\lambda_t)$. It follows from (6.13) that $\mu_t = (1 + h\lambda_t/2)/(1 - h\lambda_t/2), t = 1, 2, \ldots, m$; in particular, B must have an eigenvalue $\bar{\mu}$ such that

$$\bar{\mu} = (1 + h\bar{\lambda}/2)/(1 - h\bar{\lambda}/2). \tag{6.16}$$

Comparing (6.14) with (6.15) we see that μ_t approximates $\exp(\lambda_t h)$; note that $|\mu_t| < 1$, so that $(\mu_t)^n \to 0$ as $n \to \infty$ (which must happen, since the method is A-stable). However if $|\mathrm{Re}\,\bar{\lambda}|$ is very large and h not particularly small (and the whole idea of A-stability is not to have h excessively small), then $|h\bar{\lambda}|$ will be large, and we see from (6.16) that $\bar{\mu}$ will be close to -1. Thus the term $[\exp(h\bar{\lambda})]^n$, which tends to zero extremely rapidly as $n \to \infty$ is approximated by the term $(\bar{\mu})^n$ which tends to zero very slowly, and with alternating sign, as $n \to \infty$. We can expect a slowly damped oscillating error when we apply the Trapezoidal Rule in such circumstances.

In contrast, if we apply the Backward Euler method $y_{n+1} = y_n + hf_{n+1}$, then we obtain the difference system (6.13), but with B now given by $B = (I - hA)^{-1}$; (6.16) is then replaced by

$$\bar{\mu} = 1/(1 - h\bar{\lambda}), \tag{6.17}$$

which is close to zero when $|h\bar{\lambda}|$ is large. Thus the terms $[\exp(h\bar{\lambda})]^n$ and $(\bar{\mu})^n$ both tend to zero rapidly as $n \to \infty$, and we would not expect to see a slowly damped error.

We illustrate this phenomenon by applying both methods, with a fixed steplength, to System 2 of the preceding section. If we apply the initial conditions of Problem 2 of §6.1, which annihilated the fast transient, the phenomenon cannot be observed. However, if we change the initial conditions to $y(0) = [0, 0]^T$, the exact solution becomes

$$\begin{bmatrix} {}^1y \\ {}^2y \end{bmatrix} = \frac{-1}{999}\exp(-x)\begin{bmatrix} 1 \\ 1 \end{bmatrix} + \frac{1}{999}\exp(-1000x)\begin{bmatrix} 1 \\ -998 \end{bmatrix} + \begin{bmatrix} \sin x \\ \cos x \end{bmatrix}$$

and the fast transient is present. Figure 6.4(a) shows the results (for the component ${}^2y$ only) of applying the Trapezoidal Rule with the fixed steplength $h = 0.2$ in the interval

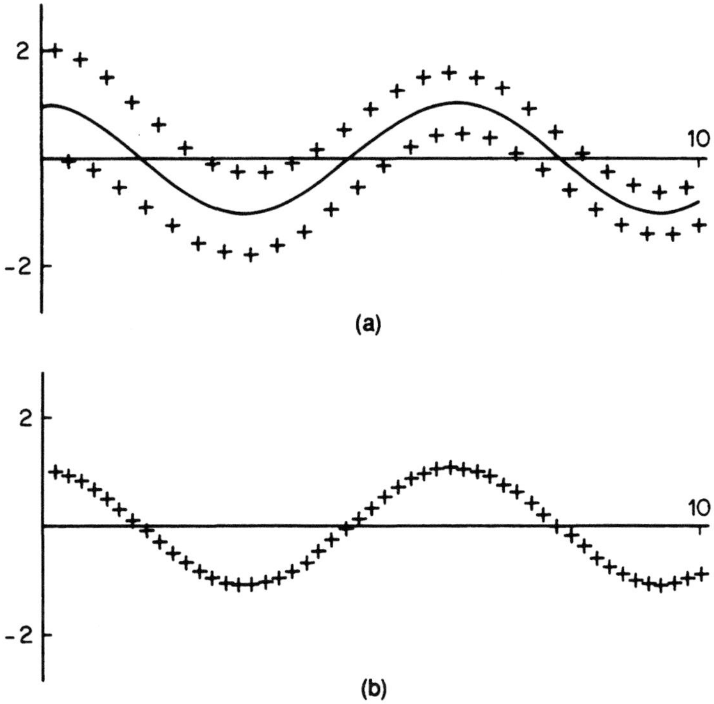

Figure 6.4 (a) Trapezoidal Rule. (b) Backward Euler method

$0 \leqslant x \leqslant 10$; the continuous line represents the exact solution and the numerical solution points are indicated by the symbol $+$. (Note that the fast transient dies so quickly, that the graph of the exact solution jumps immediately from 0 to the slow transient plus steady-state solution.) The slowly damped oscillating error in the numerical solution is clearly demonstrated. In Figure 6.4(b), the numerical solution for ${}^2y$, given by the Backward Euler method applied with the same fixed steplength, is plotted; there is clearly no slowly damped error.

In attempting to frame a definition that separates out methods like the Backward Euler method from those like the Trapezoidal Rule, we observe firstly that we can cope only with one-step methods, and secondly that, since the essential point is the difference in behaviour between (6.16) and (6.17), it would be enough to consider a scalar test problem $y' = \lambda y$. We thus arrive at the following definition (Ehle, 1969; Axelsson, 1969):

Definiton *A one-step method is said to be* **L-stable** *if it is A-stable and, in addition, when applied to the scalar test equation* $y' = \lambda y$, λ *a complex constant with* $\operatorname{Re}\lambda < 0$, *it yields* $y_{n+1} = R(h\lambda)y_n$, *where* $|R(h\lambda)| \to 0$ *as* $\operatorname{Re} h\lambda \to -\infty$.

This property is sometimes called *stiff A-stability* or *strong A-stability*; the '*L*' in *L*-stability indicates that special behaviour far to the *left* of the origin is required. Note the hierarchy

$$L\text{-stability} \Rightarrow A\text{-stability} \Rightarrow \text{stiff-stability} \Rightarrow A(\alpha)\text{-stability}$$
$$\Rightarrow A(0)\text{-stability} \Rightarrow A_0\text{-stability}.$$

In practice, the slowly-damped-error phenomenon exhibited by methods which are A-stable but not L-stable is nowhere near as disastrous as Figure 6.4(a) would suggest. In our example, the fixed steplength of 0.2 is much too large to allow the method to provide a good approximation during the phase when the fast transient in the exact solution is still alive, and a large error is introduced. This error is then damped only very slowly. A more practical setting would be to apply the Trapezoidal Rule in an automatic code which monitored error and changed the steplength. If we do this, then the code initially selects a very small steplength, but increases the steplength to a 'normal' value when the fast transient is dead. We can see this happening if we apply the automatic code based on the 2-stage Gauss method, employed in §6.1, to the above example. For the Gauss method, the function $R(h\lambda)$ appearing in the definition of L-stability is $[1 + h\lambda/2 + (h\lambda)^2/12]/[1 - h\lambda/2 + (h\lambda)^2/12]$. The method is A-stable, but not L-stable; clearly, $R(h\lambda) \to 1$ as $\operatorname{Re} h\lambda \to -\infty$, and there would be a slowly damped (but now non-oscillatory) error if the method were applied to a stiff system, using a *fixed* steplength with is not excessively small. However, when the automatic code is applied to the above problem it chooses the following initial sequence of steplengths:

$$0.0015625,\ 0.0015625,\ 0.003125,\ 0.00625,\ 0.4,\ 0.4, \ldots.$$

Thereafter the steplength stays at 0.4 except for a single step of 0.8, and the numerical solution so obtained is perfectly acceptable. The moral is clear; never compute with a fixed steplength, particularly if stiffness is around!

An alternative means of removing the slowly damped error associated with methods which are A-stable but not L-stable is to employ *smoothing*, as first advocated by Lindberg (1971) for the Trapezoidal Rule. This consists of replacing y_n by $\hat{y}_n := (y_{n-1} + 2y_n + y_{n+1})/4$, and then using the value $\hat{y}_n$ to propagate the solution. Smoothing can be carried out at the first p steps only (when the fast transients are still alive) or introduced whenever the numerical solution exhibits lack of smoothness. An analysis of the effect of smoothing on the truncation error can be found in Lindberg (1971). Let us examine the effect of applying smoothing just once to the numerical example illustrated in Figure 6.4. Using the Trapezoidal Rule with the fixed steplength of 0.2, we compute y_1 and y_2, replace y_1 by $\hat{y}_1$ and let the computation proceed from there. The graphical solution so obtained is indistinguishable from Figure 6.4(b).

Some authors appear to consider it self-evident that L-stability is to be preferred to A-stability, but this is not always so. Consider the following example:

$$y' = \begin{bmatrix} 42.2 & 50.1 & -42.1 \\ -66.1 & -58 & 58.1 \\ 26.1 & 42.1 & -34 \end{bmatrix} y, \qquad y(0) = \begin{bmatrix} 1 \\ 0 \\ 2 \end{bmatrix} \tag{6.18}$$

with exact solution

$$y(x) = \begin{bmatrix} \exp(0.1x)\sin 8x + \exp(-50x) \\ \exp(0.1x)\cos 8x - \exp(-50x) \\ \exp(0.1x)(\cos 8x + \sin 8x) + \exp(-50x) \end{bmatrix}. \tag{6.19}$$

In Figure 6.5(a), the continuous line is the graph of the exact solution for the component ${}^1y(x)$, given by (6.19), in the interval $[0, 1.5]$; after the fast transient $\exp(-50x)$ becomes

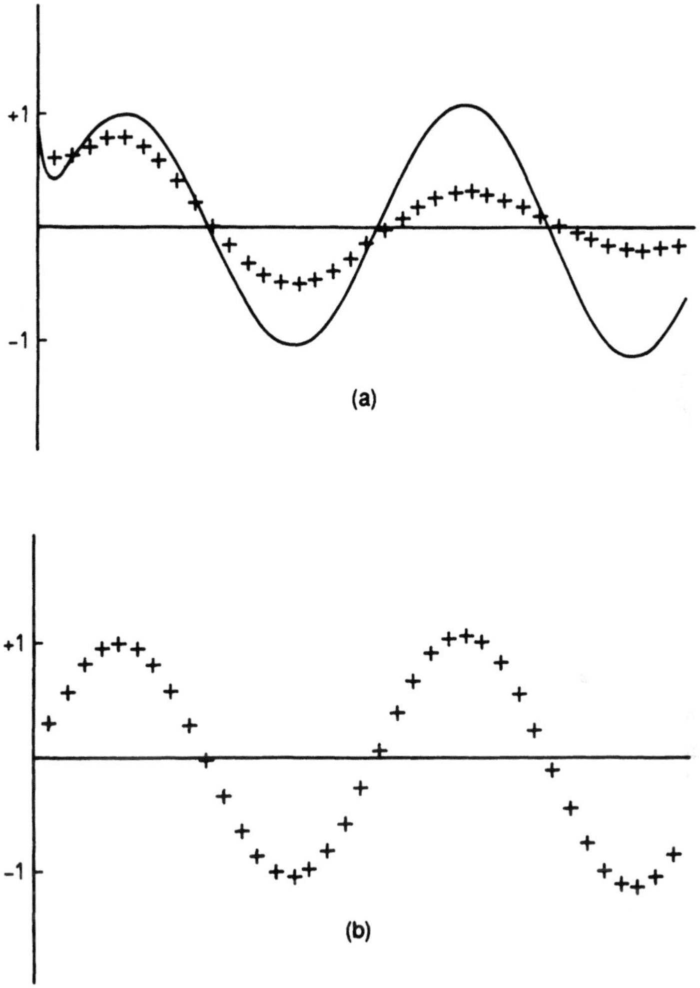

Figure 6.5 (a) Backward Euler method. (b) Trapezoidal Rule

negligible, the solution is sinusoidal, with a very slowly increasing amplitude. The numerical solution points given by the L-stable Backward Euler method with a fixed steplength of 0.04 are indicated by the symbols $+$, and clearly give an unacceptable solution. Figure 6.5(b) shows the acceptable results given by the A-stable but non-L-stable Trapezoidal Rule.

The explanation of the Backward Euler's peculiar behaviour for this problem lies in the shape of its absolute stability region. This region is the *exterior* of the circle radius 1 and centre $\operatorname{Re}\hat{h} = 1$, $\operatorname{Im}\hat{h} = 0$ (see Figure 3.4 of §3.12), and thus includes part of the right half-plane as well as the whole of the left half-plane. The eigenvalues of the system in (6.18) are -50 and $0.1 \pm 8i$, and for moderate values of h, the points $h(0.1 \pm 8i)$ lie in the region of absolute stability; in Figure 6.6 (which shows the first quadrant of the complex $\hat{h}$-plane only) the position of the point $h(0.1 + 8i)$ when $h = 0.2$ is shown in relation to the boundary of the absolute stability region.

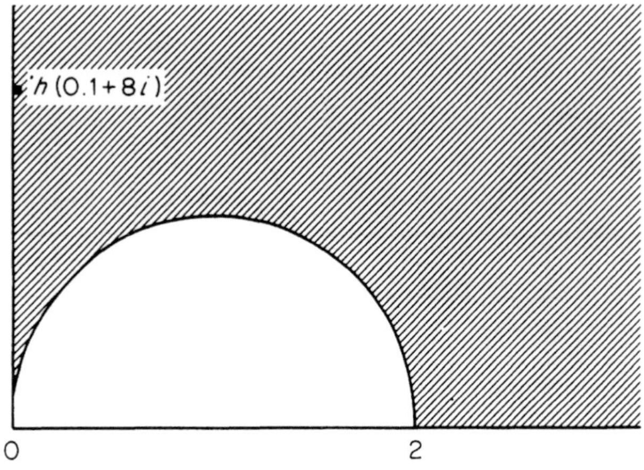

Figure 6.6

The eigenvalues $(0.1 \pm 8i)$ give rise to sinusoidals with *decaying* amplitudes in the numerical solution, whilst the corresponding eigenvalues of the system in (6.18) give rise to sinusoidals with *increasing* amplitude in the exact solution (6.19). In order to get a numerical solution which faithfully represented the exact solution, we would need to choose h sufficiently small for $h(0.1 \pm 8i)$ to lie within the region of absolute *instability*! A simple calculation shows that this implies $h < 0.0031245$.

Unlike the phenomenon of the slowly damped error produced by the Trapezoidal Rule, moving from a fixed step to an automatic code does not improve matters. We can expect nothing else; the difficulty has nothing to do with the fast transient, and therefore cannot be expected to go away when the fast transient dies. Indeed, an automatic code consisting of that used in §6.1, but with the Backward Euler method replacing the 2-stage Gauss method, runs into serious trouble when applied to (6.18). With TOL set at 0.01, it chooses an initial steplength of 0.003 125 (by chance, almost precisely the crucial value at which $h(0.1 \pm 8i)$ hits the boundary of the region of absolute stability) but then doubles the steplength to 0.006 25 when x is around 0.08. That steplength is maintained for the rest of the integration to $x = 1.5$, which is achieved at a total cost of 252 steps. The result is a convincingly smooth solution, but one in which the sinusoidals *decay* slowly in amplitude rather than increase. Wrong numerical results that look right are the most dangerous ones! If we set TOL at 0.001, then the code chooses an initial steplength of 0.000 781 25, doubles it to 0.0015 625 at around $x = 0.06$ and maintains that value for the remainder of the integration, taking a total of 999 steps. The numerical solution now correctly generates a sinusoidal solution with slowly increasing amplitude. The disturbing thing about this example is that the automatic code proves capable of generating convincing solutions which are not only inaccurate but which give incorrect *qualitative* information. Perhaps the moral is that one should use a range of tolerances; even if that range produced nothing but damped sinusoidals, one would notice that as the tolerance decreases, the sinusoidals became *less* damped, and one's suspicions would be aroused. In contrast, the same code, but with the Trapezoidal Rule replacing the Backward Euler method, encounters no difficulty; with TOL $= 0.01$, it initially chooses a steplength of 0.006 25, but once the fast transient is dead the steplength rises to 0.025, and a satisfactory solution is achieved in 66 steps.

The behaviour observed above is not specific to the backward Euler method; *all* L-stable methods are liable to behave like this when the system has some eigenvalues with positive real parts. For all conventional one-step methods, the function $R(h\lambda)$ appearing in the definition of L-stability is a rational function, and the requirement $|R(h\lambda)| \to 0$ *as* $\operatorname{Re} h\lambda \to -\infty$ implies that $|R(h\lambda)| \to 0$ *as* $\operatorname{Re} h\lambda \to +\infty$. The region of absolute stability must therefore include part of the positive half-plane, since the method is clearly absolutely stable for very large positive $h\lambda$. Behaviour of the sort illustrated above is bound to happen for a suitably chosen problem.

Summarizing the two pehnomena described above, we can say that methods which are A-stable but not L-stable will, for most problems, produce solutions with slowly damped errors, but these errors can be satisfactorily controlled by an automatic code; L-stable methods do not produce such slowly damped errors but can produce misleading results, which will not be easily detected by an automatic code, when applied to an infrequently met class of problems. On the whole, there seems to be something to be said in favour of what we might call *precisely A-stable* methods, that is, methods whose regions of absolute stability are precisely the left half-plane; the numerical solutions of the test system $y' = Ay$ then tend to zero as n tends to infinity *if and only if* the exact solutions tend to zero as x tends to infinity.

We conclude by emphasizing (yet again) that although all of the definitions in this section have been motivated by the class of linear constant coefficient systems, they are widely used in the context of variable coefficient or nonlinear systems; for much of the time, but not always, they continue to make sense.

Exercises

6.3.1. Use the result of Exercise 1.9.2 with z and w replaced by r and $\hat{h}$ respectively to prove that no explicit linear multistep method and no predictor–corrector pair in $P(EC)^{\mu}E^{1-t}$, $t = 0, 1$, where μ is finite, can be A_0-stable.

6.3.2. The Backward Euler method applied with a sequence of variable steplengths $\{h_n\}$ to the *scalar* test equation $y' = f(y)$ gives

$$y_{n+1} = y_n + h_n f(y_{n+1}). \tag{1}$$

(i) Show that for the test equation $y' = \lambda y$, λ a real negative constant, $y_n \to 0$ as $n \to \infty$ for all $h > 0$.

(ii) Consider the following novel way of applying (1). Given y_n, instead of specifying h_n and solving the implicit equation (1) for y_{n+1}, let us specify y_{n+1} and solve (1) for h_n which is now given *explicitly*. In order to specify y_{n+1} sensibly, let us use a 2-stage second order explicit Runge–Kutta method to estimate y_{n+1} from y_n, using h_{n-1} (known from the previous step) as steplength. But the sequence $\{y_n\}$ obtained by such an algorithm is precisely the same as that generated by the 2-stage explicit Runge–Kutta method on its own (though not on the same discretization), and cannot possibly have the property proved in (i)! Resolve this paradox. (If all else fails, try a numerical example.)

6.3.3*. Use the Routh–Hurwitz criterion to show that the method

$$y_{n+2} - y_{n+1} = \frac{h}{4}(f_{n+2} + 2f_{n+1} + f_n) \tag{1}$$

is A_0-stable. Use the boundary locus method to find $\partial\mathscr{R}_A$. Consider a linear constant coefficient system whose eigenvalues $\{\lambda_t\}$ are given by

$$\lambda_1 = -\mu + i, \quad \lambda_2 = -\mu - i, \quad \lambda_t \text{ real}, \quad -\mu < \lambda_t < 0, \qquad t = 3, 4, \ldots, m.$$

By considering where the infinite wedge containing these eigenvalues intersects $\partial\mathscr{R}_A$, show that the method is not $A(0)$-stable.

6.4 RATIONAL APPROXIMATIONS TO THE EXPONENTIAL; ORDER STARS

We saw in §5.12 that if a Runge–Kutta method is applied to the test equation for absolute stability, $y' = \lambda y$, $\lambda \in \mathbb{C}^1$ a constant, we obtain

$$y_{n+1} = R(\hat{h})y_n, \tag{6.20}$$

where $\hat{h} := h\lambda$ and $R(\cdot)$ is a rational function when the method is implicit or semi-implicit and a polynomial function when the method is explicit. It is clear that if any implicit linear 1-step method is applied to the test equation, we again obtain (6.20) with $R(\cdot)$ a rational function. The exact solution of the test equation is $y(x) = K\exp(\lambda x)$, K an arbitrary constant, whence

$$y(x_{n+1}) = \exp(h\lambda)y(x_n) = \exp(\hat{h})y(x_n). \tag{6.21}$$

Let the method have order p; then, using the localizing assumption $y_n = y(x_n)$, it follows from (6.20) and (6.21) and the definition of local truncation error that

$$y(x_{n+1}) - y_{n+1} = [\exp(\hat{h}) - R(\hat{h})]y(x_n) = 0(h^{p+1})$$

and hence that

$$R(\hat{h}) = \exp(\hat{h}) + 0(h^{p+1}). \tag{6.22}$$

We are thus motivated to study rational approximations to the exponential $\exp(q)$, $q \in \mathbb{C}$.

Let $q \in \mathbb{C}$, and let $R_T^S(q)$, whereas $S \geqslant 0, T \geqslant 0$, be defined by

$$R_T^S(q) = \left(\sum_{i=0}^{S} a_i q^i\right) \bigg/ \left(\sum_{j=0}^{T} b_j q^j\right), \quad a_0 = b_0 = 1, \quad a_S \neq 0, \quad b_T \neq 0 \tag{6.23}$$

where $a_i, b_j \in \mathbb{R}$, $i = 0, 1, \ldots, S$, $j = 0, 1, \ldots, T$. We say that $R_T^S(q)$ is an (S, T) *rational approximation of order p to the exponential* $\exp(q)$ if $R_T^S(q) = \exp(q) + 0(q^{p+1})$. It follows from (6.23) that we can find the order of a given rational approximation from the fact that if

$$1 + a_1 q + \cdots + a_S q^S - (1 + b_1 q + \cdots + b_T q^T)(1 + q + q^2/2! + \cdots) = 0(q^{p+1}) \tag{6.24}$$

then the approximation is of order p. For example, putting $S = T = 2$, $a_1 = 1$, $a_2 = \frac{1}{3}$, $b_1 = 0, b_2 = -\frac{1}{6}$ in (6.24), we find that $p = 3$. Thus the rational approximation

$$R_2^2(q) = \frac{1 + q + q^2/3}{1 - q^2/6}$$

Table 6.2 The Padé table

1	$1+q$	$1+q+\frac{1}{2}q^2$	$1+q+\frac{1}{2}q^2+\frac{1}{6}q^2$
$\dfrac{1}{1-q}$	$\dfrac{1+\frac{1}{2}q}{1-\frac{1}{2}q}$	$\dfrac{1+\frac{2}{3}q+\frac{1}{6}q^2}{1-\frac{1}{3}q}$	$\dfrac{1+\frac{3}{4}q+\frac{1}{4}q^2+\frac{1}{24}q^3}{1-\frac{1}{4}q}$
$\dfrac{1}{1-q+\frac{1}{2}q^2}$	$\dfrac{1+\frac{1}{3}q}{1-\frac{2}{3}q+\frac{1}{6}q^2}$	$\dfrac{1+\frac{1}{2}q+\frac{1}{12}q^2}{1-\frac{1}{2}q+\frac{1}{12}q^2}$	$\dfrac{1+\frac{3}{5}q+\frac{3}{20}q^2+\frac{1}{60}q^3}{1-\frac{2}{5}q+\frac{1}{20}q^2}$
$\dfrac{1}{1-q+\frac{1}{2}q^2-\frac{1}{6}q^3}$	$\dfrac{1+\frac{1}{4}q}{1-\frac{3}{4}q+\frac{1}{4}q^2-\frac{1}{24}q^3}$	$\dfrac{1+\frac{2}{5}q+\frac{1}{20}q^2}{1-\frac{3}{5}q+\frac{3}{20}q^2-\frac{1}{60}q^3}$	$\dfrac{1+\frac{1}{2}q+\frac{1}{10}q^2+\frac{1}{120}q^3}{1-\frac{1}{2}q+\frac{1}{10}q^2-\frac{1}{120}q^3}$

to $\exp(q)$ has order 3. As we would expect from a count of the parameters a_i, b_j in (6.24), the maximum order that a $R_T^S(q)$ approximation can attain is $S+T$. Such approximations of maximal order are known as *Padé approximations*, and we denote them by $\hat{R}_T^S(q)$. It can be shown (see, for example, Butcher (1987)) that the coefficients of $\hat{R}_T^S(q)$ are given by

$$a_i = \frac{S!}{(S+T)!}\,\frac{(S+T-i)!}{i!(S-i)!}\,i=1,2,\ldots,S, \quad b_j = (-1)^j\frac{T!}{(S+T)!}\,\frac{(S+T-j)!}{j!(T-j)!}\,j=1,2,\ldots,T$$

Using these results, we construct in Table 6.2 the so-called *Padé table* of $\hat{R}_T^S(q)$ for S, $T=0,1,2,3$. A more extensive Padé table can be found in Butcher (1987).

The linear stability properties of a one-step method which generates the difference equation (6.20) are determined by the behaviour of $R(\hat{h})$. It is convenient to use the nonmenclature of *acceptability*, introduced by Ehle (1969); we adopt the following definition:

Definition *A rational approximation* $R(q)$ *to* $\exp(q)$ *is said to be*
(*a*) **A-acceptable** *if* $|R(q)|<1$ *whenever* $\operatorname{Re} q<0$,
(*b*) **A_0-acceptable** *if* $|R(q)|<1$ *whenever* q *is real and negative, and*
(*c*) ***L*-acceptable** *if it is A-acceptable and* $|R(q)|\to 0$ *as* $\operatorname{Re} q\to -\infty$.

Clearly the method is A-, A_0- or L-stable according as $R(\hat{h})$ is A-, A_0- or L-acceptable. It is obvious that the rational approximation $R_T^S(q)$ cannot be A-acceptable if $S>T$, and that if $R_T^S(q)$ is A-acceptable and $T>S$, then $R_T^S(q)$ is also L-acceptable.

Our first two results on acceptability concern rational approximations which contain free parameters and are not, in general Padé approximations. Define

$$\left.\begin{aligned} R_1(q;\alpha) &:= \frac{1+\frac{1}{2}(1-\alpha)q}{1-\frac{1}{2}(1+\alpha)q},\\ R_2(q;\alpha,\beta) &:= \frac{1+\frac{1}{2}(1-\alpha)q+\frac{1}{4}(\beta-\alpha)q^2}{1-\frac{1}{2}(1+\alpha)q+\frac{1}{4}(\beta+\alpha)q^2} \end{aligned}\right\} \qquad (6.25)$$

(We do not label these approximations according to the notation of (6.23) since S and T depend on the values taken by the parameters α and β.) For general α, $R_1(q;\alpha)$ has order one, but has order two (Padé approximations) if $\alpha = 0$; $R_2(q;\alpha,\beta)$ has order two for general α,β, order three if $\alpha \neq 0, \beta = \frac{1}{3}$ and order four (Padé approximation) if $\alpha = 0, \beta = \frac{1}{3}$.

Theorem 6.1 (Liniger and Willoughby, 1970) Let $R_1(q;\alpha)$ and $R_2(q;\alpha,\beta)$ be defined by (6.25). Then

(a) $R_1(q;\alpha)$ is A-acceptable if and only if $\alpha \geqslant 0$ and L-acceptable if and only if $\alpha = 1$, and
(b) $R_2(q;\alpha,\beta)$ is A-acceptable if and only if $\alpha \geqslant 0$, $\beta \geqslant 0$ and L-acceptable if and only if $\alpha = \beta > 0$.

Further results on acceptability concern the Padé approximations $\hat{R}^S_T(q)$.

Theorem 6.2 (Birkhoff and Varga, 1965) If $T = S$, then $\hat{R}^S_T(q)$ is A-acceptable.

Theorem 6.3 (Varga, 1961) If $T \geqslant S$, then $\hat{R}^S_T(q)$ is A_0-acceptable.

Theorem 6.4 (Ehle, 1969) If $T = S+1$ or $T = S+2$, then $\hat{R}^S_T(q)$ is L-acceptable.

Thus, all entries on the main diagonal of the Padé table and anywhere below that diagonal are A_0-acceptable. Those on the main diagonal and on the *two* sub-diagonals below it are A-acceptable; note that any A-acceptable entry below the main diagonal is automatically L-acceptable. It is clear that there cannot be any A-acceptable entries above the main diagonal, so the key question is whether there exist A-acceptable entries in the Padé table *below* the second subdiagonal. The well-known *Ehle conjecture* asserted that there are none, so that $\hat{R}^S_T(q)$ is A-acceptable if and only if $T-2 \leqslant S \leqslant T$. This conjecture remained unresolved for many years, and was eventually proved (with a remarkable absence of any heavy analysis) by Wanner, Hairer and Nørsett (1978), using their elegant theory of order stars. In this book, we have adopted the policy of not providing proofs of theorems when such proofs do not add to an understanding of the result. We make an exception here. Order star theory appears in many contexts besides this one (see, for example, the survey paper by Wanner (1987) which uses order star theory to prove, *inter alia*, Theorem 3.1 of §3.4), and some familiarity with the ideas involved may prove helpful; besides, the proof is such a nice piece of mathematics! The following is a sketch of how the proof goes.

The region of absolute stability $\mathscr{R}_A$ of a Runge–Kutta or linear 1-step method is defined by $\mathscr{R}_A = \{q \in \mathbb{C} \,|\, |R(q)| < 1\}$, where $R(\cdot)$ is defined by (6.20). We consider, instead of $\mathscr{R}_A$, the region $\mathscr{B} = \{q \in \mathbb{C} \,|\, |R(q)| > |\exp(q)|\}$. (Note that $C(\mathscr{B})$, the complement of $\mathscr{B}$, is essentially the region of *relative* stability, according to Criterion B of §3.8.) The region $\mathscr{B}$ is called an *order star*, because of its star-like shape (see Figure 6.7). By elementary applications of classical complex variable theory, four lemmas can be established:

Lemma 1 $R(q)$ is a rational approximation to $\exp(q)$ of order p if and only if, for $q \to 0$, $\mathscr{B}$ consists of $p+1$ sectors each of angle $\pi/(p+1)$ separated by $p+1$ sectors of $C(\mathscr{B})$ each of the same angle.

(Proof follows from considering $R(q)\exp(-q) = 1 + Kq^{p+1} + 0(q^{p+2})$, $K \neq 0$ a real constant; for sufficiently small ρ, where $q = \rho\exp(i\theta)$, the condition $|R(q)| > |\exp(-q)|$

becomes $\text{Re}\{K\rho^{p+1}\exp[i(p+1)\theta]\} < 0(\rho^{p+2})$, i.e. $\cos[(p+1)\theta] > 0(\rho)$ if $K > 0$ and $\cos[(p+1)\theta] < 0(\rho)$ if $K < 0$.)

Figure 6.7(a) is a magnified view of how the order star, in the case $p = 5$, must look in a small square surrounding the origin; the region $\mathscr{B}$ is shaded and $C(\mathscr{B})$ is unshaded. Temporarily writing $q = x + iy$, it is easy to see by example that $R(q) = R(x + iy) = R(x, y^2)$, so that $\mathscr{B}$ must be symmetric about the real axis; thus the configuration in Figure 6.7(a) is the only possible one.

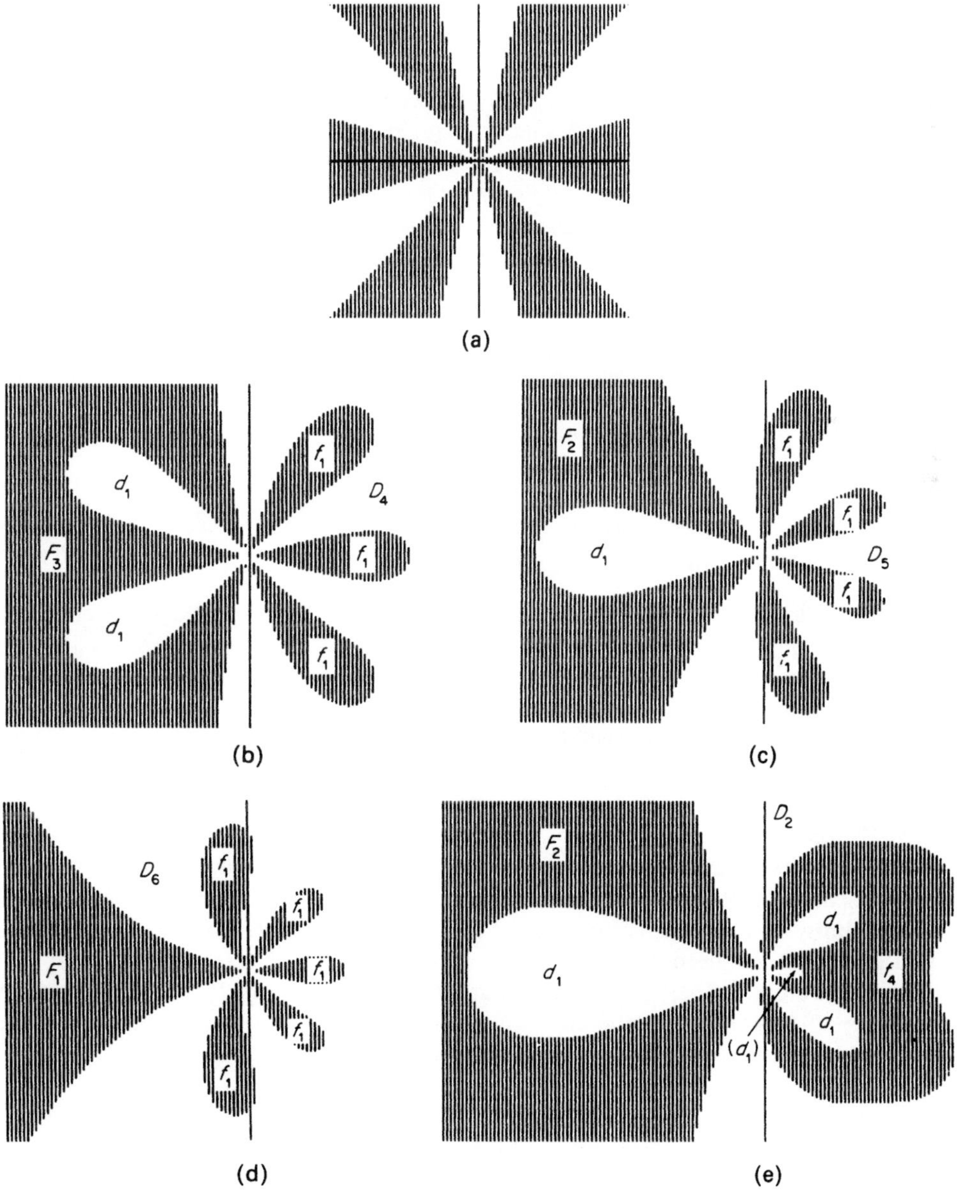

Figure 6.7

Lemma 2 *The boundary $\partial\mathscr{B}$ of $\mathscr{B}$ possesses precisely two branches which go to infinity.*

(Proof follows from considering the strong increase in $|\exp(q)|$ as $\operatorname{Re} q \to \infty$, and its strong decrease as $\operatorname{Re} q \to -\infty$.)

The boundary $\partial\mathscr{B}$ is continuous; Figures 6.7(b), (c), (d), (e) show the order stars for a selection of 5th order rational approximations to the exponential, and illustrate some of the possible configurations. These figures (produced by an obvious modification of the scanning technique described in §5.12) by no means exhaust the possibilities for the case $p = 5$; the reader is invited to sketch other configurations and also to consider the situation for other values of p. Regions of $\mathscr{B}$ arising from a single sector are called *fingers*, and are labelled f_1 in Figure 6.7 if they are bounded, and F_1 if they are unbounded. Similarly, bounded and unbounded fingers of $C(\mathscr{B})$ are called *dual fingers*, and are labelled d_1 and D_1 respectively. If a region stems from n sectors of $\mathscr{B}$ it is called a *finger of multiplicity n*, and is labelled f_n if it is bounded and F_n if it is unbounded; similar regions of $C(\mathscr{B})$ are called *dual fingers of multiplicity n* and are labelled d_n if they are bounded and D_n if they are unbounded. All regions of the q-plane can be labelled in this way; note that the sum of the subscripts in f_n, F_n, d_n and D_n always equals $2(p + 1)$.

Lemma 3 *Each bounded finger of multiplicity n contains at least n poles of $R(q)$ (a pole of multiplicity v counting as v poles), and each bounded dual finger of multiplicity n contains at least n zeros of $R(q)$ (a zero of multiplicity v counting as v zeros).*

(Proof follows from the principle of the argument and the fact that the argument of $R(q)\exp(-q)$ can be shown to decrease along $\partial\mathscr{B}$.)

Lemma 4 *$R(q)$ is A-acceptable if and only if $\mathscr{B}$ has no intersection with the imaginary axis and $R(q)$ has no poles in the half-plane* $\operatorname{Re} q < 0$.

(Proof of the 'if' part follows from the fact that $|\exp(q)| = 1$ on the imaginary axis and the maximum principle; the 'only if' part follows from the definition of $\mathscr{B}$.)

(In the following, we use the notation $[x]$ to mean the biggest integer which is less than or equal to x, and $\mathbb{C}^-$ to indicate the half-plane $\operatorname{Re} q < 0$.)

Let $R(q)$ be an A-acceptable approximation of order p. Then at least $[(p + 1)/2]$ fingers (a finger of multiplicity v counting as v fingers) start in $\mathbb{C}^-$ (Lemma 1), none of which can cross the imaginary axis (Lemma 4), and none of which are bounded (Lemmas 3 and 4). Hence these $[(p + 1)/2]$ fingers must collapse into one unbounded multiple finger (Lemma 2), and will therefore enclose $[(p + 1)/2] - 1$ bounded dual fingers in $\mathbb{C}^-$, each of which contains at least one zero of $R(q)$ (Lemma 3). For example, when $p = 5$, A-acceptability of $R(q)$ means that only the configuration (b) in Figure 6.7 is possible; note that there are indeed $2(= [(5 + 1)/2] - 1)$ bounded dual fingers in $\mathbb{C}^-$.

The *total* number of zeros of $R(q)$ is therefore at least $[(p + 1)/2] - 1$. Now suppose that $R(q) = \hat{R}^S_T(q)$, the (S, T) Padé approximation; then $p = S + T$, and $\hat{R}^S_T(q)$ has S zeros. Hence we have $S \geqslant [(p + 1)/2] - 1$ or $2S + 2 \geqslant 2[(p + 1)/2] \geqslant p$, since

$$2[(p + 1)/2] = \begin{cases} p + 1 & \text{if } p \text{ is odd} \\ p & \text{if } p \text{ is even.} \end{cases}$$

We thus have that $2S + 2 \geqslant S + T$, or $S \geqslant T - 2$. It is of course trivial to show that we cannot have A-acceptability if $S > T$, so that the Ehle conjecture is proved:

Theorem 6.5 The (S, T) Padé approximation, $\hat{R}^S_T(q)$, is A-acceptable if and only if $T-2 \leqslant S \leqslant T$.

Figure 6.7(b) is the order star of $\hat{R}^2_3(q)$ which, by Theorem 6.5, is A-acceptable; note that it satisfies the requirements of Lemma 4. Figures 6.7(c) and (d) are the order stars of $\hat{R}^1_4(q)$ and $\hat{R}^0_5(q)$ which are A-unacceptable; note that for both, the region $\mathscr{B}$ intersects with the imaginary axis. Figure 6.7(e) is the order star for a non-Padé $R^4_5(q)$ approximation of order 5; see Exercise 6.4.5.

Exercises

6.4.1. Execute the following program!

```
begin
  while PATIENCE > TOL do
    begin
      for S:= 0 to 3 do
      for T:= 0 to 3 do
      [use (6.24) to check the entries in Table 6.2];
    end;
end.
```

6.4.2. Show that the method of Exercise 5.7.4 applied to the test equation $y' = \lambda y$ generates the (2, 1) Padé approximation to $\exp(h\lambda)$ and therefore cannot be A_0-stable.

6.4.3. The following method, due to Liniger and Willoughby (1970), uses the second derivatives of y, obtained by differentiating the differential system:

$$y_{n+1} - y_n = \frac{h}{2}[(1+\alpha)y^{(1)}_{n+1} + (1-\alpha)y^{(1)}_n] - \frac{h^2}{4}[(\beta+\alpha)y^{(2)}_{n+1} - (\beta-\alpha)y^{(2)}_n].$$

By an obvious extension of the definition for a linear multistep method, show that the method has order three if $\beta = \frac{1}{3}$ and order four if, in addition, $\alpha = 0$. Find the range of values for α and β for which the method is (i) A-stable and (ii) L-stable.

6.4.4. Find the position of the poles and zeros of $\hat{R}^2_3(q)$ and check that Figure 6.7(b) is compatible with Lemma 3.

6.4.5. The approximation $R^4_5(q)$ whose order star is given by Figure 6.7(e) was constructed as follows. Choose the denominator to have zeros at $q = \frac{1}{2}, 1 \pm i, 2 \pm i$ (check that, with this choice of poles for $R^4_5(q)$, Figure 6.7(e) is compatible with Lemma 3) and choose the coefficients in the numerator so that the approximation has order 5. Find $R^4_5(q)$.

6.5 HANDLING IMPLICITNESS IN THE CONTEXT OF STIFFNESS

As we shall see presently, no explicit linear multistep or explicit Runge–Kutta method can possess any of the properties of A-, $A(\alpha)$-, A_0- or stiff-stability; neither can a predictor–corrector method in $P(EC)^{\mu}E^{1-t}$ mode, where μ is fixed and finite (an explicit process). See, for example, Figures 4.1 and 4.2 of Chapter 4. We are thus forced to use only implicit methods to solve a problem involving a stiff system. It might appear that if, for example, we want to preserve the A-stability of the Trapezoidal Rule, all we need do is use it as corrector in a predictor–corrector pair applied in the mode of correcting to convergence, when the linear stability properties of the pair will be those of the corrector alone (see §4.1). Alas, stiffness has another trick up its sleeve!

Consider the general implicit linear multistep method

$$\sum_{j=0}^{k} \alpha_j y_{n+j} = h \sum_{j=0}^{k} \beta_j f_{n+j} \tag{6.26}$$

applied to the general problem $y' = f(x, y), y(a) = \eta$. We can rewrite (6.26) in the form

$$y_{n+k} = h\beta_k f(x_{n+k}, y_{n+k}) + \Psi_n, \tag{6.27}$$

where

$$\Psi_n = \Psi(x_n, y_n, y_{n+1}, \dots, y_{n+k-1}; h) := \sum_{j=0}^{k-1} (-\alpha_j y_{n+j} + h\beta_j f_{n+j})$$

is a known function of previously computed values. If (6.26) is the corrector in a predictor–corrector pair, then the mode of correcting to convergence consists of allowing the fixed point iteration

$$y_{n+k}^{[\nu+1]} = h\beta_k f(x_{n+k}, y_{n+k}^{[\nu]}) + \Psi_n, \qquad \nu = 0, 1, \dots \tag{6.28}$$

to run until convergence is achieved. However, we recall from §4.1 that the iteration (6.28) converges if

$$h < 1/(|\beta_k| L), \tag{6.29}$$

where L is the Lipschitz constant of f with respect to y. If f is assumed differentiable with respect to y, then we may take L to be $\sup \| \partial f / \partial y \|$, and we have that

$$L = \sup \| \partial f / \partial y \| \geqslant \max_t |\lambda_t| \geqslant |\mathrm{Re}\, \bar{\lambda}|,$$

where $\lambda_t, t = 1, 2, \dots, m$ are the eigenvalues of $\partial f / \partial y$, $\bar{\lambda}$ being the eigenvalue with largest modulus real part. If the system is stiff, then $|\mathrm{Re}\, \bar{\lambda}|$ is very large, and it follows from (6.29) that h must be very small. We are thus in a Catch-22 situation. If we use an explicit method to solve a stiff system we have to use an excessively small steplength to avoid instability; if we use an implicit method with an absolute stability region large enough to impose no stability restriction, we can choose a steplength as large as we please, but we will not be able to solve the implicit equation by the iteration (6.28) unless the steplength is excessively small! To see how nasty this Catch-22 can be in practice, consider the nonlinear example (6.12) of §6.2. Recall that this problem did not exhibit stiffness until x reached the value 1.6, but that thereafter it became increasingly stiff, as witnessed by Table 6.1, which showed the actual steplength (h(ACT)) used by the explicit RKF45 code. Suppose we were to solve the problem by the A-stable Trapezoidal Rule. In order

Table 6.3

x	1.0	1.2	1.4	1.6	1.8	2.0	2.2
h(ACT)	0.01	0.01	0.01	0.01	0.005	0.0013	0.000 16
h(TR)	0.25	0.11	0.04	0.01	0.003	0.0007	0.000 13

to compute the bound (6.29) on the steplength, we take as generous a view as possible and choose L to be not $\sup\|\partial f/\partial y\|$ but the local value at x of $\|\partial f/\partial y\|$ (using the L_2-norm) obtained by evaluating the Jacobian on the known exact solution. We denote by h(TR) the resulting maximum steplength that could be used with the Trapezoidal Rule if the convergence condition (6.29) is to be satisfied. Table 6.3 compares h(TR) with h(ACT).

We see that, once stiffness manifests itself, the convergence condition (6.29) forces the Trapezoidal Rule to use a steplength *smaller* than that which stabilit forces the RKF45 code to use!

The only way out of this difficulty is to abandon fixed point iteration in favour of Newton iteration (see §1.8); in order to save on LU decompositions, modified Newton iteration (see §1.8, equation (1.27)) is almost invariably employed. Applying this to (6.27) gives

$$\left[I - h\beta_k \frac{\partial f}{\partial y}(x_{n+k}, y^{[0]}_{n+k})\right]\tilde{\Delta} y^{[\nu]}_{n+k} = -y^{[\nu]}_{n+k} + h\beta_k f(x_{n+k}, y^{[\nu]}_{n+k}) + \Psi_n \qquad \nu = 0, 1, \ldots \tag{6.30}$$

where we recall that $\tilde{\Delta} y^{[\nu]}_{n+k} = y^{[\nu+1]}_{n+k} - y^{[\nu]}_{n+k}$. If (6.30) converges, then it does so to the solution of (6.27); accordingly, it is quite common to keep the matrix $I - h\beta_k \partial f/\partial y$ constant not only throughout the iteration, but to use it for the next one or two integration steps we well. The matrix is up-dated, and a new LU decomposition computed, only when the iteration fails to converge.

If the system is linear, then $f(x, y) = A(x)y + \varphi(x)$ and (6.27) becomes

$$[I - h\beta_k A(x_{n+k})]y_{n+k} = h\beta_k \varphi(x_{n+k}) + \Psi_n \tag{6.31}$$

and the computational cost per step is one LU decomposition and back substitution. If $A(x) = A$, independent of x, then the same LU decomposition is used throughout the interval of integration. Note that if we put $f(x, y) = A(x)y + \varphi(x)$ in (6.30) we obtain

$$\begin{aligned}[I - h\beta_k A(x_{n+k})]\tilde{\Delta} y^{[\nu]}_{n+k} &= -y^{[\nu]}_{n+k} + h\beta_k[A(x_{n+k})y^{[\nu]}_{n+k} + \varphi(x_{n+k})] + \Psi_n \\ &= -[I - h\beta_k A(x_{n+k})]y^{[\nu]}_{n+k} + h\beta_k \varphi(x_{n+k}) + \Psi_n, \qquad \nu = 0, 1, \ldots\end{aligned}$$

whence

$$[I - h\beta_k A(x_{n+k})]y^{[\nu+1]}_{n+k} = h\beta_k \varphi(x_{n+k}) + \Psi_n, \qquad \nu = 0, 1, \ldots$$

which states that the iterates $y^{[\nu]}_{n+k}, \nu = 1, 2, \ldots$ are all equal to the solution for y_{n+k} given by (6.31). In other words, Newton iteration (now identical with modified Newton iteration) converges in just one step when the problem is linear.

Similar difficulties arise if we attempt to use fixed point iteration to implement an implicit Runge–Kutta method; the notation is now a little more complicated. Consider an implicit s-stage Runge–Kutta method specified by the Butcher array

$$\begin{array}{c|c} c & A \\ \hline & b^{\mathrm{T}} \end{array}$$

It is marginally easier in this context to use the alternative form (5.6) of the method:

$$\left.\begin{aligned} y_{n+1} &= y_n + h\sum_{i=1}^{s} b_i f(x_n + c_i h, Y_i) \\ Y_i &= y_n + h\sum_{j=1}^{s} a_{ij} f(x_n + c_j h, Y_j), \qquad i = 1, 2, \ldots, s. \end{aligned}\right\} \tag{6.32}$$

The second of (6.32) may be written as

$$Y = F(Y), \tag{6.33}$$

where

$$Y := [Y_1^{\mathrm{T}}, Y_2^{\mathrm{T}}, \ldots, Y_s^{\mathrm{T}}]^{\mathrm{T}} \in \mathbb{R}^{ms}, \quad F(Y) := [F_1^{\mathrm{T}}, F_2^{\mathrm{T}}, \ldots, F_s^{\mathrm{T}}]^{\mathrm{T}} \in \mathbb{R}^{ms}$$

and

$$F_i = F_i(Y) = y_n + h\sum_{j=1}^{s} a_{ij} f(x_n + c_j h, Y_j), \qquad i = 1, 2, \ldots, s.$$

The fixed point iteration

$$Y^{[\nu+1]} = F(Y^{[\nu]}), \qquad \nu = 0, 1, \ldots \tag{6.34}$$

will converge if $0 \leqslant M < 1$, where M is the Lipschitz constant of $F(Y)$ with respect to Y; assuming differentiability we take M to be $\sup \| \partial F/\partial Y \|$. Now

$$\partial F/\partial Y = h\begin{bmatrix} a_{11}\partial f/\partial y & a_{12}\partial f/\partial y & \cdots & a_{1s}\partial f/\partial y \\ a_{21}\partial f/\partial y & a_{22}\partial f/\partial y & \cdots & a_{2s}\partial f/\partial y \\ \vdots & \vdots & \vdots & \vdots \\ a_{s1}\partial f/\partial y & a_{s2}\partial f/\partial y & \cdots & a_{ss}\partial f/\partial y \end{bmatrix}.$$

Using the notation of direct products, described in §1.11, we can write

$$\partial F/\partial Y = hA \otimes \partial f/\partial y. \tag{6.35}$$

By Property (3) of §1.11, the eigenvalues of $\partial F/\partial Y$ are $h\mu_i\lambda_t$, $i = 1, 2, \ldots, s$, $t = 1, 2, \ldots, m$, where the μ_i are the eigenvalues of A and the λ_t those of $\partial f/\partial y$. The eigenvalues μ_i depend only on the method, and are not particularly small in practice, and if the system is stiff then $\{\lambda_t\}$ contains an eigenvalue $\bar{\lambda}$ where $|\mathrm{Re}\,\bar{\lambda}|$ is large. It follows from an argument analogous to that used for implicit linear multistep methods that $\partial F/\partial Y$ has eigenvalues with large modulus, and that the iteration (6.34) will converge only for

Table 6.4

x	1.0	1.2	1.4	1.6	1.8	2.0	2.2
h(TR)	0.25	0.11	0.04	0.01	0.003	0.0007	0.000 13
h(2G)	0.29	0.13	0.05	0.01	0.004	0.0008	0.000 14

excessively small steplengths. As an illustration, consider the 2-stage Gauss method applied to the nonlinear problem (6.12). A calculation similar to that which led to the results displayed in Table 6.3 can be used to establish the maximum steplength h(2G) which can be employed with the 2-stage Gauss method if the iteration (6.34) is to converge; Table 6.4 shows that the restrictions that convergence of the fixed point iteration imposes on the 2-step Gauss method are comparable with those it imposes on the Trapezoidal Rule.

Once again, we are forced to replace the fixed point iteration (6.34) by Newton or modified Newton iteration; applying the latter to (6.33) and using (6.35) gives

$$(I_{ms} - hA \otimes J)\tilde{\Delta} Y^{[\nu]} = \Gamma^{[\nu]}, \qquad \nu = 0, 1, \ldots \tag{6.36(a)}$$

where

$$J = \frac{\partial f}{\partial y}(Y^{[0]}), \qquad \tilde{\Delta} Y^{[\nu]} = Y^{[\nu+1]} - Y^{[\nu]}, \qquad \Gamma^{[\nu]} = -Y^{[\nu]} + F(Y^{[\nu]}) \tag{6.36(b)}$$

and I_{ms} is the $ms \times ms$ unit matrix. Note that (6.36(a)) requires us to compute, at each integration step, a single LU decomposition of an $ms \times ms$ matrix, and a new back substitution for each call of the iteration. The LU decomposition is the expensive part, particularly since the dimension of the matrix may be quite large. It is possible, in some circumstances, to reduce this computational effort by utilizing, as follows, a transformation due to Butcher (1976).

For any $s \times s$ matrix A there exists a nonsingular matrix H such that

$$M := H^{-1}AH = \begin{bmatrix} \mu_1 & \omega_1 & 0 & 0 & \cdots & 0 \\ 0 & \mu_2 & \omega_2 & 0 & & 0 \\ \vdots & & & & & \vdots \\ 0 & & 0 & 0 & \mu_{s-1} & \omega_{s-1} \\ 0 & \cdots & 0 & 0 & 0 & \mu_s \end{bmatrix} \tag{6.37}$$

where $\mu_i, i = 1, 2, \ldots, s$ are the eigenvalues of A and the ω_j are all either 0 or 1; this is the Jordan canonical form of A (see, for example, Gourlay and Watson (1973)). If the μ_i are all distinct, then $\omega_j = 0, j = 1, 2, \ldots, s-1$, and M is a diagonal matrix whose elements are the eigenvalues of A. Consider the transformation from $Y^{[\nu]}, \Gamma^{[\nu]}$ to $\hat{Y}^{[\nu]}, \hat{\Gamma}^{[\nu]}$ given by

$$\hat{Y}^{[\nu]} := (H^{-1} \otimes I_m) Y^{[\nu]}, \qquad \hat{\Gamma}^{[\nu]} := (H^{-1} \otimes I_m)\Gamma^{[\nu]}, \tag{6.38(a)}$$

where I_m is the $m \times m$ unit matrix. By Property (2) of §1.11, the inverse transformations are

$$Y^{[\nu]} := (H \otimes I_m)\hat{Y}^{[\nu]}, \qquad \Gamma^{[\nu]} := (H \otimes I_m)\hat{\Gamma}^{[\nu]}. \tag{6.38(b)}$$

Defining $\tilde{\Delta}\hat{Y}^{[\nu]} := \hat{Y}^{[\nu+1]} - \hat{Y}^{[\nu]}$, (6.36) becomes

$$(I_{ms} - hA \otimes J)(H \otimes I_m)\tilde{\Delta}\hat{Y}^{[\nu]} = (H \otimes I_m)\hat{\Gamma}^{[\nu]}$$

or

$$[I_{ms} - h(H \otimes I_m)^{-1}(A \otimes J)(H \otimes I_m)]\tilde{\Delta}\hat{Y}^{[\nu]} = \hat{\Gamma}^{[\nu]}.$$

Now, using Properties (1) and (2) §1.11 and (6.37), we obtain

$$(H \otimes I_m)^{-1}(A \otimes J)(H \otimes I_m) = (H^{-1} \otimes I_m)(AH \otimes J) = H^{-1}AH \otimes J = M \otimes J.$$

We thus have that

$$(I_{ms} - hM \otimes J)\tilde{\Delta}\hat{Y}^{[\nu]} = \hat{\Gamma}^{[\nu]}, \qquad \nu = 0, 1, \dots \tag{6.39}$$

which is the modified Newton iteration now couched in terms of the transformed variables $\hat{Y}^{[\nu]}, \hat{\Gamma}^{[\nu]}$. The crucial difference between (6.39) and (6.36(a)) is that whilst the latter required the LU decomposition of an $ms \times ms$ matrix, (6.39) requires s LU decompositions of the blocks $I_m - h\mu_t J$; the computational effort is proportional to sm^3 instead of $(sm)^3$, a considerable saving. The saving is even greater if the Runge–Kutta method is one of the class of singly-implicit (SIRK) methods described in §5.11. For such methods, the μ_t are all equal, and there is only one LU decomposition to be performed at each integration step, giving SIRK methods a level of efficiency approaching that of DIRK methods. Note that the procedure described here is less attractive if the matrix A has complex eigenvalues; in that case, the transforming matrix H in (6.38) will also be complex, as will be the transformed variables $\hat{Y}^{[\nu]}, \hat{\Gamma}^{[\nu]}$. We conclude with two illustrations.

Illustration 1 Consider a 2-stage implicit Runge–Kutta method for which the matrix A has real distinct eigenvalues μ_1 and μ_2. Then

$$M = \begin{bmatrix} \mu_1 & \cdot \\ \cdot & \mu_2 \end{bmatrix}, \qquad I_{2m} - hM \otimes J = \begin{bmatrix} I_m - h\mu_1 J & 0 \\ 0 & I_m - h\mu_2 J \end{bmatrix}.$$

We compute two LU decompositions of $m \times m$ matrices, namely $I - h\mu_1 J = L_1 U_1$, $I_m - h\mu_2 J = L_2 U_2$; then

$$I_{2m} - hM \otimes J = \begin{bmatrix} L_1 U_1 & 0 \\ 0 & L_2 U_2 \end{bmatrix} = LU,$$

where

$$L = \begin{bmatrix} L_1 & 0 \\ 0 & L_2 \end{bmatrix}, \quad U = \begin{bmatrix} U_1 & 0 \\ 0 & U_2 \end{bmatrix}.$$

Illustration 2 Consider a 2-stage singly implicit method; the matrix A has an eigenvalue μ of multiplicity 2. Then

$$M = \begin{bmatrix} \mu & 1 \\ \cdot & \mu \end{bmatrix}, \qquad I_{2m} - hM \otimes J = \begin{bmatrix} I_m - h\mu J & -hJ \\ 0 & I_m - h\mu J \end{bmatrix}.$$

We compute the single LU decomposition $I_m - h\mu J = L_1 U_1$; then

$$I_{2m} - hM \otimes J = \begin{bmatrix} L_1 U_1 & -hJ \\ 0 & L_1 U_1 \end{bmatrix} = LU,$$

where

$$L = \begin{bmatrix} L_1 & 0 \\ 0 & L_1 \end{bmatrix}, \qquad U = \begin{bmatrix} U_1 & -hL_1^{-1}J \\ 0 & U_1 \end{bmatrix}.$$

Exercise

6.5.1. Illustrate the use of the Butcher transformation by computing one step, of length 0.1, of the numerical solution of the problem $u' = v, v' = 5(1 - u^2)v - u, u(0) = 2, v(0) = 0$ by the Runge–Kutta method with Butcher array

$$\begin{array}{c|cc} 2-\sqrt{2} & 1-\dfrac{\sqrt{2}}{4} & 1-\dfrac{3\sqrt{2}}{4} \\[2ex] 2+\sqrt{2} & 1+\dfrac{3\sqrt{2}}{4} & 1+\dfrac{\sqrt{2}}{4} \\[2ex] \hline & \dfrac{1}{2}+\dfrac{3\sqrt{2}}{8} & \dfrac{1}{2}-\dfrac{3\sqrt{2}}{8} \end{array}$$

6.6 LINEAR MULTISTEP METHODS FOR STIFF SYSTEMS

If one is looking for methods with any of the linear stability properties defined in §6.3, the class of linear multistep methods does not provide a particularly good hunting ground; happily there is one notable exception (the BDF) to this statement. Just how difficult it is for linear multistep methods to achieve A-stability is spelled out in the following theorem:

Theorem 6.6 (Dahlquist, 1963) (i) An explicit linear multistep method cannot be A-stable. (ii) The order of an A-stable linear multistep method cannot exceed two. (iii) The second-order A-stable linear multistep method with smallest error constant is the Trapezoidal Rule.

In other words, if we insist on full A-stability, the Trapezoidal Rule (which, earlier in this chapter, may have seemed to be just a conveniently simple example) is the most accurate linear multistep method we have! Theorem 6.6 is often known as the *second Dahlquist barrier*. (Recall the first Dahlquist barrier, Theorem 3.1 of §3.4.) It is of interest to note that order star theory (see §6.4) can be used to provide a proof of this theorem (Wanner, 1987). Implementations of the Trapezoidal Rule for stiff problems usually incorporate the smoothing procedure described in §6.3 and an extrapolation technique which effectively raises the order to 4; local error estimation is usually achieved by Richardson extrapolation (see §5.10).

A linear 1-step method which can be of value in the context of stiffness is the *Theta method*,

$$y_{n+1} - y_n = h[(1-\theta)f_{n+1} + \theta f_n]. \tag{6.40}$$

It has order only 1 in general, and order 2 if $\theta = \frac{1}{2}$, when it becomes the Trapezoidal Rule. Applying (6.40) to the test equation $y' = \lambda y$ gives

$$y_{n+1} = \frac{1 + \hat{h}\theta}{1 - (1-\theta)\hat{h}} y_n = R_1(\hat{h}; 1 - 2\theta) y_n, \qquad \hat{h} = h\lambda, \tag{6.41}$$

where $R_1(\cdot;\cdot)$ is defined by (6.25) of §6.4. It follows from Theorem 6.1 of §6.4 that the Theta method is A-stable if and only if $\theta \leqslant \frac{1}{2}$. One way in which the free parameter θ in (6.40) can be gainfully employed is in the technique of *exponential fitting* (Liniger and Willoughby, 1970). A method is said to be exponentially fitted at a (real) value λ_0 if, when the method is applied to the scalar test problem $y' = \lambda_0 y$, $y(x_0) = y_0$, it yields the exact solution. If the Theta method is applied to this test problem, we clearly get $y_n = [R_1(h\lambda_0; 1 - 2\theta)]^n y_0$, which coincides with the exact solution $y(x_n) = y_0 \exp(\lambda_0(x_n - x_0))$ if $\exp(\lambda_0 h) = R_1(h\lambda_0; 1 - 2\theta)$. From (6.41), this is equivalent to choosing

$$\theta = -\frac{1}{h\lambda_0} - \frac{\exp(h\lambda_0)}{1 - \exp(h\lambda_0)}. \tag{6.42}$$

For all $h\lambda_0 < 0$, the value of θ given by (6.42) satisfies $\theta \leqslant \frac{1}{2}$, so that A-stability is preserved. If, for the linear constant coefficient system $y' = Ay$, stiffness is caused by an isolated real negative eigenvalue of A, exponential fitting to that eigenvalue (estimated by the power method) gives good results. If the stiffness is caused by a cluster of such eigenvalues, then it can be beneficial to fit exponentially to some mean of the eigenvalues. For variable coefficient linear and nonlinear systems (subject as always to our reservations about 'frozen Jacobians'), the same technique can be applied, with periodic updating of the estimate of the dominant eigenvalue.

Turning to linear multistep methods with less than full A-stability, we meet more barriers, but less fearsome ones:

Theorem 6.7 (*Widlund, 1967*) (*i*) *An explicit linear multistep method cannot be $A(0)$-stable.* (*ii*) *There is only one $A(0)$-stable linear k-step method whose order exceeds k, namely the Trapezoidal Rule.* (*iii*) *For all $\alpha \in [0, \pi/2)$ there exist $A(\alpha)$-stable linear k-step methods of order p for which $k = p = 3$, $k = p = 4$.*

Theorem 6.8 (*Cryer, 1973*) (*i*) *An explicit linear multistep method cannot be A_0-stable.* (*ii*) *There exist A_0-stable linear multistep methods of arbitrary order.*

Table 6.5

k	1	2	3	4	5	6
$\alpha_{\max}$	90°	90°	88°	73°	52°	19°
$a_{\min}$	0	0	0.1	0.7	2.4	6.1

If, in Theorem 6.7(iii), we relax the requirement that methods be $A(\alpha)$-stable for all $\alpha\in[0,\pi/2)$, then we can easily find k-step methods of order k for $k>4$ that are $A(\alpha)$-stable for specific values of α. The most notable class of such methods is that consisting of the k-step backward differentiation formulae (BDF), $k=1,2,\ldots,6$, derived in §3.12. (Recall that these methods are zero-stable only for $k\leqslant 6$.) From the regions of absolute stability for the BDF, shown in Figure 3.4 of §3.12, we see that it is more natural to consider stiff-stability rather than $A(\alpha)$-stability; indeed the definition of stiff stability given in §6.3 is virtually tailor-made for the BDF. All of the BDF, $k\leqslant 6$, are stiffly stable (and therefore $A(\alpha)$-stable); Table 6.5 lists the maximum value of α and the minimum value of a, the parameter appearing in the definition of stiff stability.

One can see by the following informal argument that the BDF are likely to be well-suited to dealing with stiffness. Recall from §3.8 the stability polynomial $\pi(r,\hat{h}):=\rho(r)-\hat{h}\sigma(r)$; for a stiff system $|\hat{h}|$ will be large, and $\pi(r,\hat{h})$ will be dominated by $\hat{h}\sigma(r)$. It would therefore seem a good idea to choose $\sigma(r)$ so that its roots are all safely inside the unit circle, and where safer than at the centre of the unit circle? Thus we are led to the choice $\sigma(r)=r^k$, which (together with the implicitness that Theorem 6.7 demands and the requirement that the order be as high as possible) defines the BDF.

The BDF are central to the construction of efficient algorithms for handling stiff systems; they play the same role in stiff problems as the Adams methods do in non-stiff ones. Although the use of the BDF for stiff problems goes back to Curtiss and Hirschfelder (1952), their implementation in VSVO form stems from the work of Gear (1969). Such implementations have essentially the same structures as those we described in Chapter 4 for non-stiff problems. The basic differences are that the Adams–Moulton correctors are replaced by the BDF and the (truncated) fixed point iteration of the corrector is replaced by modified Newton iteration (as described in §6.5), pursued to convergence. Newton iteration (unlike fixed-point iteration) does not converge for an arbitrary starting value, but only for one sufficiently close to the solution, so that it is essential—as well as desirable from the point of view of efficiency—that an accurate starting value be provided. Robertson and Williams (1975) showed that if one attempts to obtain such a starting value by means of a predictor which involves any of the previously computed f values, then traces of the fast transient solution are liable to pollute the predicted value and lead to loss of accuracy (another manifestation of the fact that when a system is stiff $f(y)$ is an ill-conditioned function). Prediction by extrapolation of previously calculated y-values only is preferable. To be more specific, consider the k-step kth-order BDF in standard form (see (3.115) of §3.12),

$$\sum_{j=0}^{k}\alpha_j y_{n+j}=h\beta_k f_{n+k},\qquad \alpha_k=1. \tag{6.43}$$

The order is k and the error constant C_{k+1}. If the predicted value for y_{n+k} is to have order k then it will be necessary to interpolate the $k+1$ points (x_{n+j}, y_{n+j}), $j = -1, 0, 1, \ldots, k-1$. From (1.30) of §1.10, the appropriate interpolant is

$$I_k(x_{n+k-1} + rh) =: P_k(r) = \sum_{i=0}^{k} (-1)^i \binom{-r}{i} \nabla^i y_{n+k-1}.$$

Putting $r = 1$, we take $y_{n+k}^{[0]}$ to be $I_k(x_{n+k})$, whence

$$y_{n+k}^{[0]} = \sum_{i=0}^{k} (-1)^i \binom{-1}{i} \nabla^i y_{n+k-1} = \sum_{i=0}^{k} \nabla^i y_{n+k-1}. \tag{6.44}$$

It is still possible to apply Milne's estimate for the local truncation error. If we regard (6.44) as a linear multistep method then by (1.31) of §1.10 it has order k and error constant

$$C_{k+1}^* = (-1)^{k+1} \binom{-1}{k+1} = 1.$$

From (4.11) of §4.3 we have that

$$\text{PLTE} = \frac{C_{k+1}}{1 - C_{k+1}} (y_{n+k} - y_{n+k}^{[0]}), \tag{6.45}$$

where y_{n+k} is the solution of (6.43) obtained by modified Newton iteration.

Thus, the building blocks for a VSVO algorithm are all available, and a development parallel to that we have discussed in detail in Chapter 4 for ABM methods is possible. In particular, two aspects of the BDF can be exploited; firstly, the BDF can be conveniently expressed in backward difference form (see (3.117) of §3.12) and, secondly, the kth-order BDF is equivalent to setting f_{n+k} equal to the slope of the interpolant through (x_{n+j}, y_{n+j}), $j = 0, 1, \ldots, k$ (again, see §3.12). Further details can be found in Prothero (1976) and Brayton, Gustavson and Hachtel (1972). Finally, we mention two aspects of VSVO implementations of the BDF which do not arise in similar implementations of the ABM. If the system is large and complicated, then finding analytically the elements of the Jacobian matrix can be an onerous task (but a symbolic manipulator can help); accordingly some codes estimate the derivatives $\partial^i f/\partial^j y$ by differencing. Secondly, there can be a choice of strategy in the initial phase depending on whether the user wants to see an accurate representation of the fast transients or is content merely to see the solution after such transients are dead.

The codes DIFSUB, GEAR and EPISODE, referenced in §4.11, all have stiff options. (It is possible for codes to determine automatically whether or not to use stiff options; see, for example, Shampine (1982).) There also exist variants of GEAR adapted for particular circumstances. Thus GEARB (Hindmarsh, 1975) is intended for problems where the Jacobian is large and banded; one important instance of this situation arises when a partial differential equation is semi-discretized—a topic we shall touch upon in §6.9. GEARS (Spellman and Hindmarsh, 1975) is appropriate for problems where the Jacobian is large and sparse. Finally we mention the code FACSIMILE (Curtis and Sweetenham, 1985) which deals only with stiff problems; a survey of this code can be found in Curtis (1987).

Exercises

6.6.1. Show that the Trapezoidal Rule is exponentially fitted at 0 and that the Backward Euler method (and indeed *any* one-step *L*-stable method) is exponentially fitted at $-\infty$.

6.6.2. By subtracting $y_{n+k}^{[0]}$, given by (6.44), from y_{n+k} and using the fact that $y_{n+k-1} = (1-\nabla)y_{n+k}$, find an alternative derivation of (6.45).

6.6.3. The Implicit Midpoint Rule

$$y_{n+1} - y_n = hf(\tfrac{1}{2}(x_n + x_{n+1}), \tfrac{1}{2}(y_n + y_{n+1})) \tag{1}$$

is a close relative of the Trapezoidal Rule

$$y_{n+1} - y_n = \tfrac{1}{2}h[f(x_{n+1}, y_{n+1}) + f(x_n, y_n)]. \tag{2}$$

(It is indeed its *one-leg twin*; see §7.4.) Show that both methods give the same result when applied to the scalar test equation $y' = \lambda y$, and deduce that (1) is *A*-stable. Show, however, that when both methods are applied to the scalar test equation $y' = \lambda(x)y$, $\lambda(x) < 0$ for all x, all of whose solutions tend to zero as x tends to infinity, then all solutions of (1) satisfy $y_n \to 0$ as $n \to \infty$ for all $h > 0$, but those of (2) do so only if h satisfies a condition of the form $0 < h < H(x_n, x_{n+1})$. Find $H(x_n, x_{n+1})$. Devise and carry out a numerical experiment to illustrate this result.

6.7 RUNGE–KUTTA METHODS FOR STIFF SYSTEMS

It is much easier to find implicit Runge–Kutta methods—as opposed to linear multistep methods—with the linear stability properties defined in §6.3; of course, as we have seen, implicit Runge–Kutta methods are more expensive to implement than their linear multistep counterparts. We saw in §5.12 that when an *s*-stage Runge–Kutta method with Butcher array

$$\begin{array}{c|c} c & A \\ \hline & b^{\mathrm{T}} \end{array}$$

is applied to the usual scalar test equation $y' = \lambda y$, $\lambda \in \mathbb{C}$, it yields the difference equation $y_{n+1} = R(\hat{h})y_n$, where $R(\hat{h})$, a rational function of $\hat{h}$, is the stability function and $\hat{h} = h\lambda$. In §5.12 we derived two alternative forms (5.86) and (5.87) of $R(\hat{h})$; for convenience, we reiterate them here:

$$R(\hat{h}) = 1 + \hat{h}b^{\mathrm{T}}(I - \hat{h}A)^{-1}e \tag{6.46i}$$

$$R(\hat{h}) = \frac{\det[I - \hat{h}(A - eb^{\mathrm{T}})]}{\det(I - \hat{h}A)}, \tag{6.46ii}$$

where $e = [1, 1, \ldots, 1]^{\mathrm{T}} \in \mathbb{R}^s$. The method will be *A*-, A_0- or *L*-stable according as $R(\hat{h})$ is *A*-, A_0- or *L*-acceptable. We have already observed that when the method is explicit, (6.46) implies that $R(\hat{h})$ is a polynomial in $\hat{h}$, and there is no possibility of the method having any of these stability properties. In §5.11 we provided several examples of implicit and semi-implicit methods, and all we need do to establish their stability properties is to use (6.46) to determine $R(\hat{h})$ and then use the results of §6.4 on the acceptability of rational approximations to the exponential. However, it can be heavy work applying either of (6.46) to specific examples, and we can obtain the results we want much more easily, and in greater generality, by employing an argument used in §5.12.

Consider first the class of Gauss methods described in §5.11. The s-stage Gauss method has order $2s$, which means that when applied to the test equation $y' = \lambda y$ it will produce $y_{n+1} = R(\hat{h})y_n$, where $R(\hat{h}) = \exp(\hat{h}) + O(h^{2s+1})$; that is, $R(\hat{h})$ is a rational approximation of order $2s$ to the exponential $\exp(\hat{h})$. By (6.46ii), the numerator and denominator in $R(\hat{h})$ are polynomials in $\hat{h}$ of degree at most s. Since there exists a unique (s, s) rational approximation of order $2s$ to $\exp(\hat{h})$, namely the Padé approximation $\hat{R}_s^s(\hat{h})$, it follows that $R(\hat{h}) = \hat{R}_s^s(\hat{h})$, which, by Theorem 6.2 of §6.4, is A-acceptable. Thus all Gauss methods are A-stable. We note that this implies the existence of A-stable implicit Runge–Kutta methods of arbitrarily high order.

For the remaining classes of fully implicit methods described in §5.11 we use an approach due to Dekker and Verwer (1984). The essential point is the structure of the matrix $A - eb^{\mathrm{T}}$ appearing in (6.46ii). We shall establish this structure merely by observation of the examples quoted in §5.11, so that our stability results will be formally proven only for these examples. The extension to the general case can be found in Dekker and Verwer; alternative proofs can be found in Butcher (1987). First let us consider a tempting argument, which is false. Take, for example, an s-stage Radau method, which has order $2s - 1$. We could attempt to show that the denominator in $R(\hat{h})$ has degree s, and then deduce from an order argument that $R(\hat{h})$ was the $(s - 1, s)$ Padé approximation; the flaw is that it could equally be an (s, s) non-Padé approximation of order $2s - 1$. The valid argument goes as follows.

From the examples of §5.11, we observe that the matrices $A - eb^{\mathrm{T}}$ have special forms, from which follow the corresponding forms of $I - \hat{h}(A - eb^{\mathrm{T}})$; we then apply (6.46ii). All matrices are $s \times s$; the symbol + denotes a constant element, independent of h, and * one which is of degree at most 1 in $\hat{h}$.

Radau IA

$$A - eb^{\mathrm{T}} = \begin{bmatrix} 0 & + & + & \cdots & + \\ & & & & \\ 0 & + & + & \cdots & + \\ \vdots & & & & \\ 0 & + & + & \cdots & + \end{bmatrix}, \qquad I - \hat{h}(A - eb^{\mathrm{T}}) = \begin{bmatrix} 1 & * & * & \cdots & * \\ & & & & \\ 0 & * & * & \cdots & * \\ \vdots & & & & \\ 0 & * & * & \cdots & * \end{bmatrix}.$$

Expanding $\det[I - \hat{h}(A - eb^{\mathrm{T}})]$ by the elements of its first column, we see that it is a polynomial in $\hat{h}$ of degree at most $s - 1$. Since $\det(I - \hat{h}A)$ has degree at most s in $\hat{h}$ and the approximation has order $2s - 1$, $R(\hat{h})$ must be the $(s - 1, s)$ Padé approximation. By Theorem 6.4 of §6.4, the methods are all A-stable, indeed L-stable.

Radau IIA

$$A - eb^{\mathrm{T}} = \begin{bmatrix} + & + & \cdots & + & + \\ + & + & \cdots & + & + \\ \vdots & & & & \\ + & + & \cdots & + & + \\ 0 & 0 & \cdots & 0 & 0 \end{bmatrix}, \qquad I - \hat{h}(A - eb^{\mathrm{T}}) = \begin{bmatrix} * & * & \cdots & * & * \\ * & * & \cdots & * & * \\ \vdots & & & & \\ * & * & \cdots & * & * \\ 0 & 0 & \cdots & 0 & 1 \end{bmatrix}.$$

Expanding $\det[I - \hat{h}(A - eb^{\mathrm{T}})]$ by the elements of its last row and arguing as in the case of the Radau IA, we see that $R(\hat{h})$ is again the $(s-1, s)$ Padé approximation, and the methods are L-stable.

Lobatto IIIA

$$A - eb^{\mathrm{T}} = \begin{bmatrix} + & + & \cdots & + & + \\ + & + & \cdots & + & + \\ \vdots & & & & \\ + & + & \cdots & + & + \\ 0 & 0 & \cdots & 0 & 0 \end{bmatrix}, \qquad I - \hat{h}(A - eb^{\mathrm{T}}) = \begin{bmatrix} * & * & \cdots & * & * \\ * & * & \cdots & * & * \\ \vdots & & & & \\ * & * & \cdots & * & * \\ 0 & 0 & \cdots & 0 & 1 \end{bmatrix}.$$

On expanding, we see that $\det[I - \hat{h}(A - eb^{\mathrm{T}})]$ has degree at most $s-1$ in $\hat{h}$. Moreover, $\det(I - \hat{h}A)$ now has degree at most $s-1$, and, since the order is $2s-2$, $R(\hat{h})$ must be the $(s-1, s-1)$ Padé approximation; by Theorem 6.2 of §6.4, the methods are A-stable.

Lobatto IIIB

$$A - eb^{\mathrm{T}} = \begin{bmatrix} 0 & + & \cdots & + & + \\ 0 & + & \cdots & + & + \\ \vdots & & & & \\ 0 & + & \cdots & + & + \\ 0 & + & \cdots & + & + \end{bmatrix}, \qquad I - \hat{h}(A - eb^{\mathrm{T}}) = \begin{bmatrix} 1 & * & \cdots & * & * \\ 0 & * & \cdots & * & * \\ \vdots & & & & \\ 0 & * & \cdots & * & * \\ 0 & * & \cdots & * & * \end{bmatrix}.$$

Again, $\det[I - \hat{h}(A - eb^{\mathrm{T}})]$ and $\det(I - \hat{h}A)$ both have degree at most $s-1$ in $\hat{h}$, and $R(\hat{h})$ must be the $(s-1, s-1)$ Padé approximation and the methods are A-stable.

Lobatto IIIC

$$A - eb^{\mathrm{T}} = \begin{bmatrix} 0 & + & \cdots & + & + \\ 0 & + & \cdots & + & + \\ \vdots & & & & \\ 0 & + & \cdots & + & + \\ 0 & 0 & \cdots & 0 & 0 \end{bmatrix}, \qquad I - \hat{h}(A - eb^{\mathrm{T}}) = \begin{bmatrix} 1 & * & \cdots & * & * \\ 0 & * & \cdots & * & * \\ \vdots & & & & \\ 0 & * & \cdots & * & * \\ 0 & * & \cdots & * & * \end{bmatrix}.$$

On expanding, we see that $\det[I - \hat{h}(A - eb^{\mathrm{T}})]$ has degree at most $s-2$ in $\hat{h}$, while $\det(I - hA)$ has degree at most s. Hence $R(\hat{h})$ must be the $(s-2, s)$ Padé approximation and by Theorem 6.4 of §6.4 the methods are L-stable.

The above results are summarized in Table 6.6.

Turning to semi-implicit methods, a straightforward calculation using either of (6.46) shows that for the pair of 2-stage DIRK methods of order 3 given by (5.76) of §5.11,

$$R(\hat{h}) = \frac{1 \mp \sqrt{3}\hat{h}/3 - (1 \pm \sqrt{3})\hat{h}^2/6}{1 - (3 \pm \sqrt{3})\hat{h}/3 + (2 \pm \sqrt{3})\hat{h}^2/6}.$$

Recalling the general (2, 2) non-Padé approximation $R_2(\hat{h}; \alpha, \beta)$ given by (6.25) of §6.4,

Table 6.6

s-stage RK method	Order	Stability function $R(\hat{h})$	Linear stability property
Gauss	2s	$\hat{R}_s^s(\hat{h})$	A-stability
Radau IA, IIA	2s − 1	$\hat{R}_s^{s-1}(\hat{h})$	L-stability
Lobatto IIIA, IIIB	2s − 2	$\hat{R}_{s-1}^{s-1}(\hat{h})$	A-stability
Lobatto IIIC	2s − 2	$\hat{R}_s^{s-2}(\hat{h})$	L-stability

we see that $R(\hat{h}) = R_2(\hat{h}; \alpha, \beta)$ if $\alpha = 1 \pm 2\sqrt{3}/3$ and $\beta = 1/3$. It follows from Theorem 6.1 of the same section that $R(\hat{h})$ is A-acceptable only for $\alpha = 1 + 2\sqrt{3}/3$ and that (5.76) is A-stable only if we take the upper of the alternative signs. For the 3-stage DIRK methods of order 4 given by (5.77), $R(\hat{h})$ is a (3, 3) non-Padé approximation which is not so easily handled, since we have no result similar to Theorem 6.1 for (3, 3) approximations. However, it can be shown (see, for example, Butcher (1987)) that only one of the three methods, that given by the choice $\nu = (2/\sqrt{3})\cos(10°)$, is A-stable.

Finally, consider the 2-stage SIRK method (5.78) containing a parameter μ; it has order 2 in general and order 3 if $\mu = (3 \pm \sqrt{3})/6$. It turns out to be easier to use (6.46i) rather than (6.46ii) to establish $R(\hat{h})$, which is given by

$$R(\hat{h}) = \frac{1 + (1 - 2\mu)\hat{h} + (\mu^2 - 2\mu + \frac{1}{2})\hat{h}^2}{1 - 2\mu\hat{h} + \mu^2\hat{h}^2}$$

which is identical with $R_2(\hat{h}; \alpha, \beta)$ if $\alpha = 4\mu - 1$, $\beta = (2\mu - 1)^2$. It follows from Theorem 6.1 that $R(\hat{h})$ is A-acceptable for $\mu \geqslant \frac{1}{4}$; the method can be A-stable and have order 3 only if we choose $\mu = (3 + \sqrt{3})/6$. It also follows from Theorem 6.1 that the method will be L-stable if $\mu = 1 \pm \sqrt{2}/2$, but the order is then ony 2.

It is clear that we have no difficulty in finding implicit or semi-implicit Runge–Kutta methods which are A- or L-stable. Any of these methods could, with no especially heavy programming effort, be made into an automatic algorithm. Step-changing is no problem, and estimation of the local truncation error can be done either by Richardson extrapolation or by embedding (see, for example, Burrage (1978a)). For algorithms based on explicit Runge–Kutta methods, using embedded methods for error estimation was considerably more efficient (in terms of the number of function calls per step) than using Richardson extrapolation. The advantages of embedding for algorithms based on implicit methods are much less significant, since the major computational costs arise from the handling of the implicitness, discussed in §6.5.

Algorithms constructed in this way will work—and often be very robust—but, in terms of efficiency, they will fall well short of the BDF-based VSVO codes described in the preceding section. To make implicit Runge–Kutta algorithms competitive one must cut down as much as possible the computational effort of handling the implicitness. In particular, use of the Butcher transformation in conjunction with SIRK methods, described in §6.5, reduces the costs to a level comparable with that of VSVO codes.

STRIDE (Burrage, Butcher and Chipman, 1979) is a variable order code which uses this approach.

Exercises

(Use both forms of (6.46) in the following exercises.)

6.7.1. Show that the method in Exercise 5.7.4 is not A-stable.

6.7.2. Find the range of β for which the method in Exercise 5.7.7 is A-stable, and show that there are two values of β for which it is L-stable.

6.7.3. Investigate the stability of the method in Exercise 6.5.1.

6.7.4. Find the order of the implicit method

$$k_1 = f\left(x_n, y_n + \frac{h}{4}k_1 - \frac{h}{4}k_2\right), \quad k_2 = f\left(x_n + \frac{2h}{3}, y_n + \frac{h}{4}k_1 + \frac{5h}{12}k_2\right)$$

$$y_{n+1} - y_n = \frac{h}{4}(k_1 + 3k_2)$$

and show that it is L-stable.

6.8 METHODS INVOLVING THE JACOBIAN

Although stiffness was known to users from a much earlier data, interest in the topic amongst numerical analysts stems from the seminal paper of Dahlquist (1963), which first defined A-stability. Since then, there has appeared in the literature a large number of suggested techniques, ranging from the ingenious to the bizarre, for dealing with stiffness. It is not practicable to survey all of these here, and we restrict our attentions to classes of methods, of general applicability in the context of stiffness, which have a common theme: the major classes we have studied, namely linear multistep and Runge–Kutta methods are modified so that they directly involve the Jacobian of the system, and are thus *adaptive*. The motivation for this is obvious. Stiffness requires that implicit equations be solved by Newton iteration, which in turn demands that we evaluate the Jacobian of the system; so why not try to use the Jacobian in the method itself?

The first such class we look at constitutes a sub-class of the so-called *Obrechkoff methods*, which are methods of linear multistep form but which involve higher derivatives of y. Such derivatives can be obtained by repeatedly differentiating the system of differential equations, as we did at the start of §5.4. We saw there that this procedure can soon get out of hand, particularly if the system is large, so we restrict our attention to the case when only first and second derivatives of y are involved. As we saw in §1.4, there is no loss of generality in assuming that the system is autonomous, so that we take $y' = f(y)$, whence $y'' = (\partial f/\partial y)f =: f^{(1)}(y)$. The general k-step Obrechkoff method containing up to second derivatives is given by

$$\sum_{j=0}^{k} \alpha_j y_{n+j} = h \sum_{j=0}^{k} \beta_j f_{n+j} + h^2 \sum_{j=0}^{k} \gamma_j f^{(1)}_{n+j}, \qquad \alpha_k = 1, \ |\alpha_0| + |\beta_0| + |\gamma_0| \neq 0. \quad (6.47)$$

Order is defined in an obvious manner analogous to that for linear multistep methods. The stability polynomial is clearly $\pi(r;\hat{h}) := \rho(r) - \hat{h}\sigma(r) - \hat{h}^2\omega(r)$ where $\rho(r)$, $\sigma(r)$ and $\omega(r)$ are polynomials of degree k in r with coefficients α_j, β_j and γ_j respectively. Enright (1974a) advocates the use for stiff systems of the sub-class of (6.47) defined by

$$y_{n+k} - y_{n+k-1} = h\sum_{j=0}^{k} \beta_j f_{n+j} + h^2\gamma_k f^{(1)}_{n+k} \tag{6.48}$$

which have order $p = k + 2$. The choice $\rho(r) = r^k - r^{k-1}$ ensures, as for Adams methods, that the spurious roots are situated at the origin, and the argument we used in §6.6 to motivate the BDF equally motivates the choice $\omega(r) = r^k$. Indeed, Enright shows that the class (6.48) is stiffly stable for $k \leqslant 7$ and A-stable for $k = 1, 2$. Tables of the coefficients appearing in (6.48) together with the angle α of $A(\alpha)$-stability and the parameter a appearing in the definition of stiff stability can be found in Enright's paper. There is an additional bonus in using Obrechkoff methods, namely that the error constants are considerably smaller than those of linear multistep methods. Table 6.7 compares the error constants of Enright's methods with those of the BDF and the Adams–Moulton methods of orders 3, 4 and 5.

Handling the implicitness presents some new difficulties. Modified Newton iteration is of course necessary, but now the implicit equation to be solved (analogous to (6.27) of §6.5 for linear multistep methods) is

$$y_{n+k} = h\beta_k f(y_{n+k}) + h^2\gamma_k f^{(1)}(y_{n+k}) + \Psi_n.$$

The formula for modified Newton iteration is an obvious modification of (6.30), the key difference being that the matrix multiplying the increment $\tilde{\Delta} y_{n+k}$ becomes

$$I - h\beta_k \frac{\partial f}{\partial y} - h^2\gamma_k \frac{\partial}{\partial y} f^{(1)} \tag{6.49}$$

evaluated at y_{n+k}. Since $f^{(1)} = (\partial f/\partial y)f$, the second derivative of the components of f with respect to those of y will appear in (6.49). Following a suggestion of Liniger and Willoughby (1970), who had earlier studied a more restricted set of second derivative methods, Enright proposes that the terms $\partial/\partial y[(\partial f/\partial y)f]$ in (6.49) be replaced by $(\partial f/\partial y)^2$, and only first derivatives are involved. For large systems, the evaluation of $(\partial f/\partial y)^2$ still represents a considerable amount of computation, and Enright (1974b) proposes a modification of (6.48) which leads to a more efficient implementation; the price is that the order drops to $k + 1$. These methods are implemented in VSVO format in the codes SDBASIC (see Enright, Hull and Lindberg (1975) and SECDER (Addison, 1979)).

Table 6.7

	$p = 3$	$p = 4$	$p = 5$
Enright	$\frac{1}{72}$	$\frac{7}{1440}$	$\frac{-5}{4320}$
BDF	$\frac{-3}{22}$	$\frac{-12}{125}$	$\frac{-10}{137}$
Adams–Moulton	$\frac{-1}{24}$	$\frac{-19}{720}$	$\frac{-3}{160}$

An alternative means of introducing the Jacobian directly into linear multistep like methods consists of the *variable coefficient multistep methods or VCMM*; see Brunner (1967), Lambert and Sigurdsson (1972) and Sanz-Serna (1981). These methods are essentially linear multistep methods whose coefficients are functions of a variable matrix Q_n which, in practice, is taken to be an approximation to the negative Jacobian $-\partial f/\partial y$ evaluated at (x_n, y_n). The class is defined by

$$\sum_{j=0}^{k}\left[\sum_{s=0}^{S} a_j^{(s)} h^s Q_n^s\right] y_{n+j} = h\sum_{j=0}^{k}\left[\sum_{s=0}^{S-1} b_j^{(s)} h^s Q_n^s\right] f_{n+j}, \tag{6.50}$$

where Q_n is a variable $m \times m$ matrix such that $\|Q_n\|$ is bounded for all n, I is the $m \times m$ unit matrix, and Q_n^s is interpreted to be I when $s = 0$. In general, (6.50) is fully implicit and requires the solution at each step of a nonlinear system of dimension m. If, however, $b_k^{(s)} = 0$, $s = 0, 1, \ldots, S-1$, then the method is said to be *linearly implicit*; at each step it is necessary to solve one linear system of dimension m, so that linear implicitness is in a sense part way between implicitness and explicitness. Note that the computational cost per step of a linearly implicit method is the same as that which would arise if we solved an implicit linear multistep method by Newton iteration and terminated the iteration after one step. The order of (6.50), again defined by an obvious extension of the definition for linear multistep methods, is independent of Q_n, so that one can afford to have inaccurate representations of the negative Jacobian without affecting the accuracy of the method.

A-, $A(\alpha)$- and stiff-stability are investigated by applying (6.50) to the test equation $y' = Ay$, A an $m \times m$ matrix with all its eigenvalues in the left half plane, and setting $Q_n = -A$. The maximum order that an A- stable method of the type (6.50) can attain is $2S$. An example of a 2-step linearly implicit VCCM (which we shall use in the next section) has order $p = 2$ and $S = 1$, and is given by

$$(I + \tfrac{1}{2}hQ_n)y_{n+2} - [(1+\alpha)I + hQ_n]y_{n+1} + (\alpha I + \tfrac{1}{2}hQ_n)y_n = h[\tfrac{1}{2}(3-\alpha)f_{n+1} - \tfrac{1}{2}(1+\alpha)f_n]. \tag{6.51}$$

It has local truncation error

$$\mathrm{TE} = h^3\left[\frac{5+\alpha}{12} y^{(3)}(x_n) + \tfrac{1}{2}Q_n y^{(2)}(x_n)\right] + 0(h^4). \tag{6.52}$$

This method is A-stable if $-1 < \alpha < 1$. A-stable linearly implicit VCMM with $S = 2$, $p = k = 3$ and $S = 2, p = k = 4$ can be found in Lambert and Sigurdsson (1972). Sanz-Serna (1981) proved the interesting result that to every convergent linear k-step method of order k (the k-step BDF is the most significant example) there corresponds a linearly implicit k-step VCMM (with $S = 1$) of order k such that both methods generate the same numerical solution when applied to $y' = Ay$ (and Q_n is chosen to be $-A$).

Another interpretation of VCMM presents itself if we simply gather the terms in (6.50) in a different way; (6.50) can be written as

$$\sum_{s=0}^{S-1} h^s Q^s \sum_{j=0}^{k} [a_j^{(s)} y_{n+j} - h b_j^{(s)} f_{n+j}] + \sum_{j=0}^{k} a_j^{(S)} y_{n+j} = 0. \tag{6.53}$$

Thus a VCMM can be interpreted as a linear combination of linear multistep methods, the coefficients in the combination being powers of hQ_n. If the VCMM has order p, then the linear multistep method

$$\sum_{j=0}^{k} [a_j^{(s)} y_{n+j} - hb_j^{(s)} f_{n+j}] = 0$$

has order $p-s$. Combinations such as (6.53) are christened *blended linear multistep methods* by Skeel and Kong (1977), who develop a VSVO algorithm based on blends of the Adams–Moulton methods and the BDF.

Finally, the Jacobian can also be injected directly into the coefficients of a Runge–Kutta method, an idea first proposed by Rosenbrock (1963). Rosenbrock methods have been extensively developed in recent years, and various forms have been studied; that most usually considered is

$$\left.\begin{aligned} y_{n+1} &= y_n + h\sum_{i=1}^{s} b_i k_i \\ \text{where}\quad (I-\gamma hJ)k_i &= f\left(y_n + h\sum_{j=0}^{i-1} a_{ij}k_j\right) + hJ\sum_{j=0}^{i-1}\gamma_{ij}k_j, \qquad i=1,2,\ldots,s \end{aligned}\right\} \tag{6.54}$$

and we have assumed that the differential system is in autonomous form $y' = f(y)$. The matrix J is usually taken to be the Jacobian $\partial f/\partial y$ evaluated at y_n. We see that $(I-\gamma hJ)k_i$ is given explicitly in terms of previously computed k_j, so that a linear system for k_i has to be solved at each stage of the s-stage method. The method is thus linearly implicit. One can regard (6.54) as either a modification of an explicit Runge–Kutta method or a linearization of a semi-implicit Runge–Kutta method. A-stable (or nearly A-stable) methods of the form (6.54) of orders up to 6 can be found (Kaps and Wanner, 1981), and embedded Rosenbrock methods yielding error estimates have been derived by Kaps and Rentrop (1979). For further information on Rosenbrock methods the interested reader is referred to Verwer (1982), where a list of relevant references can be found.

Exercise

6.8.1. For the cases $k=1,2$, find the coefficients in the methods defined by (6.48), assumed to have order $k+2$, and show that the methods are A-stable.

6.9 CORRELATION WITH FINITE DIFFERENCE METHODS FOR PARTIAL DIFFERENTIAL EQUATIONS

This section is by way of a diversion. Most readers will have had some exposure to finite difference (FD) schemes for partial differential equations (PDEs); those who have not are referred to Mitchell and Griffiths (1980). Here we take a very superficial look at such methods, with the sole aim of seeing how the ideas we have developed in this chapter for ordinary differential equations (ODEs) correlate with those that have evolved in the study of FD schemes for PDEs. The two subject areas have developed independently, with the consequence that the nomenclatures are different and sometimes contradictory. All the early work on FD schemes concerned only scalar linear constant coefficient PDEs. The counterpart in ODEs would be the scalar equation $y' = \lambda y$ and is of course trivial; that corresponding equations in PDEs are far from trivial arises from the fact that the region in which the PDE holds plays a major role. As a vehicle for our discussions, we take the simplest possible PDE and region, and consider the parabolic equation

$$\frac{\partial}{\partial t} u(x,t) = \frac{\partial^2}{\partial x^2} u(x,t), \tag{6.55i}$$

with the initial/boundary conditions

$$u(x,0) = \varphi(x), \qquad 0 \leqslant x \leqslant 1; \qquad u(0,t) = u(1,t) = 0, \qquad t \geqslant 0. \tag{6.55ii}$$

To apply a FD scheme, we first construct a rectangular mesh in the semi-infinite rectangle $0 \leqslant x \leqslant 1$, $t \geqslant 0$ by drawing lines parallel to the x- and t-axes with mesh spacings Δx and Δt, and seek approximate solutions to (6.55) at the mesh points (x_m, t_n), $m = 1, 2, \ldots, M$, $t = 1, 2, \ldots$, where $x_m = m\Delta x$, $t_n = n\Delta t$ and $(M+1)\Delta x = 1$. We denote such an approximate solution by $U_m^n \approx u(m\Delta x, n\Delta t)$. A FD scheme consists of a linear relationship between this approximation and similar ones at neighbouring mesh points, and involves the *mesh ratio* $r := \Delta t/(\Delta x)^2$, which plays a role similar to that played by the steplength in a linear multistep method. We shall assume that r is *fixed*. If the relationship involves approximations at k successive levels of t, it is called a *k-level* method. If it gives the value of U_m^n at the newest level of t explicitly, without the need to solve a linear system, it is said to be *explicit*; otherwise it is *implicit*. If a FD scheme is 'stable' only for a certain interval of the mesh ratio r, it is said to be *conditionally stable*, and if it is 'stable' for all positive r it is said to be *unconditionally stable*. Definitions of these properties (which can be found in Mitchell and Griffiths (1980)) are not particularly relevant to our discussion here; our aim is merely to interpret conditional and unconditional stability in terms of the linear stability definitions developed for ODEs in §6.3. We quote below four well-known FD methods for problem (6.55). The local truncation errors, defined as the residuals when the approximate solution is replaced by the exact, are found by Taylor expansions. It is notationally convenient to make use of the central difference operator δ_x^2, defined by $\delta_x^2 U_m^n := U_{m+1}^n - 2U_m^n + U_{m-1}^n$.

The elementary explicit scheme

$$U_m^{n+1} = (1 + r\delta_x^2) U_m^n \tag{6.56}$$

$$\text{LTE} = 0((\Delta t)^2) + 0(\Delta t(\Delta x)^2).$$

The method is explicit, two-level and is stable if and only if $r \leqslant \frac{1}{2}$.

The Crank–Nicolson scheme

$$(1 - \tfrac{1}{2} r\delta_x^2) U_m^{n+1} = (1 + \tfrac{1}{2} r\delta_x^2) U_m^n \tag{6.57}$$

$$\text{LTE} = 0((\Delta t)^3) + 0(\Delta t(\Delta x)^2).$$

The method is implicit, two-level and unconditionally stable.

The Douglas scheme

$$[1-\tfrac{1}{2}(r-\tfrac{1}{6})\delta_x^2]U_m^{n+1}=[1+\tfrac{1}{2}(r+\tfrac{1}{6})\delta_x^2]U_m^n \qquad (6.58)$$

$$\text{LTE}=0((\Delta t)^3)+0(\Delta t(\Delta x)^4).$$

The method is implicit, two-level and unconditionally stable. Note that it has higher accuracy than the Crank–Nicolson method.

The Du Fort–Frankel scheme

$$(1+2r)U_m^{n+1}=2r(U_{m+1}^n+U_{m-1}^n)+(1-2r)U_m^{n-1} \qquad (6.59)$$

$$\text{LTE}=0((\Delta t)^2)+0(\Delta t(\Delta x)^2).$$

The method is explicit, three-level and unconditionally stable.

We note with interest that (6.59) is unconditionally stable *and* explicit; in the context of ODEs, we never found an explicit method with an infinite region of absolute stability! There is, however, a difficulty about the convergence of the Du Fort–Frankel scheme, namely that (6.59) converges to the exact solution only if the so-called *consistency restraint* is satisfied; see, for example, Richtmyer and Morton (1967). This requires that

$$\Delta t/\Delta x\to 0 \qquad \text{as } \Delta t,\ \Delta x\to 0. \qquad (6.60)$$

Let us now attempt to interpret these methods in terms of methods for ODEs. Instead of discretizing the problem (6.55) completely, we *semi-discretize* it by leaving t as a continuous variable and discretizing only the x variable. This is equivalent to replacing the rectangular mesh by a sequence of lines parallel to the t-axis; the process of semi-discretization is thus sometimes called the *method of lines*. Define $u(t):=[u(x_1,t),u(x_2,t),\ldots,u(x_M,t)]^T$ and replace $\partial^2u(x_m,t)/\partial x^2$ by $[u(x_{m+1},t)-2u(x_m,t)+u(x_{m-1},t)]/(\Delta x)^2$, $m=1,2,\ldots,M$. Let us denote by $U(t)\in\mathbb{R}^M$ the exact solution of the resulting semi-discrete problem

$$\left.\begin{aligned}
&\mathrm{d}U(t)/\mathrm{d}t=BU(t),\qquad U(0)=\varphi\\
&\text{where}\\
&U(t)=[{}^1U(t),{}^2U(t),\ldots,{}^MU(t)]^T,\\
&\varphi=[\varphi(x_1),\varphi(x_2),\ldots,\varphi(x_M)]^T,\\
&\text{and}\\
&B=\frac{1}{(\Delta x)^2}\begin{bmatrix}-2 & 1 & 0 & 0 & \cdots & 0\\ 1 & -2 & 1 & 0 & & 0\\ \vdots & & & & & \vdots\\ 0 & & & 1 & -2 & 1\\ 0 & \cdot & \cdot & 0 & 1 & -2\end{bmatrix}.
\end{aligned}\right\} \qquad (6.61)$$

The initial value problem (6.61) can now be solved numerically by any appropriate numerical method for ODEs. In order to apply such a method, we make the discretization $t_n = n\Delta t$, $n = 0, 1, \ldots$, and denote the numerical solution so obtained by $U_n := [{}^1U_n, {}^2U_n, \ldots, {}^MU_n]^T$. The eigenvalues of the matrix B are known to be $\lambda_j = [-2 + 2\cos(j\pi/(M+1))]/(\Delta x)^2$, $j = 1, 2, \ldots, M$; they are real and lie in the interval $(-4/(\Delta x)^2, 0)$ of the negative real axis. We observe that, if Δx is small, the system in (6.61) is stiff. If $\mathscr{R}_A$ is the region of absolute stability of the numerical method employed, then we will achieve absolute stability if the steplength Δt is such that

$$(-4\Delta t/(\Delta x)^2, 0) \equiv (-4r, 0) \subseteq \mathscr{R}_A. \tag{6.62}$$

Note that this procedure (which we are developing here only as a means of examining the correlation between PDE and ODE methods) is a viable numerical method in its own right for solving (6.55), and one that is frequently used. A suitable means of solving (6.61) would be the GEARB code mentioned in §6.6.

Let us first solve (6.61) by Euler's Rule, $U_{n+1} - U_n = \Delta t B U_n$. The mth component of this equation is

$$ {}^mU_{n+1} - {}^mU_n = \frac{\Delta t}{(\Delta x)^2}[{}^{m+1}U_n - 2\,{}^mU_n + {}^{m-1}U_n] \tag{6.63}$$

and on identifying ${}^mU_n$ with U_m^n, we see that the one-step method (6.63) is precisely the two-level elementary explicit FD scheme (6.56). The order is 1 so that the PLTE $= 0((\Delta t)^2)$, consistent with the first term in the LTE given in (6.56); the second term arises from the truncation error of the semi-discretization process. The interval of absolute stability of Euler's Rule is $(-2, 0)$, so that the stability condition (6.62) is satisfied if and only if $r \leqslant \frac{1}{2}$, suggesting that we may interpret conditional stability of a FD scheme as being equivalent to absolute stability of an ODE method.

In an exactly analogous way, we find that applying the Trapezoidal Rule to (6.61) is equivalent to the Crank–Nicolson FD scheme (6.57). The A-stability of the Trapezoidal Rule corresponds to the unconditional stability of the Crank–Nicolson scheme. Recalling that the eigenvalues of B are real, it is clear that a *sufficient* condition for an FD scheme to be unconditionally stable is that the equivalent numerical method for (6.61) be A_0-stable. At this stage we might hazard a guess that the condition is also *necessary*—but that guess would be wrong! Consider the Theta method (6.40), $U_{n+1} - U_n = \Delta t B[(1-\theta)U_{n+1} + \theta U_n]$, applied to (6.61). A straightforward (but painful) manipulation with Taylor series establishes that, taking into account the truncation error associated with the semi-discretization, there is a cancellation of terms in the LTE of the equivalent FD scheme if we choose

$$\theta = \frac{1}{2} + \frac{1}{12r}. \tag{6.64}$$

The equivalent FD scheme is then the Douglas scheme (6.58). It is all too easy to jump to the conclusion that the unconditional stability of the Douglas scheme follows from the fact that the Theta method is A-stable for a certain range of θ (and all the easier in view of the fact that some authors quote the Theta method with θ and $1-\theta$ interchanged!). The necessary and sufficient condition for the Theta method to be A-stable is that $\theta \leqslant \frac{1}{2}$ (see §6.6), a condition clearly *not* satisfied by the choice (6.64)! It is straightforward to establish that when $\theta > \frac{1}{2}$, the region of absolute stability of the Theta method is a circle on $(2/(1-2\theta), 0)$ as diameter. On substituting from (6.64), this becomes the circle on $(-12r, 0)$ as diameter, and the stability condition (6.62) is satisfied. Thus the Douglas method is unconditionally stable, but the equivalent ODE method is not A_0-stable. (Note, however, that with the choice (6.64), the Theta method is no longer a linear multistep method, since its coefficients depend on the steplength Δt.)

A similar phenomenon explains why it is possible for the explicit Du Fort–Frankel scheme to be unconditionally stable. This time the equivalent ODE method is derived from the linearly implicit VCMM (6.51) of §6.8, which, when applied to (6.61), gives

$$\begin{aligned}(I+\tfrac{1}{2}\Delta t Q_n)U_{n+2}-[(1+\alpha)I+\Delta t Q_n]U_{n+1}+(\alpha I+\tfrac{1}{2}\Delta t Q_n)U_n\\ =\Delta t B[\tfrac{1}{2}(3-\alpha)U_{n+1}-\tfrac{1}{2}(1+\alpha)U_n].\end{aligned} \tag{6.65}$$

If $-1<\alpha<1$, the VCMM is A-stable, provided we take $Q_n=-B$. However, the fact that (6.65) is linearly implicit means that the equivalent FD scheme will be implicit. In an attempt to force explicitness, let us choose $Q_n=qI$, where q is a scalar and I the unit matrix. The coefficient of U_{n+2} in (6.65) is now scalar and the equivalent FD scheme is explicit. However, the price that has to be paid for forcing explicitness is that (6.65) is no longer A-stable. Sigurdsson (1973) shows that when $Q_n=qI$, the region of absolute stability is a simple *closed* region of the negative half-plane which intercepts the real axis in the interval $[-1-\alpha-q\Delta t, 0]$. The stability criterion (6.62) is thus satisfied if we choose α and q such that

$$1+\alpha+q\Delta t\geqslant 4r. \tag{6.66}$$

A simple choice which satisfies (6.66) is $\alpha=-1$, $q=4/(\Delta x)^2$; for this choice the region of absolute stability is now an ellipse with axis $(-4r, 0)$. Note that $-Q_n$ now has an eigenvalue $-4/(\Delta x)^2$ of multiplicity M, while the eigenvalues of B all lie in $(-4/(\Delta x)^2, 0)$. With the above choice of parameters, the FD scheme equivalent to (6.65) is the Du Fort–Frankel scheme (6.59). This equivalence not only explains how an explicit scheme can be unconditionally stable, but also affords an alternative interpretation of the consistency restraint (6.60). Qualitatively, we can see that there is going to be trouble as we let Δt and Δx approach zero; since $Q_n=qI=-4I/(\Delta x)^2$, the condition (stated in §6.8) that $\|Q_n\|$ be bounded will be violated. Quantitatively, it follows from (6.52) of §6.8 that

$$\mathrm{LTE}=\left[\frac{5+\alpha}{12}U^{(3)}(t_n)+\tfrac{1}{2}Q_nU^{(2)}(t_n)\right](\Delta t)^3+0((\Delta t)^4).$$

Since $Q_n=0(1/(\Delta x)^2)$, the second term in the bracket is

$$0((\Delta t)^3/(\Delta x)^2)=0\left(\left(\frac{\Delta t}{\Delta x}\right)^2\right)0(\Delta t).$$

If $\Delta t/\Delta x$ were to tend to a non-zero constant as Δt, $\Delta x\to 0$, then the method would effectively have order only zero, and would be inconsistent. Thus the restraint (6.60) is necessary for consistency.

Other VCMMs can be shown to be equivalent to *splitting techniques* in FD schemes (see Exercise 6.9.1); further details can be found in Lambert (1975), where fuller details of the work of this section can also be found.

Exercise

6.9.1*. The following fully implicit VCMM for the system $y' = f(x, y)$ has order two and is A-stable (choosing $Q_n = -A$) for $a \geqslant 0$, $2(b + c) \geqslant a$:

$$[I + ahQ_n + b(hQ_n)^2](y_{n+1} - y_n) = h[(\tfrac{1}{2}I + chQ_n)f_{n+1} + [\tfrac{1}{2}I + (a - c)hQ_n]f_n].$$

Apply this method to $y' = Ay$, A a symmetric matrix, with the choice $Q_n = -C$, where C is a triangular matrix such that $C + C^{\mathrm{T}} = A$. Show that the resulting difference system simplifies considerably if we choose $a = 0$, $b = -c$. Make the further choice $c = \frac{1}{4}$ to obtain a difference equation which can be split into a two-stage form involving an intermediate value (call it $y_{n+1/2}$). Apply the resulting method to (6.61) to obtain an effectively explicit unconditionally stable finite difference method for the problem (6.55). Show that this method (known in the PDE literature as Saul'ev's method) suffers the same consistency restraint as does the DuFort–Frankel method.

7 Stiffness: Nonlinear Stability Theory

7.1 THE SHORTCOMINGS OF LINEAR STABILITY THEORY

At various stages in this book we have pointed out the inadequacies of linear stability theory when applied to nonlinear or even linear variable coefficient systems. Over recent years there has emerged an alternative theory, which suffers none of the shortcomings of linear stability theory. A full development of this *nonlinear stability theory* is beyond the scope of this book, and our intention in this chapter is merely to give a flavour of the work and present some of the more significant results. The reader who wishes to see a rigorous account of the theory is referred to the excellent treatise by Dekker and Verwer (1984), whose general approach we follow here.

In §3.8, where we considered the linear stability of linear multistep methods, we produced a popular (but false) argument which seeks to extend the applicability of the linear theory by deriving the linearized error equation (3.78),

$$\sum_{j=0}^{k} [\alpha_j I - h\beta_j J]E_{n+j} = T, \tag{7.1}$$

where the α_j and β_j are the coefficients of the linear multistep methods, E_{n+j} is the global error at x_{n+j}, T is the local truncation error (assumed constant) and J is the Jacobian of the system, also assumed constant (or 'frozen'). We showed by example that it could happen that the solutions of (7.1) did not correctly represent, even in a qualitative manner, the behaviour of the global error; this could happen even if the Jacobian were taken to be 'piecewise frozen' (that is, the constant value assumed for the Jacobian is re-computed from time to time as the computation of the numerical solution progresses).

In the context of stiffness, a false argument analogous to that which produced (7.1) would go as follows. Consider the general initial value problem

$$y' = f(x, y), \qquad y(a) = \eta, \qquad f: \mathbb{R} \times \mathbb{R}^m \to \mathbb{R}^m, \tag{7.2}$$

where $f(x, y)$ satisfies a Lipschitz condition with respect to y, so that there exists a unique solution $y(x)$. In some neighbourhood of this exact solution, $y(x)$ can be well represented by a solution of

$$y' = f((x, y(x)) + \frac{\partial f}{\partial y}(x, y(x))[y - y(x)] \tag{7.3}$$

the so-called *variational equation.* Now assume that the Jacobian $\partial f/\partial y$ can be locally 'frozen'. Then (7.3) takes the form $y' = Ay + \varphi$, where $\varphi = \varphi(x, y(x))$ does not depend on y. Since stability essentially depends only on A, we ignore φ and arrive at the conclusion that the behaviour of the solutions of the equation $y' = Ay$, where A is a 'piecewise frozen' value of the Jacobian, in some way locally represents the behaviour of the solutions of (7.2), thus justifying the use of the linear test equation $y' = Ay$ in a nonlinear context. We can see more clearly what is being asserted if we restrict ourselves to the homogeneous linear variable coefficient case (for which the argument remains false) by setting $f(x, y) = A(x)y$ in (7.2); then equations (7.3) and (7.2) both become

$$y' = A(x)y. \tag{7.4}$$

Let x^* be some *fixed* value of x; then the 'piecewise frozen' Jacobian argument would assert that in some neighbourhood of x^*, the solutions of (7.4) behave like those of

$$y' = A(x^*)y. \tag{7.5}$$

Since $A(x^*)$ is constant, the general solutions of (7.5) has the form

$$y(x) = \sum_{t=1}^{m} \varkappa_t \exp(\lambda_t^* x)c_t,$$

where λ_t^*, $t = 1, 2, \dots, m$ are the eigenvalues (assumed distinct) of $A(x^*)$. The 'frozen' Jacobian argument would assert that if these eigenvalues were complex we would expect (7.4) to have oscillatory solutions; if they had negative real part we would expect (7.4) to have decaying solutions. The following simple example shows that the first of these assertions is false:

Example 1

$$y' = A(x)y = \frac{1}{1+x^2}\begin{bmatrix} 0 & 1 \\ -1 & 0 \end{bmatrix} y. \tag{7.6}$$

The eigenvalues of $A(x)$ are $\lambda_1, \lambda_2 = \pm i/(1 + x^2)$, and are purely imaginary for all values of x; yet the general solution of (7.6),

$$y(x) = \varkappa_1(1 + x^2)^{-1/2}\begin{bmatrix} 1 \\ -x \end{bmatrix} + \varkappa_2(1 + x^2)^{-1/2}\begin{bmatrix} x \\ 1 \end{bmatrix},$$

is not oscillatory. In fact, we hardly need an example to see that it is impossible for the 'piecewise frozen' Jacobian argument to predict oscillatory solutions. The question we are investigating is whether the solutions of (7.5) mimic those of (7.4) *in some neighbourhood of* x^*; we are thus looking for an indication of *local* behaviour. Oscillatory behaviour is a *global* phenomenon, and it does not make sense to talk of a solution being oscillatory in some neighbourhood of x^*.

More important is the question of whether negativity of the real parts of the eigenvalues of $A(x^*)$ imply that (7.4) has decaying solutions in a neighbourhood of x^*. The only interpretation of 'decaying solutions' that makes sense for a nonlinear or a linear variable coefficient system is to take the phrase to mean that for *any* solution $y(x)$ of (7.4), $\| y(x) \|$

is monotonic decreasing in a neighbourhood of x^*. Perhaps the most striking example which shows that such a property does not follow from the negativity of the eigenvalues of the 'piecewise frozen' Jacobian is one due to Vinograd (1952) (a generalization of which can be found in Dekker and Verwer (1984):

Example 2

$$y' = A(x)y = \begin{bmatrix} -1-9\cos^2 6x + 6\sin 12x & 12\cos^2 6x + 4.5\sin 12x \\ -12\sin^2 6x + 4.5\sin 12x & -1-9\sin^2 6x - 6\sin 12x \end{bmatrix} y. \tag{7.7}$$

The remarkable thing about this example is that the eigenvalues of $A(x)$ are $\lambda_1 = -1$, $\lambda_2 = -10$, and are independent of x. The general solution of (7.7) is

$$y(x) = \varkappa_1 e^{2x} \begin{bmatrix} \cos 6x + 2\sin 6x \\ 2\cos 6x - \sin 6x \end{bmatrix} + \varkappa_2 e^{-13x} \begin{bmatrix} \sin 6x - 2\cos 6x \\ 2\sin 6x + \cos 6x \end{bmatrix}.$$

The general solution is certainly not monotonic decreasing for any x.

The following is an example going the other way, where the eigenvalues of $A(x)$ have positive real part but there exists a solution which is monotonic decreasing:

Example 3

$$y' = A(x)y = \begin{bmatrix} 0 & 1 \\ \dfrac{1-x}{x} & \dfrac{1-2x}{x} \end{bmatrix} y, \qquad x > 0. \tag{7.8}$$

The eigenvalues of $A(x)$ are $\lambda_1 = -1$, $\lambda_2 = (1-x)/x$, so that for $x \in (0, 1)$, $\lambda_2 > 0$. The general solution of (7.8) is

$$y(x) = \varkappa_1 e^{-x} \begin{bmatrix} 1 \\ -1 \end{bmatrix} + \varkappa_2 e^{-x} \begin{bmatrix} x^2 \\ x(2-x) \end{bmatrix} \tag{7.9}$$

and the solution given by taking $\varkappa_2 = 0$ is certainly not monotonic increasing for $x \in (0, 1)$.

We consider one further example, which we shall use again in a later section:

Example 4

$$y' = A(x)y = \begin{bmatrix} \dfrac{-1}{2x} & \dfrac{2}{x^3} \\ \dfrac{-x}{2} & \dfrac{-1}{2x} \end{bmatrix} y, \qquad x \geqslant 1. \tag{7.10}$$

The eigenvalues of $A(x)$ are $\lambda_1, \lambda_2 = (-1 \pm 2i)/(2x)$, so that both eigenvalues have negative real parts for the indicated interval $x \geqslant 1$. The general solution of (7.10) is

$$y(x) = \varkappa_1 \begin{bmatrix} x^{-3/2} \\ -\tfrac{1}{2}x^{1/2} \end{bmatrix} + \varkappa_2 \begin{bmatrix} 2x^{-3/2}\ln x \\ x^{1/2}(1-\ln x) \end{bmatrix}. \tag{7.11}$$

For the solution given by $\varkappa_1 = 1$, $\varkappa_2 = 0$,

$$\| y(x) \|_2 := \sqrt{[y^{\mathrm{T}}(x)y(x)]} = \sqrt{(x^{-3} + x/4)}$$

and $\| y(x) \|_2$ is monotonic increasing for $x > (12)^{1/4} \approx 1.86$.

Of course, no one seriously believes the 'frozen' Jacobian argument; but we have become so used to the application of linear stability theory to stiff systems, that it is all too easy to find ourselves making statements like 'The eigenvalues have negative real parts and are close to the imaginary axis, so the solutions will be slowly damped oscillations'. Such statements are strictly valid only for the linear constant coefficient system $y' = Ay$; for general systems they will sometimes be true and sometimes false.

Exercises

7.1.1. Find the eigenvalues of the 2×2 linear system $y' = A(x)y$, where

$$A(x) = \begin{bmatrix} 0 & 1 \\ \dfrac{\cos x - \sin x}{2 + \sin x + \cos x} & \dfrac{-2(1 + \sin x)}{2 + \sin x + \cos x} \end{bmatrix}.$$

Show that $y(x) = [2 + \sin x, \cos x]^{\mathrm{T}}$ is a solution of the system and find (by guessing a bit) another solution. Conclude that a system with oscillatory solutions can have real eigenvalues.

7.1.2. Find the eigenvalues of the 2×2 linear system $y' = A(x)y$, where

$$A(x) = \begin{bmatrix} 0 & 1 \\ -1/(16x^2) & -1/2x \end{bmatrix}, \qquad x > 0.$$

Show that $[4x^{1/4}, x^{-3/4}]^{\mathrm{T}}$ is a solution of the system, and deduce that this example backs up the conclusions we drew from Example 2.

7.1.3. We reveal here how we found Example 3. Consider the system

$$\begin{bmatrix} u' \\ v' \end{bmatrix} = \begin{bmatrix} 0 & 1 \\ \varphi(x) & \varphi(x) - 1 \end{bmatrix} \begin{bmatrix} u \\ v \end{bmatrix}.$$

Find the eigenvalues and the general solution of the system.
(*Hint*: Eliminate v from the system and set $w = u + u'$.)

Use your results to construct more examples like Example 3 which will confirm that there is no relationship between decaying solutions and negativity of the real parts of the eigenvalues of the system.

7.1.4. The nonlinear system

$$u' = \beta[\exp(3x) - u^3] + v, \qquad u(0) = 1$$
$$v' = \exp(3x) - v^3 + \exp(x), \qquad v(0) = 1$$

has solution $u(x) = v(x) = \exp(x)$ (independent of β); the solution clearly increases with x. Show that the eigenvalues λ_1, λ_2 of the Jacobian are always real and *negative*, and that by choosing β appropriately we can make the system apparently as stiff as we like (in the sense that $\lambda_1 \ll \lambda_2 < 0$).

7.1.5. Here is a device for constructing linear variable coefficient and nonlinear *initial value problems* with known solutions. (It will not establish general solutions.) The linear variable coefficient problem

$$y' = A(x)[z(x) - y] + z'(x), \quad y(a) = z(a) = \eta \tag{1}$$

has solution $y(x) = z(x)$. Likewise, the nonlinear problem

$$y' = f(x, y) = \varphi(x, y) - \varphi(x, z(x)) + z'(x), \quad y(a) = z(a) = \eta \tag{2}$$

has solution $y(x) = z(x)$. (An important difference between these two is that *all* linear problems can be put in the form (1), whereas not all nonlinear problems can be put in the form (2).)

Use these constructions to devise problems which support the general conclusions of §7.1.

7.2 CONTRACTIVITY

The examples of the preceding section should not only convince us that the 'frozen' Jacobian argument gives the wrong answer to the question 'When do the solutions of a general system decay?', but also suggest that we are asking the wrong question. For the linear constant coefficient system $y' = Ay$, negativity of the real parts of the eigenvalues of A implies that $\| y(x) \|$ decreases, but also implies that neighbouring solution curves get closer together as x increases. It turns out to be much more fruitful to seek generalizations of this second property. We are thus motivated to make the following definition:

Definition *Let $y(x)$ and $\tilde{y}(x)$ be any two solutions of the system $y' = f(x, y)$ satisfying initial conditions $y(a) = \eta$, $\tilde{y}(a) = \tilde{\eta}$, $\eta \neq \tilde{\eta}$. Then if*

$$\| y(x_2) - \tilde{y}(x_2) \| \leqslant \| y(x_1 - \tilde{y}(x_1) \| \tag{7.12i}$$

for all x_1, x_2 such that

$$a \leqslant x_1 \leqslant x_2 \leqslant b, \tag{7.12ii}$$

the solutions of the system are said to be **contractive** *in $[a, b]$.*

We see at once the possibility of an analogous definition for numerical solutions. For a k-step method define $Y_n, \tilde{Y}_n \in \mathbb{R}^{mk}$ by

$$\left.\begin{aligned} Y_n &:= [y_{n+k-1}^{\mathrm{T}}, y_{n+k-2}^{\mathrm{T}}, \ldots, y_n^{\mathrm{T}}]^{\mathrm{T}}, \; Y_0 = z_0 \\ \tilde{Y}_n &:= [\tilde{y}_{n+k-1}^{\mathrm{T}}, \tilde{y}_{n+k-2}^{\mathrm{T}}, \ldots, \tilde{y}_n^{\mathrm{T}}]^{\mathrm{T}}, \; \tilde{Y}_0 = \tilde{z}_0 \end{aligned}\right\} z_0 \neq \tilde{z}_0, \tag{7.13}$$

where $\{y_n\}$ and $\{\tilde{y}_n\}$ are two numerical solutions generated by the method with different starting values. (Note that for a one-step method $Y_n = y_n$.)

Definition *Let $\{Y_n\}$ and $\{\tilde{Y}_n\}$ be defined by (7.13). Then if*

$$\| Y_{n+1} - \tilde{Y}_{n+1} \| \leqslant \| Y_n - \tilde{Y}_n \|, \qquad 0 \leqslant n \leqslant N \tag{7.14}$$

the numerical solutions and the method are said to be **contractive** *for $n \in [0, N]$.*

The requirement (7.14) makes a lot of practical sense. The inevitable introduction of discretization errors in a numerical solution can be thought of as being equivalent to jumping on to a neighbouring solution curve; if we demand that the numerical solutions be contractive whenever the exact solutions are, then we are ensuring that the numerical solution cannot wander away from the exact solution. We are thus led to a new breed of stability definition with the syntax diagram (see §2.6).

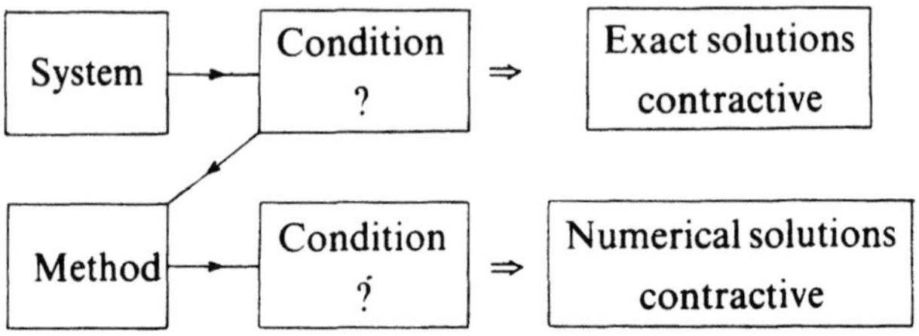

The first task is to find an appropriate condition for the middle box of the top line.

7.3 THE ONE-SIDED LIPSCHITZ CONSTANT AND THE LOGARITHMIC NORM

Recall the Lipschitz condition of the system $y' = f(x, y)$, defined by

$$\|f(y) - f(\tilde{y})\| \leqslant L\|y - \tilde{y}\|. \tag{7.15}$$

where L is the Lipschitz constant. In our search for a sufficient condition for the solutions of $y' = f(x, y)$ to be contractive, it is clear that the Lipschitz condition is not going to be subtle enough. To see this we need only look at the scalar equations $y' = -y$ and $y' = y$, both of which have the same Lipschitz constant of $+1$; the solutions of the first are contractive while those of the second are not. We consider instead the so-called *one-sided Lipschitz condition.* Let $\langle \cdot, \cdot \rangle$ be an inner product and $\|\cdot\|$ the corresponding inner product norm defined by $\|u\|^2 := \langle u, u \rangle$. The theory holds for any inner product, but we shall normally use only the inner product $\langle u, v \rangle_2 := u^T v$, $u, v \in \mathbb{R}^m$, for which the corresponding norm is the L_2-norm defined by $\|u\|_2 = (u^T u)^{1/2} = (\sum_{t=1}^{m} {}^t u^2)^{1/2}$.

Definition The function $f(x, y)$ and the system $y' = f(x, y)$ are said to satisfy a **one-sided Lipschitz condition** *if*

$$\langle f(x, y) - f(x, \tilde{y}), y - \tilde{y} \rangle \leqslant \nu(x)\|y - \tilde{y}\|^2 \tag{7.16}$$

holds for all $y, \tilde{y} \in M_x$ and for $a \leqslant x \leqslant b$. The function $\nu(x)$ is called a **one-sided Lipschitz constant.**

The convex region $M_x \in \mathbb{R}^m$ is the domain of the function $f(x, y)$, regarded as a function of y; clearly, if $f(x, y) = A(x)y$, M_x can be taken to be the whole of $\mathbb{R}^m$. Note that a one-sided Lipschitz constant is, in general, a function of x; it is only constant as far as y is concerned. Condition (7.16) is less demanding than (7.15). To see this, we make use of *Schwarz's inequality*, $\langle u, v \rangle \leqslant \|u\|.\|v\|$, from which it follows that

$$\langle f(x,y) - f(x,\tilde{y}), y - \tilde{y} \rangle \leqslant \|f(x,y) - f(x,\tilde{y})\|.\|y - \tilde{y}\| \leqslant L\|y - \tilde{y}\|^2$$

if we assume that (7.15) holds. Thus if $f(x, y)$ satisfies a Lipschitz condition, then it satisfies a one-sided Lipschitz condition.

Condition (7.16), unlike (7.15), does succeed in separating the trivial scalar examples we used above. For $y' = -y$, (7.16) reads

$$\langle -y + \tilde{y}, y - \tilde{y} \rangle = -\|y - \tilde{y}\|^2 \leqslant v(x)\|y - \tilde{y}\|^2$$

and we can take $v(x) = -1$. (Note that a one-sided Lipschitz constant can be negative.) For $y' = y$, (7.16) reads

$$\langle y - \tilde{y}, y - \tilde{y} \rangle = \|y - \tilde{y}\|^2 \leqslant v(x)\|y - \tilde{y}\|^2$$

and we take $v(x) = +1$. For these simple examples, contractivity appears to be associated with negativity of the one-sided Lipschitz constant, a result we shall now show holds in general.

Let $y(x)$ and $\tilde{y}(x)$ be two solutions of $y' = f(x, y)$ satisfying initial conditions $y(a) = \eta$, $\tilde{y}(a) = \tilde{\eta}$, where $\eta \neq \tilde{\eta}$, define $\Omega(x) := \|y(x) - \tilde{y}(x)\|^2$, and assume that (7.16) holds. Then

$$\begin{aligned}\Omega'(x) &= \frac{\mathrm{d}}{\mathrm{d}x}\langle y(x) - \tilde{y}(x), y(x) - \tilde{y}(x)\rangle \\ &= 2\langle y'(x) - \tilde{y}'(x), y(x) - \tilde{y}(x)\rangle \\ &= 2\langle f(x, y(x)) - f(x, \tilde{y}(x)), y(x) - \tilde{y}(x)\rangle \leqslant 2v(x)\Omega(x),\end{aligned}$$

by (7.16). The differential inequality $\Omega'(x) \leqslant 2v(x)\Omega(x)$ can be handled in the same way as the corresponding differential equation; defining the integrating factor $\omega(x) := \exp(-2\int_0^x v(\xi)\,\mathrm{d}\xi)$, we obtain $(\mathrm{d}/\mathrm{d}x)[\omega(x)\Omega(x)] \leqslant 0$, which means that $\omega(x)\Omega(x)$ is monotonic non-increasing for all x in $[a, b]$. Since $\omega(x)$ is always positive, it follows that $\Omega(x_2) \leqslant \Omega(x_1)\omega(x_1)/\omega(x_2)$, for $a \leqslant x_1 \leqslant x_2 \leqslant b$. Now

$$\omega(x_1)/\omega(x_2) = \exp\left(2\int_{x_1}^{x_2} v(\xi)\,\mathrm{d}\xi\right) = \left(\exp\left(\int_{x_1}^{x_2} v(\xi)\,\mathrm{d}\xi\right)\right)^2,$$

whence we have that

$$\|y(x_2) - \tilde{y}(x_2)\| \leqslant \exp\left(\int_{x_1}^{x_2} v(\xi)\,\mathrm{d}\xi\right)\|y(x_1) - \tilde{y}(x_1)\|, \qquad a \leqslant x_1 \leqslant x_2 \leqslant b. \qquad (7.17)$$

It follows from (7.17) that if $v(x) \leqslant 0$ for all $x \in [a, b]$ then

$$\|y(x_2) - \tilde{y}(x_2)\| \leqslant \|y(x_1 - \tilde{y}(x_1)\|, \qquad a \leqslant x_1 \leqslant x_2 \leqslant b \qquad (7.18)$$

and we have contractivity. In particular, it follows that if (7.16) holds with $v(x) \equiv 0$ then (7.18) follows, thus motivating another definition:

Definition *The system* $y' = f(x, y)$ *is said to be* **dissipative** *in* $[a, b]$ *if*

$$\langle f(x,y) - f(x,\tilde{y}), y - \tilde{y} \rangle \leqslant 0 \tag{7.19}$$

holds for all $y, \tilde{y} \in M_x$ *and for all* $x \in [a, b]$.

Clearly the solutions of a dissipative system are contractive.

We have not yet tackled the question of how to find a one-sided Lipschitz constant. The answer lies in the *logarithmic norm*, defined by Dahlquist (1959).

Definition *The* **logarithmic norm** $\mu[A]$ *of a square matrix* A *is defined by*

$$\mu[A] := \lim_{\delta \to 0^+} (\|I + \delta A\| - 1)/\delta, \tag{7.20}$$

where I *is the unit matrix and* $\delta \in \mathbb{R}$.

The name 'logarithmic norm' is a little misleading. Although in some ways it behaves like a norm, $\mu[A]$ is not a norm; in particular, it can be negative. Note that $\mu[A]$ is norm-dependent; if the norm $\|\cdot\|$ on the right side of (7.20) is the L_2-norm $\|\cdot\|_2$, we shall denote the corresponding logarithmic norm by $\mu_2[\cdot]$.

Properties of the logarithmic norm

1. (See Dahlquist, 1959; Coppel, 1965.) Let the eigenvalues of A be $\lambda_t, t = 1, 2, \dots, m$; then

$$\max_t \operatorname{Re} \lambda_t \leqslant \mu[A] \leqslant \|A\|. \tag{7.21}$$

In particular, if $\mu[A] < 0$ then all the eigenvalues of A lie in the left half-plane; the converse is not true.

2. (See Dekker and Verwer, 1984.) If, in (7.20), $\|\cdot\|$ is an inner product norm, then

$$\mu[A] = \max_z \frac{\langle Az, z \rangle}{\|z\|^2}. \tag{7.22}$$

3. (See Dekker and Verwer, 1984.) Let $\sigma_t, t = 1, 2, \dots, m$ be the eigenvalues of $\frac{1}{2}(A + A^T)$; note that they are necessarily real. If, in (7.20), $\|\cdot\|$ is the L_2-norm, given by $\|u\|_2^2 := \langle u, u \rangle_2 = u^T u$, then

$$\mu_2[A] = \max_t \sigma_t. \tag{7.23}$$

Let us consider the general homogeneous linear system $y' = A(x)y$. Then in the left side of the one-sided Lipschitz condition (7.16), $\langle f(x,y) - f(x,\tilde{y}), y - \tilde{y} \rangle$ becomes $\langle A(x)(y - \tilde{y}), y - \tilde{y} \rangle$ and it follows from (7.22) that

$$\langle A(x)(y - \tilde{y}), y - \tilde{y} \rangle \leqslant \mu[A(x)] \|y - \tilde{y}\|^2$$

and we may thus take $\mu[A(x)]$ to be the one-sided Lipschitz constant; it is clear from (7.22) that $\mu[A(x)]$ is indeed the smallest possible one-sided Lipschitz constant. It follows immediately from (7.17) that we have found a sufficient condition for the solutions of $y' = A(x)y$ to be contractive for $x \in [a, b]$, namely $\mu[A(x)] \leqslant 0$ for $x \in [a, b]$.

We now come to the key result of this section, namely that the above result actually generalizes to the full nonlinear system $y' = f(x, y)$ (and indeed holds for an arbitrary norm); no dubious linearizing or 'freezing' arguments are involved, and the result can be stated precisely as a theorem:

Theorem 7.1 Let $\|\cdot\|$ be a given norm and let $v(x)$ be a piecewise continuous function such that

$$\mu\left[\frac{\partial f}{\partial y}(x, y)\right] \leqslant v(x) \text{ for all } x \in [a, b], y \in M_x.$$

Then, for any two solutions $y(x), \tilde{y}(x)$ of $y' = f(x, y)$ satisfying initial conditions $y(a) = \eta, \tilde{y}(a) = \tilde{\eta}, \eta \neq \tilde{\eta}$,

$$\| y(x_2) - \tilde{y}(x_2) \| \leqslant \exp\left(\int_{x^1}^{x_2} v(\xi) \mathrm{d}\xi\right) \| y(x_1) - \tilde{y}(x_1) \|,$$

for all x_1, x_2 satisfying $a \leqslant x_1 \leqslant x_2 \leqslant b$.

This theorem goes back a long way. It was first proved by Dahlquist (1959); a more accessible reference where a proof can be found is Dekker and Verwer (1984).

We now have a sufficient condition for contractivity of the solutions of a general system, namely that $\mu[\partial f / \partial y]$ be non-positive in some convex region enclosing the solution we are interested in; moreover, in the case when $\|\cdot\| = \|\cdot\|_2$, we have, from (7.23), a practical means of testing whether this condition is satisfied.

Let us try this out by conducting an experiment on Example 4 of §7.1. Recall the system (7.10)

$$y' = A(x)y = \begin{bmatrix} \dfrac{-1}{2x} & \dfrac{2}{x^3} \\ \dfrac{-x}{2} & \dfrac{-1}{2x} \end{bmatrix} y, \qquad x \geqslant 1$$

with general solution, given by (7.11),

$$y(x) = \kappa_1 \begin{bmatrix} x^{-3/2} \\ -\frac{1}{2}x^{1/2} \end{bmatrix} + \kappa_2 \begin{bmatrix} 2x^{-3/2} \ln x \\ x^{1/2}(1 - \ln x) \end{bmatrix}.$$

The eigenvalues σ_1, σ_2 of $[A(x) + A^{\mathrm{T}}(x)]/2$ are readily found to be given by

$$\sigma_1, \sigma_2 = \frac{-1}{2x} \pm \left(\frac{1}{x^3} - \frac{x}{4}\right)$$

and a straightforward calculation shows that $\mu_2[A(x)] = \max(\sigma_1, \sigma_2) \leqslant 0$ if and only if $\underline{x} \leqslant x \leqslant \bar{x}$, where $\underline{x} = \sqrt{(\sqrt{5} - 1)} \approx 1.112$ and $\bar{x} = \sqrt{(\sqrt{5} + 1)} \approx 1.799$. Now, for a linear system $y' = A(x)y$, we may take M_x to be the whole of $\mathbb{R}^m$, and we can choose $\tilde{y}(x) \equiv 0$,

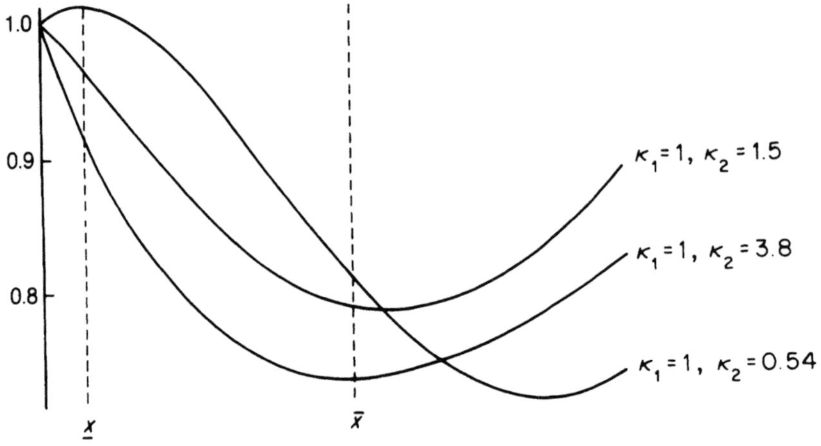

Figure 7.1

so that contractivity in $[\underline{x}, \bar{x}]$ implies that $\| y(x_2)\|_2 \leqslant \| y(x_1)\|_2$ for all x_1, x_2 satisfying $\underline{x} \leqslant x_1 \leqslant x_2 \leqslant \bar{x}$. That is, $\| y(x)\|_2$ should be monotonic non-increasing in $[\underline{x}, \bar{x}]$. Let us test this by a numerical search over the parameter space (κ_1, κ_2). In order to keep a uniform scale in the graphs, we normalize $\| y(x)\|_2$ by dividing by $\| y(1)\|_2$, and look at plots of $Y(x) := \| y(x)\|_2 / \| y(1)\|_2$. Since $Y(x)$ is clearly a function of κ_2/κ_1, we have to search only in a one-dimensional parameter space. Figure 7.1 shows some plots of $Y(x)$ against x in the interval $1 \leqslant x \leqslant 2.5$ for various values of κ_1, κ_2. The curve for $\kappa_1 = 1, \kappa_2 = 1.5$ (an arbitrary choice) is certainly monotonic non-increasing in $(\underline{x}, \bar{x})$, and indeed this turns out to be the case for all values of κ_1 and κ_2 tested. However, for $\kappa_1 = 1, \kappa_2 = 0.54$, $Y(x)$ has a maximum at $x = \underline{x}$, and so is monotonic increasing to the left of $\underline{x}$. Similarly, we find that for $\kappa_1 = 1$, $\kappa_2 = 3.8$, $Y(x)$ has a minimum at $\bar{x}$ and is thus monotonic increasing to the right of $\bar{x}$. This experiment suggests that $Y(x)$ is monotonic non-increasing for all κ_1, κ_2 if and only if $x \in [\underline{x}, \bar{x}]$, precisely the interval in which $\mu[A(x)]$ is non-positive.

We do not always get results as sharp as this. If we repeat the above calculations for Example 3 of §7.1 we find from (7.9) that

$$\| y(x)\|_2 = 2e^{-2x}\{[\kappa_1 + \kappa_2 x(x-1)]^2 + \kappa_2 x^2\},$$

and it is clear that for all finite κ_1, κ_2, $\| y(x)\|_2$ is monotonic non-increasing for all sufficiently large x. However, from (7.8) and (7.9), the eigenvalues of $[A(x) + A^{\mathrm{T}}(x)]/2$ are given by

$$\sigma_1, \sigma_2 = \frac{1 - 2x \pm \sqrt{[(1-2x)^2 + 1]}}{2x}$$

from which it is clear that $\mu[A(x)] = \max(\sigma_1, \sigma_2) > 0$ for all $x > 0$. Thus Theorem 7.1 (which, of course, gives only a *sufficient* condition for contractivity) declines to tell us whether or not the solutions are contractive for large x.

We are now able to fill in the middle box on the top line of the syntax of the new stability definitions, given at the end of the preceding section. The most suitable condition to use is dissipativity, defined by (7.19), giving the syntax diagram shown below.

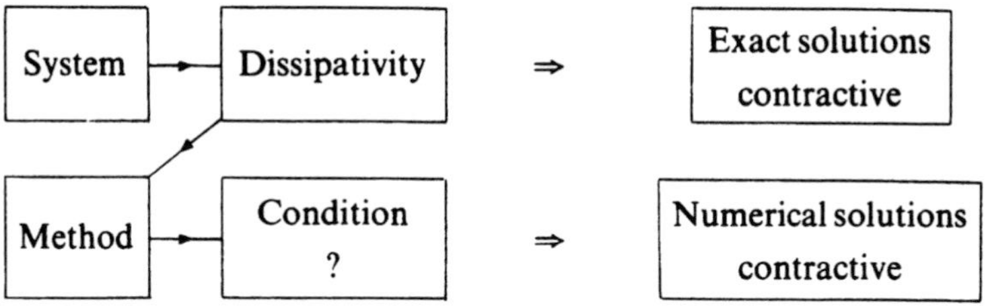

The next two sections of this chapter will be devoted to filling in the middle box of the bottom line of this syntax diagram.

Exercise

7.3.1. We have shown above that for Example 3 of §7.1, the logarithmic norm of $A(x)$ is positive for all positive x. Generalize this result by showing that the same is true for the system given in Exercise 7.1.3.

7.4 G-STABILITY

The earliest work on constructing conditions for a numerical method to be contractive is due to Dahlquist (1975, 1976). There is no loss of generality in assuming the system to be autonomous, and we assume the system to be $y' = f(y)$, satisfying the one-sided Lipschitz condition

$$\left.\begin{aligned} &\langle f(y) - f(\tilde{y}), y - \tilde{y}\rangle \leqslant \nu(x)\|y - \tilde{y}\|^2 \\ \text{with} \quad &\nu(x) \leqslant 0 \text{ for all } x \in [a, b]. \end{aligned}\right\} \tag{7.24}$$

The system is thus assumed dissipative. The definition to be developed applies not to a linear multistep method, but to a close relative. Let a linear k-step method be defined in operator notation (see (3.5) of §3.1) by

$$\rho(E)y_n = h\sigma(E)f(y_n). \tag{7.25}$$

Then the *one-leg twin* of (7.25) is defined by

$$\rho(E)y_n = hf(\sigma(E)y_n). \tag{7.26}$$

For example, the one-leg twin of the Trapezoidal Rule, $y_{n+1} - y_n = \frac{1}{2}h(f_{n+1} + f_n)$ is the Implicit Midpoint Rule, $y_{n+1} - y_n = hf(\frac{1}{2}(y_{n+1} + y_n))$ (see (5.68) of §5.11). Not surprisingly, there exists a relationship between the solutions of a linear multistep method and those of its one-leg twin. Let $\{y_n\}$ be a solution of (7.26), and define $\hat{y}_n = \sigma(E)y_n$. Then, since $\rho(E)$ and $\sigma(E)$ commute,

$$\rho(E)\sigma(E)y_n = \sigma(E)\rho(E)y_n = h\sigma(E)f(\sigma(E)y_n)$$

or

$$\rho(E)\hat{y}_n = h\sigma(E)f(\hat{y}_n)$$

and $\{\hat{y}_n\}$ is a solution of the linear multistep method (7.25). This relationship between the solutions of (7.25) and (7.26) allows results for the one-leg twin to be translated into (admittedly more complicated) results for the linear multistep method. The one-leg twin is not just a device to make the analysis of this section work; there is some evidence (Nevanlinna and Liniger, 1978, 1979) that the one-leg twin is to be preferred in variable steplength applications.

We can now state the first of our nonlinear stability definitions, due to Dahlquist (1975).

Definition *Let* $w_0, w_1, \ldots, w_k$ *be any real numbers, and define the vectors* $W_0, W_1 \in \mathbb{R}^k$ *by* $W_0 = [w_0, w_1, \ldots, w_{k-1}]^T$, $W_1 = [w_1, w_2, \ldots, w_k]^T$. *Then the k-step method* (7.26) *is said to be* **G-stable** *if there exists a real symmetric positive definite matrix G such that*

$$W_1^T G W_1 - W_0^T G W_0 \leqslant 2[\sigma(E)w_0][\rho(E)w_0]/\sigma^2(1) \tag{7.27}$$

for all such W_0, W_1.

(Note that some authors normalize the standard linear multistep method by requiring that $\sigma(1) = 1$, whereas we chose in §3.1 to normalize by requiring that $\alpha_k = 1$; thus the reader will find in the quoted references that the divisor $\sigma^2(1)$ does not appear on the right side of (7.27).)

In what can be interpreted as a vector analogue of the structure inherent in the above, it is possible to define a norm, the *G-norm*, of a vector $Z_n \in \mathbb{R}^{mk}$, defined by $Z_n := [z_{n+k-1}^T, z_{n+k-2}^T, \ldots, z_n^T]^T$, where $z_{n+j} \in \mathbb{R}^n$, $j = 0, 1, \ldots, k-1$. The G-norm $\|\cdot\|_G$ is defined by

$$\|Z_n\|_G^2 := \sum_{i=1}^{k} \sum_{j=1}^{k} g_{ij} \langle z_{n+k-i}, z_{n+k-j} \rangle, \tag{7.28}$$

where g_{ij} is the (i,j)th element of G, and the inner product is the one used in (7.24). It can be shown (Dahlquist, 1976) that if the method (7.26) is G-stable, then

$$\|Z_{n+1}\|_G^2 - \|Z_n\|_G^2 \leqslant 2\langle \sigma(E)z_n, \rho(E)z_n \rangle / \sigma^2(1) \tag{7.29}$$

for any vectors $z_n, z_{n+1}, \ldots, z_{n+k}$. Equation (7.29) can be seen as a vector extension of (7.27), and indeed can be taken as an alternative definition of G-stability; of course, (7.27) is easier to apply in practice.

Let $\{y_n\}$ and $\{\tilde{y}_n\}$ be two solutions of $y' = f(y)$, given by (7.26) with different starting values, and assume that (7.24) is satisfied. Further, define Y_n, $\tilde{Y}_n \in \mathbb{R}^{mk}$ by

$$Y_n := [y_{n+k-1}^T, y_{n+k-2}^T, \ldots, y_n^T]^T, \; \tilde{Y}_n := [\tilde{y}_{n+k-1}^T, \tilde{y}_{n+k-2}^T, \ldots, \tilde{y}_n^T]^T$$

If the method is G-stable, it follows from (7.29) that

$$\begin{aligned}
\|Y_{n+1} - \tilde{Y}_{n+1}\|_G^2 - \|Y_n - \tilde{Y}_n\|_G^2 &\leqslant 2\langle \sigma(E)(y_n - \tilde{y}_n), \rho(E)(y_n - \tilde{y}_n) \rangle / \sigma^2(1) \\
&\leqslant 2\langle \sigma(E)(y_n - \tilde{y}_n), hf(\sigma(E)y_n) - hf(\sigma(E)\tilde{y}_n) \rangle / \sigma^2(1) \\
&\leqslant 2h\nu(x)\|\sigma(E)(y_n - \tilde{y}_n)\|^2 / \sigma^2(1),
\end{aligned}$$

by (7.24), and, since $v(x) \leqslant 0$ for all $x \in [a, b]$, the condition (7.14) of §7.2 is satisfied, and we have contractivity of the numerical solution. We thus have the syntax diagram for G-stability shown below.

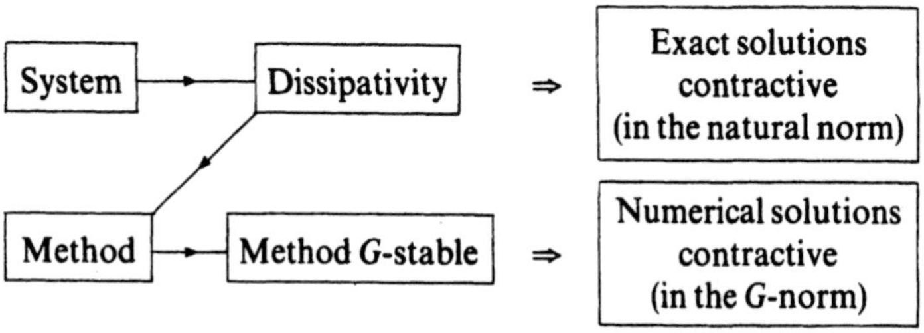

By 'the natural norm', we mean, of course, the norm appearing in (7.24), namely the norm associated with the inner product appearing on the right sides of (7.28) and (7.29).

It is not hard to show that G-stability implies A-stability; however, Dahlquist (1978) proved the unexpected result that A-stability implies G-stability, so that *G-stability and A-stability are equivalent.* This is a remarkable result; it means that for any method which is A-stable (a result based on a linear constant coefficient test system), there exists a norm (the G-norm) in which the numerical solutions are contractive whenever the general *nonlinear* system being solved is dissipative. It is not, of course, the result we want. We would like to have a stability condition which assured contractivity of the numerical solution in the norm corresponding to the inner product in which the system is dissipative; the exact and numerical solutions would then be contractive in the *same* norm. G-stability, despite the fact that it does not give us what we want, played an important role in the development of nonlinear stability theory; moreover, useful bounds on the G-norm of the error can be established (see Dahlquist, 1975, 1976).

Exercises

7.4.1. Show that the Trapezoidal Rule satisfies the G-stability condition (7.27) with $G = 1$.

7.4.2. Dahlquist (1976) gives a construction for finding the matrix G for any particular one-leg twin and quotes, as an example, that for the 2-step BDF and its twin (why both?) we may take

$$G = \begin{bmatrix} \frac{5}{2} & -1 \\ -1 & \frac{1}{2} \end{bmatrix}.$$

Check that G is positive definite and show that the G-stability condition (7.27) is satisfied. (See Table 3.3 of §3.12 for the coefficients of the method.)

7.5 NONLINEAR STABILITY OF IMPLICIT RUNGE–KUTTA METHODS

We saw in §6.7 that it was much easier to find A-stable implicit Runge–Kutta methods than it was to find A-stable linear multistep methods. It is thus no surprise that implicit Runge–Kutta methods turn out to be the best class for which to seek nonlinear stability properties. We remind the reader of the discussion of various sub-classes of implicit and semi-implicit Runge–Kutta methods given in §5.11. We shall assume that the method has s stages and is defined by the Butcher array

$$\begin{array}{c|c} c & A \\ \hline & b^{\mathrm{T}} \end{array} \tag{7.30}$$

Further, we assume that the system $y' = f(x, y)$ is dissipative; that is, we assume that

$$\langle f(x, y(x)) - f(x, \tilde{y}(x)), y(x) - \tilde{y}(x) \rangle \leqslant 0 \tag{7.31}$$

holds for any two solutions $\{y(x)\}$ and $\{\tilde{y}(x)\}$ satisfying different initial conditions. It follows from §7.3 that the exact solutions are contractive in the norm corresponding to the inner product in (7.31). We shall assume this norm throughout this section; in particular, contractivity of the method and of the numerical solutions (in the sense of §7.2) will mean contractivity in this norm.

Definition (Butcher, 1975); Burrage and Butcher, 1979; Crouziex, 1979) If a Runge–Kutta method applied, with any steplength, to an autonomous system satisfying (7.31) generates contractive numerical solutions, then the method is said to be **B-stable**; *if the same is true when the method is applied to a non-autonomous system satisfying (7.31), the method is said to be* **BN-stable**.

Butcher (1975) proved the following sufficient condition for B-stability; a proof can also be found in Dekker and Verwer (1984). Let B and Q be $s \times s$ matrices defined by

$$B := \operatorname{diag}(b_1, b_2, \ldots, b_s) \qquad Q := BA^{-1} + A^{-\mathrm{T}}B - A^{-\mathrm{T}}bb^{\mathrm{T}}A^{-1}. \tag{7.32}$$

Then the sufficient condition for B-stability is

$$B \text{ and } Q \text{ non-negative definite.} \tag{7.33}$$

We thus have the syntax diagram shown below for B-stability.

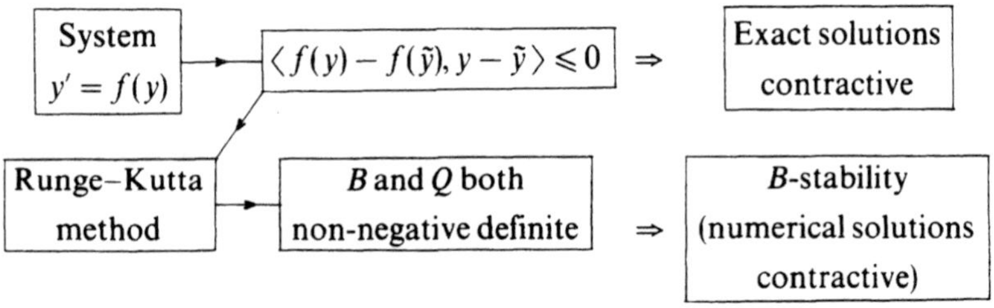

Example

For the 2-stage Gauss method,

$$A = \frac{1}{12}\begin{bmatrix} 3 & 3 - 2\sqrt{3} \\ 3 + 2\sqrt{3} & 3 \end{bmatrix}, \qquad b = \begin{bmatrix} \frac{1}{2} \\ \frac{1}{2} \end{bmatrix}, \quad B = \tfrac{1}{2}I$$

whence we find that

$$BA^{-1}+A^{-\mathsf{T}}B=\begin{bmatrix}3 & -3\\ -3 & 3\end{bmatrix}, \qquad b^{\mathsf{T}}A^{-1}=\sqrt{3}[-1,1].$$

It follows that

$$A^{-\mathsf{T}}bb^{\mathsf{T}}A^{-1}=(b^{\mathsf{T}}A^{-1})^{\mathsf{T}}b^{\mathsf{T}}A^{-1}=3\begin{bmatrix}-1\\ 1\end{bmatrix}[-1,1]=BA^{-1}+A^{-\mathsf{T}}B$$

and $Q=0$. Since B is clearly positive definite, condition (7.33) is satisfied and the method is B-stable. Indeed, it can be shown that (7.33) is satisfied for all Gauss methods, as indeed it is for all Radau IA, Radau IIA and Lobatto IIIC methods. Clearly it is not satisfied for Lobatto IIIA and IIIB methods, since A is singular for these methods.

The above condition is awkward to apply. (If the reader doubts this, let him repeat the above working for the 3-stage Gauss method!) A much more easily applied condition (which involves no matrix inversions) was discovered by Burrage and Butcher (1979) and Crouziex (1979). Let B be defined as in (7.32) and define the $s\times s$ matrix M by

$$M:=BA+A^{\mathsf{T}}B-bb^{\mathsf{T}}. \tag{7.34}$$

Definition *A Runge–Kutta method is said to be* **algebraically stable** *if the matrices B and M defined by (7.32) and (7.34) are both non-negative definite.*

Algebraic stability can be shown to be sufficient not only for B-stability, but also for BN-stability. We have the syntax diagram shown below for algebraic stability.

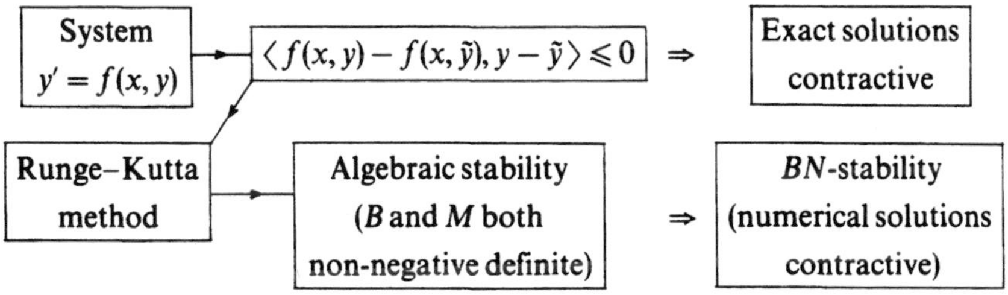

Example 1 Since no matrix inversions are involved, we are prepared to use a 3-stage example this time. Let us choose the 3-stage Lobatto IIIC for which

$$A=\tfrac{1}{12}\begin{bmatrix}2 & -4 & 2\\ 2 & 5 & -1\\ 2 & 8 & 2\end{bmatrix}, \qquad b=\tfrac{1}{6}\begin{bmatrix}1\\ 4\\ 1\end{bmatrix}, \qquad B=\tfrac{1}{6}\begin{bmatrix}1 & \cdot & \cdot\\ \cdot & 4 & \cdot\\ \cdot & \cdot & 1\end{bmatrix}.$$

A simple calculation shows that

$$M=\tfrac{1}{36}\begin{bmatrix}1 & -2 & 1\\ -2 & 4 & -2\\ 1 & -2 & 1\end{bmatrix}.$$

The eigenvalues of M are $0, 0, \frac{1}{6}$, so that M is non-negative definite; clearly so is B, and the method is algebraically stable. As for condition (7.33), it can be shown that the Gauss, Radau IA, Radau IIA and Lobatto IIIC are all algebraically stable, whereas the Lobatto IIIA and IIIB are not.

Example 2 For the one-parameter family of semi-implicit methods (5.75) of §5.11, we have

$$A = \begin{bmatrix} \dfrac{3\mu - 1}{6\mu} & \cdot \\ \mu & \dfrac{1-\mu}{2} \end{bmatrix}, \qquad b = \frac{1}{3\mu^2 + 1}\begin{bmatrix} 3\mu^2 \\ 1 \end{bmatrix}, \qquad B = \frac{1}{3\mu^2 + 1}\begin{bmatrix} 3\mu^2 & \cdot \\ \cdot & 1 \end{bmatrix}$$

and we find that $M = \omega(\mu)\tilde{M}$, where

$$\omega(\mu) = \frac{\mu(3\mu^2 - 3\mu + 1)}{(3\mu^2 + 1)^2}, \qquad \tilde{M} = \begin{bmatrix} -1 & 1 \\ 1 & -1 \end{bmatrix}.$$

The eigenvalues of $\tilde{M}$ are 0 and -2; since $3\mu^2 - 3\mu + 1$ is positive for all μ, it follows that M is non-negative definite for all $\mu \leqslant 0$. Thus, of the pair of DIRK methods given by (6.76) only the one given by choosing $\mu = -\sqrt{3}/3$ (the one we showed in §6.7 to be A-stable) is algebraically stable.

Example 3 Recall the 3-stage DIRK method (5.77), for which

$$A = \begin{bmatrix} \dfrac{1+\nu}{2} & \cdot & \cdot \\ -\dfrac{\nu}{2} & \dfrac{1+\nu}{2} & \cdot \\ 1+\nu & -1-2\nu & \dfrac{1+\nu}{2} \end{bmatrix}, \qquad b = \frac{1}{6\nu^2}\begin{bmatrix} 1 \\ 6\nu^2 - 2 \\ 1 \end{bmatrix}, \qquad B = \frac{1}{6\nu^2}\begin{bmatrix} 1 & \cdot & \cdot \\ \cdot & 6\nu^2 - 2 & \cdot \\ \cdot & \cdot & 1 \end{bmatrix},$$

where ν is one of the three real roots of

$$3\nu^3 - 3\nu - 1 = 0. \tag{7.35}$$

To construct M and find its eigenvalues for *general* ν is a horrendous task. One is reminded of the lawyer and mathematician Viète (1540–1603), who, on declining to perform a similarly tedious piece of manipulation, described it as 'work not fit for a Christian gentleman' (Perhaps we have here the basis for a definition of a symbolic manipulator?) However, by repeatedly using (7.35), the manipulation becomes tolerable, and we find that $M = \omega(\nu)\tilde{M}$, where

$$\omega(\nu) = \frac{6\nu^2 + 6\nu + 1}{36\nu^4}, \qquad \tilde{M} = \begin{bmatrix} 1 & -2 & 1 \\ -2 & 4 & -2 \\ 1 & -2 & 1 \end{bmatrix}.$$

The eigenvalues of $\tilde{M}$ are $0, 0, 6$, and $6v^2 + 6v + 1$ is positive only for one root, $v = v_1 = (2/\sqrt{3})\cos(10°)$. Thus (5.77) is algebraically stable only for $v = v_1$, the value for which we indicate in §6.7 the method is A-stable.

Hundsdorfer and Spijker (1983a) take a different approach. As we have seen, the test equation for A-stability is essentially the scalar equation $y' = \lambda y$, $\lambda \in \mathbb{C}$. This is replaced by the scalar test equation

$$y' = \lambda(x)y, \quad \lambda(x) \in \mathbb{C}. \tag{7.36}$$

Clearly any exact solution of (7.36) satisfies

$$y(x+h)/y(x) = \exp\left(\int_x^{x+h} \lambda(x)\mathrm{d}x\right)$$

and it follows that if $\operatorname{Re}\lambda(x) \leqslant 0$ for all $x \in [a, b]$, then

$$y(x+h) = Ky(x), \qquad |K| \leqslant 1 \tag{7.37}$$

for any $x \in [a, b]$ and any $h > 0$. A stability definition for a one-step method can be framed by requiring that the numerical solution of (7.36) mimics (7.37); that is, we shall demand that $y_{n+1} = \tilde{K}y_n$, $|\tilde{K}| \leqslant 1$ holds for all positive h, whenever $\operatorname{Re}\lambda(x) \leqslant 0$. The general s-stage Runge–Kutta method (written in the alternative form (5.6) of §5.1) applied to (7.36) gives

$$\left.\begin{aligned} y_{n+1} &= y_n + h\sum_{i=1}^{s} b_i\lambda(x_n + c_ih)Y_i \\ \text{where} \qquad & \\ Y_i &= y_n + h\sum_{j=1}^{s} a_{ij}\lambda(x_n + c_jh)Y_j. \end{aligned}\right\} \tag{7.38}$$

Introducing the notation

$$h\lambda(x_n + c_ih) = \gamma_i, \qquad i = 1, 2, \ldots, s, \qquad \Gamma := \operatorname{diag}(\gamma_1, \gamma_2, \ldots, \gamma_s),$$

$$Y := [Y_1, Y_2, \ldots, Y_s]^{\mathrm{T}}, \qquad e := [1, 1, \ldots, 1]^{\mathrm{T}} \in \mathbb{R}^s,$$

(7.38) can be written in the form

$$y_{n+1} = y_n + b^{\mathrm{T}}\Gamma Y, \qquad Y = y_ne + A\Gamma Y$$

whence

$$y_{n+1}/y_n =: R(\Gamma) = 1 + b^{\mathrm{T}}\Gamma(I - A\Gamma)^{-1}e. \tag{7.39}$$

Note that Γ is a function of x_n. Note also that if in (7.36) we put $\lambda(x) = \lambda$, constant, then $\Gamma = \hat{h}I$, where $\hat{h} = h\lambda$, and (7.39) reduces to

$$y_{n+1}/y_n = R(\hat{h}) = 1 + \hat{h}b^{\mathrm{T}}(I - \hat{h}A)^{-1}e,$$

which is just the stability function of the method (see (5.86) of §5.12). We are now able to frame a definition.

Definition *Let* $\gamma_i \in \mathbb{C}$ *be such that* $\operatorname{Re}\gamma_i \leqslant 0$, $i = 1, 2, \dots, s$, *with* $\gamma_i = \gamma_j$ *if* $c_i = c_j$, *and let* $\Gamma = \operatorname{diag}(\gamma_1, \gamma_2, \dots, \gamma_s)$. *Then the Runge–Kutta method is said to be* **AN-stable** *if, for all such* γ_i, $I - A\Gamma$ *is non-singular and* $R(\Gamma)$, *defined by* (7.39), *satisfies* $|R(\Gamma)| \leqslant 1$.

(The 'N' in AN-stability denotes that we are using a non-autonomous form $y' = \lambda(x)y$ of the standard linear test equation $y' = \lambda y$ for A-stability.) Obviously, AN-stability$\Rightarrow$ A-stability; the converse is not true.

The syntax diagram for AN-stability is therefore

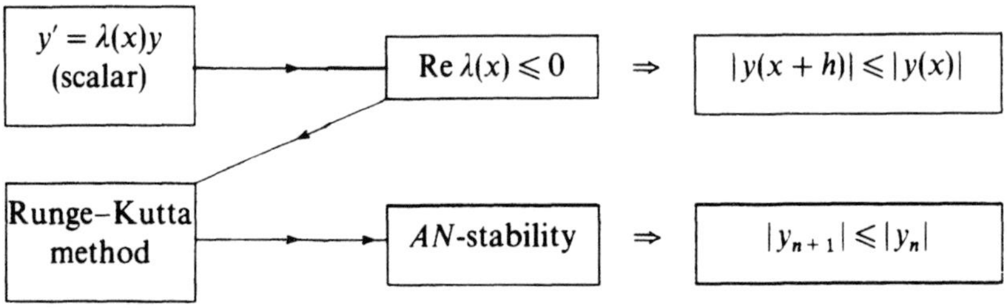

Examples For the 2-stage Lobatto IIIA method (the Trapezoidal Rule), we easily find that

$$R(\Gamma) = \frac{1 + \gamma_1/2}{1 - \gamma_2/2}.$$

To see that the condition $|R(\Gamma)| \leqslant 1$ for all $\operatorname{Re}\gamma_i \leqslant 0$, $i = 1, 2$, is *not* satisfied, all we need do is choose γ_1 and γ_2 to be real and negative, with $\gamma_1 < \gamma_2 - 4$, whereupon $R(\Gamma) < -1$. In contrast, for the 1-stage Gauss method (the Implicit Mid-point Rule, the one-leg twin of the Trapezoidal Rule), we find that

$$R(\Gamma) = \frac{1 + \gamma_1/2}{1 - \gamma_1/2}$$

and the conditions for AN-stability are clearly satisfied.

It is not difficult to find $R(\Gamma)$ for any particular Runge–Kutta method, but it can be quite difficult to determine whether or not $|R(\Gamma)| \leqslant 1$. However, we seldom need to apply this test, since in most cases, AN-stability is equivalent to algebraic stability which is much easier to test. Before we can state these relationships, we need a further definition.

Definition *A Runge–Kutta method is said to be* **nonconfluent** *if all of the* c_i, $i = 1, 2, \dots, s$, *are distinct.*

It can be shown (see, for example, Dekker and Verwer (1984)) that the various stability properties we have discussed in this section are related as follows:

For general Runge–Kutta methods

$$\text{Algebraic stability} \Rightarrow \begin{cases} BN\text{-stability} \Rightarrow B\text{-stability} \\ \qquad\qquad\qquad\qquad \Downarrow \\ AN\text{-stability} \Rightarrow A\text{-stability}. \end{cases}$$

Table 7.1

s-stage RK method	Order	Linear Stability property	Nonlinear stability property
Gauss	$2s$	A-stability	Algebraic stability
Radau IA, IIA	$2s-1$	L-stability	Algebraic stability
Lobatto IIIA, IIIB	$2s-2$	A-stability	No algebraic stability
Lobatto IIIC	$2s-2$	L-stability	Algebraic stability

For nonconfluent Runge–Kutta methods

$$\left.\begin{array}{c}\text{Algebraic stability}\\ \Updownarrow \\ BN\text{-stability}\\ \Updownarrow \\ AN\text{-stability}\end{array}\right\} \Rightarrow B\text{-stability} \Rightarrow A\text{-stability}.$$

It is worth reflecting for a moment on the import of these equivalences. We started out with a discussion of the full *nonlinear system*, and sought a criterion for some sort of 'controlled behaviour' of the solutions, leading to the notion of contractivity. Algebraic stability is a sufficient condition for the numerical solutions to behave in a similarly controlled manner. Yet, with the not-all-that-important exception of confluent methods, we can guarantee the same contractive behaviour of the numerical solutions by imposing a condition based on the *scalar linear* test equation $y' = \lambda(x)y$. In other words, we need only move a little bit away from the over-restrictive test equation $y' = \lambda y$ to be able to predict contractivity of the numerical solutions of a fully nonlinear system!

Finally, in Table 7.1, we update Table 6.6 of §6.7 to include nonlinear stability properties.

In particular, we note that the Trapezoidal Rule (the 2-stage Lobatto IIIA) is not algebraically stable, whereas the Implicit Mid-point Rule (the 1-stage Gauss) and the Backward Euler method (the 2-stage Radau IIA, the one-step BDF) are algebraically stable.

Exercises

7.5.1. Show that the 2-stage Radau IA, Radau IIA and Lobatto IIIC methods (listed in §5.11) satisfy the condition (7.33).

7.5.2. Make a selection of the Gauss, Radau and Lobatto methods listed in §5.11, and demonstrate that Gauss, Radau IA, Radau IIA and Lobatto IIIC methods are algebraically stable, while Lobatto IIIA and IIIB methods are not.

7.5.3. Derive the matrix M quoted for Example 3. (Apologies to any 'Christian gentlemen' amongst the readership.)

7.5.4. Show that the 2-stage Lobatto IIIB method is not AN-stable.

7.5.5. This nice example is due to Hundsdorfer and Spijker (1981b). Consider the Runge–Kutta method

$$\begin{array}{c|cc} \frac{1}{4} & \frac{1}{8} & \frac{1}{8} \\ \frac{3}{4} & \frac{3}{8} & \frac{3}{8} \\ \hline & \frac{1}{2} & \frac{1}{2} \end{array}$$

Show that $R(\Gamma)$, as defined by (7.39), is given by $R(\Gamma)=(8+3\gamma_1+\gamma_2)/(8-\gamma_1-3\gamma_2)$, and deduce the stability function $R(\hat{h})$. Hence show that the method is A-stable. By considering the values $\gamma_1=-3\mu i, \gamma_2=\mu i$, show that the method is not AN-stable. Show (independently) that it is not algebraically stable.

7.6 B-CONVERGENCE

Finally, we comment very briefly on the subject of *B-convergence*. There exists a substantial body of theory on this topic, which is beyond the scope of this book; we mention it here solely because the reader, when solving stiff systems numerically, may on occasion find results which are a little puzzling, and might appreciate knowing that the observed behaviour is to be expected. The phenomenon we are about to discuss was first described by Prothero and Robinson (1974), who gleaned much insight from considering the family of scalar equations

$$y'=\lambda y+g'(x)-\lambda g(x), \qquad \lambda\in\mathbb{C}. \tag{7.40}$$

The subsequent theory of B-convergence for the full nonlinear system was developed by Frank, Schneid and Ueberhuber (1981, 1985a, 1985b); an excellent review of this work can be found in Dekker and Verwer (1984).

When we solve a stiff problem numerically, we expect to have to use a very small steplength in the interval in which the fast transients are still alive. If we use a method with an appropriate stability property (such as A-stability or algebraic stability) then, once the fast transients are dead, we expect to be able to use a steplength which is not restricted by stability constraints, and this is indeed the case. However, the *accuracy* we achieve in this phase of the solution is often rather less than the order of the method would lead us to expect.

Consider the scalar equation (7.40) in the case when λ is real. The general solution is $y(x)=\kappa\exp(\lambda x)+g(x)$, where κ is an arbitrary constant. If $g(x)$ is smooth (that is, the higher derivatives are not large) and we choose λ such that $\lambda\ll 0$, then this scalar equation exhibits stiffness. Further, if we choose the initial condition $y(x_0)=g(x_0)$, then the solution is $y(x)=g(x)$ and the fast transient does not appear in the solution (exactly the situation we had in Problem 2 of §6.1). Following Seinfeld, Lapidus and Hwang (1970), we choose $g(x)=10-(10+x)\exp(-x)$, giving the initial value problem

$$y'=\lambda y+[9+10\lambda+(1+\lambda)x]\exp(-x)-10\lambda, \qquad y(0)=0 \tag{7.41}$$

with exact solution

$$y(x)=10-(10+x)\exp(-x).$$

Table 7.2

	$\lambda = -50$		$\lambda = -5 \times 10^3$		$\lambda = -5 \times 10^5$		$\lambda = -5 \times 10^7$	
h	$E(h)$	$R(h)$	$E(h)$	$R(h)$	$E(h)$	$R(h)$	$E(h)$	$R(h)$
0.2	6.4×10^{-4}		5.0×10^{-3}		5.2×10^{-3}		5.2×10^{-3}	
		17.9		4.4		4.0		4.0
0.1	3.6×10^{-5}		1.1×10^{-3}		1.3×10^{-3}		1.3×10^{-3}	
		16.4		5.8		4.0		4.0
0.05	2.2×10^{-6}		2.0×10^{-4}		3.2×10^{-4}		3.3×10^{-4}	
		15.8		12.0		4.1		4.0
0.025	1.4×10^{-7}		1.6×10^{-5}		8.0×10^{-5}		8.1×10^{-5}	

We solve this, for a range of values for λ, by the algebraically stable 2-stage Gauss method with steplength h. Naturally, no stability difficulties arise, but the pattern of the global truncation errors at $x = 1.0$ as we vary both λ and h, as displayed in Table 7.2, is not what we might expect. As $|\lambda|$ increases (and the problem becomes stiffer) the global errors increase, but flatten out for sufficiently large $|\lambda|$.

To see if these global errors are consistent with the order of the method, we have also computed the ratios $R(h) := E(h)/E(h/2)$, where $E(h)$ is the global error when the steplength is h. If a method has order p then the global error is $0(h^p)$ and the ratio $R(h)$ should be approximately 2^p; we are thus able to make a numerical estimate of the effective order of the method. From the ratios quoted in Table 7.2, we see that for $\lambda = -50$, the effective order does appear to be 4, but as $|\lambda|$ increases this order appears to decrease progressively until, for $\lambda \leqslant -5 \times 10^5$, it appears to be 2.

What is happening is that, although the exact solution is independent of the stiffness parameter λ (and this would also be the case in the steady-state phase for a problem with general initial conditions), the coefficient of h^{p+1} in the principal local truncation error (recall that this is a linear combination of elementary differentials) does depend on λ, and so therefore does the global error. When $|\lambda|$ becomes very large, the multiplier of h^p in the global error becomes so large that it robs the superscript p of any meaning. What is interesting is that, for this example at least, the effective order does not just degenerate in a random way, but appears to change progressively from 4 to precisely 2.

We turn now to the general problem for a nonlinear system; since our treatment of this topic is necessarily superficial, we shall eschew any formal definitions. A Runge–Kutta method is said to be **B-consistent** of order q if the local truncation error T_{n+1} satisfies a bound $\| T_{n+1} \| \leqslant kh^{q+1}$ for all $h \in (0, h_1]$, where the real constants k and h_1 are *independent of the stiffness of the problem*; they can depend on the one-sided Lipschitz constant of the system and on the smoothness of the solution (after the fast transient is dead), but they cannot depend on higher derivatives of the solution which are influenced by stiffness. In a similar way, the method is said to be **B-convergent** of order q if the global error E_{n+1} satisfies a bound $\| E_{n+1} \| \leqslant Kh^q$ for all $h \in (0, h_2]$, where, again, K and h_2 are independent of the stiffness of the problem. Our numerical experiment would seem to suggest that the 2-stage Gauss method is B-convergent of order only 2 *for the class of problems* (7.40). The catch is in the italicized phrase; unfortunately the order of B-consistency does depend on the problem being solved, and is thus not a property of

Table 7.3

s-stage RK method	Order p	Stage order $\tilde{p}$
Gauss	$2s$	s
Radau IA	$2s-1$	$s-1$
Radau IIA	$2s-1$	s
Lobatto IIIA	$2s-2$	s
Lobatto IIIB	$2s-2$	$s-2$
Lobatto IIIC	$2s-2$	$s-1$

the method alone. It is necessary to define the order of B-consistency and B-convergence relative to a class of problems.

It is, however, possible to find bounds for the order of B-consistency. Clearly, we always have $q \leqslant p$, where p is the order of the method defined in the conventional way. A lower bound is found as follows. Let $y(x)$ be a sufficiently smooth function and, for the general s-stage Runge–Kutta method (in the alternative form (5.6)), define the residuals $t_i, i = 1, 2, \dots, s+1$ by

$$\left.\begin{aligned} t_i &= y(x_n + c_i h) - y(x_n) - h\sum_{j=1}^{s} a_{ij} y'(x_n + c_j h), \qquad i = 1, 2, \dots, s \\ t_{s+1} &= y(x_{n+1}) - y(x_n) - h\sum_{i=1}^{s} b_i y'(x_n + c_i h). \end{aligned}\right\} \tag{7.42}$$

Essentially these residuals are the local truncation errors of each stage of the Runge–Kutta method (the $(s+1)$th stage being regarded as the statement $y_{n+1} = y_n + h\sum_{i=1}^{s} b_i f(x_n + c_i h, Y_i)$) when each stage is regarded as a method in its own right. (Compare with (5.99) of §5.13.) If, for $i = 1, 2, \dots, s+1$, $t_i = 0(h^{p_i+1})$, then $\tilde{p} := \min(p_1, p_2, \dots, p_{s+1})$ is called the **stage order** of the method. It is assumed that the method satisfies, for a given class of problems, yet another stability criterion (**BS-stability**), which requires that if each stage of the Runge–Kutta method is perturbed by $\varepsilon_i, i = 1, 2, \dots, s+1$, then the perturbation in the numerical solution given by the method must be bounded by $K \max_i \|\varepsilon_i\|$, where K is independent of the stiffness. For that class of problem, the order of B-consistency is at least the stage order. We thus have that the method is B-consistent of order q, with $\tilde{p} \leqslant q \leqslant p$. The stage order $\tilde{p}$ for the classes of implicit Runge–Kutta methods discussed in §5.11 are shown in Table 7.3.

Our numerical experiment suggested that for problem (7.41) the 2-stage Gauss method is B-consistent of order 2; note that this is within the bounds given by Table 7.3.

Exercises

7.6.1. Referring to (7.42), show that for $i = 1, 2, \dots, s$ the conditions for $t_i = 0(h^{p_i+1})$ to hold are that

$$\sum_{j=1}^{s} a_{ij} c_j^{q-1} = c^q/q, \qquad q = 1, 2, \dots, p_i \tag{1}$$

and that the condition for $t_{s+1} = 0(h^{p_{s+1}+1})$ to hold is that

$$\sum_{i=1}^{s} b_i c_i^{q-1} = 1/q, \qquad q = 1, 2, \ldots, p_{s+1}. \tag{2}$$

Using (2) and the order conditions, show that $p_{s+1} \geqslant p$, the order of the method.

Compare the conditions (1) with the collocation conditions (5.74) of §5.11.

7.6.2. Using the result of the preceding exercise, find the stage orders of the s-stage Gauss, Radau IA and IIA methods for $s = 1, 2$, and of the s-stage Lobatto IIIA, IIIB and IIIC methods for $s = 2, 3$, and check that the results are consistent with the entries in Table 7.3. (See §5.11 for the coefficients of these methods.)

7.7 CONCLUSIONS

So, what does one make of this theory of nonlinear stability? From the theoretical point of view, one can have no complaints. It removes, in a rigorous and comprehensive manner, all of the doubts that linear stability theory, with its dubious 'frozen' Jacobian arguments, raises. Moreover, it turns out that one does not need particularly powerful tools to cope with the full nonlinear system; AN-stability, based on a fairly mild extension of the scalar test equation of linear stability theory, is equivalent to algebraic stability for most Runge–Kutta methods. Further, testing for algebraic stability is not difficult (indeed sometimes easier than testing for A-stability). Finally, B-convergence gives a convincing and useful explanation of an observed phenomenon. Really, we could not have asked for more!

But what impact has the theory had on the way in which stiff systems are solved in practice? As the dates of the references show, the nonlinear theory has been around for some time, yet most real-life stiff problems continue to be solved by highly-tuned codes based on the BDF, which give excellent results for the vast majority of problems. One reason for this becomes clear when we recall the shortcomings of the linear theory. To put it somewhat fancifully, the bogus extension of the linear theory to nonlinear problems sometimes tells us fairy-tales about how the exact solutions might be expected to behave. We do not believe these fairy-tales but, significantly, neither necessarily do the methods! We saw an example of this in the latter part of §3.8, where the linear theory predicted disaster, but the method continued to behave normally.

On the other hand, it is possible that variable-order Runge–Kutta codes based on algebraically stable methods will be developed to the point where they are competitive with BDF codes (and of course the ever-decreasing cost of computer time biases users towards robustness rather than efficiency). Should that happen, it would make sense to prefer methods whose stability is so well understood.

References

Note: An extensive bibliography on numerical methods for ordinary differential equations can be found in Butcher (1987).

Addison, C. A. (1979) Implementing a stiff method based upon the second derivative formulas, *Dept. of Comp. Sci. Tech. Rept. No. 130/79, University of Toronto.*

Albrecht, P. (1985) Numerical treatment of ODEs: the theory of A-methods, *Num. Math.*, **47**, 59–87.

Albrecht, P. (1987) The extension of the theory of A-methods to RK-methods: in *Numerical Treatment of Differential Equations, Proc. 4th Seminar NUMDIFF-4*, ed. K. Strehmel, Teubner-Texte zur Mathematik, Leipzig, pp. 8–18.

Albrecht, P. (1989) Elements of a general theory of composite integration methods, *Appl. Math. Comp.*, **31**, 1–17.

Alexander, R. (1977) Diagonally implicit Runge–Kutta methods for stiff ODEs, *Siam J. Numer. Anal.*, **14**, 1006–1021.

Atkinson, L. V. and Harley, P. J. (1983) *An Introduction to Numerical Methods with Pascal*, Addison-Wesley, London.

Axelsson, O. (1969) A class of A-stable methods, *BIT*, **9**, 185–199.

Bashforth, F. and Adams, J. C. (1883) *An Attempt to Test the Theories of Capillary Action by Comparing the Theoretical and Measured Forms of Drops of Fluid, with an Explanation of the Method of Integration Employed in Constructing the Tables which Give the Theoretical Forms of Such Drops*, Cambridge University Press, Cambridge.

Birkhoff, G. and Varga, R. S. (1965) Discretization errors for well-set Cauchy problems: I, *J. Math. and Phys.*, **44**, 1–23.

Brayton, R. K., Gustavson, F. G. and Hachtel, G. D. (1972) A new efficient algorithm for solving differential-algebraic systems using implicit backward differentiation formulas, *Proc. IEEE*, **60**, 98–108.

Brunner, H. (1967) Stabilization of optimal difference operators, *Z. Angew. Math. Phys.*, **18**, 438–444.

Burrage, K. (1978a) A special family of Runge–Kutta methods for solving stiff differential equations, *BIT*, **18**, 22–41.

Burrage, K. (1978b) High order algebraically stable Runge–Kutta methods, *BIT*, **18**, 373–383.

Burrage, K. (1982) Efficiently implementable algebraically stable Runge–Kutta methods, *SIAM J. Numer. Anal.*, **19**, 245–258.

Burrage, K. and Butcher, J. C. (1979) Stability criteria for implicit Runge–Kutta methods, *SIAM J. Numer. Anal.*, **16**, pp 46–57.

Burrage, K., Butcher, J. C. and Chipman, F. H. (1979) STRIDE: Stable Runge–Kutta integrator for differential equations, *Report Series No. 150, Dept. of Mathematics, University of Auckland.*

Burrage, K., Butcher, J. C. and Chipman, F. H. (1980) An implementation of singly-implicit Runge–Kutta methods, *BIT*, **20**, 452–465.

Butcher, J. C. (1964a) Implicit Runge–Kutta processes, *Math. Comp.*, **18**, 233–244.

Butcher, J. C. (1964b) Integration processes based on Radau quadrature formulas, *Math. Comp.*, **18**, 233–244.

Butcher, J. C. (1966) On the convergence of numerical solutions to ordinary differential equations, *Math. Comp.*, **20**, 1–10.

Butcher, J. C. (1972) The numerical solution of ordinary differential equations, *Lecture notes, North British Differential Equations Symposium.*

Butcher, J. C. (1975) A stability property of implicit Runge–Kutta methods, *BIT*, **15**, 358–361.

Butcher, J. C. (1976) On the implementation of implicit Runge–Kutta methods, *BIT*, **16**, 237–240.

Butcher, J. C. (1987) *The Numerical Analysis of Ordinary Differential Equations: Runge–Kutta and General Linear Methods*, Wiley, Chichester.

Byrne, G. D. and Hindmarsh, A. C. (1975) A polyalgorithm for the numerical solution of ordinary differential equations, *ACM Trans. Math. Software*, **1**, 71–96.

Ceschino, F. (1961) Modification de la longueur du pas dans l'intégration numérique par les méthodes à liés, *Chiffres*, **4**, 101–106.

Chartres, B. A. and Stepleman, R. S. (1972) A general theory of convergence for numerical methods, *SIAM J. Numer. Anal.*, **9**, 476–492.

Chipman, F. H. (1971) A-stable Runge–Kutta processes, *BIT*, **11**, 384–388.

Coppel, W. (1965) *Stability and Asymptotic Behavior of Differential Equations*, D. C. Heath, Boston.

Crouzeix, M. (1976) Sur les méthodes de Runge–Kutta pour l'approximation des problèmes d'évolution: in *Computing Methods in Applied Science and Engineering: Second Internat. Symp., 1975*, eds R. Glowinski and J. L. Lions, Lecture Notes in Econom. and Math. Systems No. 134, Springer-Verlag, Berlin, pp. 206–233.

Crouzeix, M. (1979) Sur la B-stabilité des méthodes de Runge–Kutta, *Numer. Math.*, **32**, 75–82.

Cryer, C. W. (1972) On the instability of high order backward-difference multistep methods, *BIT*, **12**, 17–25.

Cryer, C. W. (1973) A new class of highly stable methods: A_0-stable methods, *BIT*, **13**, 153–159.

Curtis, A. R. (1987) Stiff ODE initial value problems and their solution: in *The State of the Art in Numerical Analysis*, eds A. Iserles and M. J. D. Powell, Clarendon Press, Oxford, pp. 433–450.

Curtis, A. R. and Sweetenham, W. P. (1985) FACSIMILE Release H, user's manual, *Report AERE-R 11771, Harwell Laboratory.*

Curtiss, C. F. and Hirschfelder, J. O. (1952) Integration of stiff equations, *Proc. Nat. Acad. Sci. USA*, **38**, 235–243.

Dahlquist, G. (1956) Convergence and stability in the numerical integration of ordinary differential equations, *Math. Scand.*, **4**, 33–53.

Dahlquist, G. (1959) Stability and error bounds in the numerical solution of ordinary differential equations, Thesis, in *Trans. Royal Inst. of Technology, No 130*, Stockholm.

Dahlquist, G. (1963) A special stability problem for linear multistep methods, *BIT*, **3**, 27–43.

Dahlquist, G. (1975) On stability and error analysis for stiff nonlinear problems, *Report No. TRITA-NA-7508*, Dept. of Information Processing, Computer Science, Royal Inst. of Technology, Stockholm.

Dahlquist, G. (1976) Error analysis for a class of methods for stiff nonlinear initial value problems: in *Lecture Notes in Math, No 506*, ed. G. A. Watson, Springer-Verlag, Berlin.

Dahlquist, G. (1978) G-stability is equivalent to A-stability, *BIT* **18**, 384–401.

Dekker, K. and Verwer, J. G. (1984) *Stability of Runge–Kutta Methods for Stiff Nonlinear Differential Equations*, North-Holland, Amsterdam.

Dormand, J. R. and Prince, P. J. (1980) A family of embedded Runge–Kutta formulae, *J. Comp. and Appl. Math.*, **6** 19–26.

Ehle, B. L. (1969) On Padé approximations to the exponential function and A-stable methods for the numerical solution of initial value problems, *Research Rep. CSRR 2010, Dept. AACS, University of Waterloo.*

England, R. (1969) Error estimates for Runge–Kutta type solutions to systems of ordinary differential equations, *Comput. J.*, **12**, 166–170.

Enright, W. H. (1974a) Second derivative multistep methods for stiff ordinary differential equations, *SIAM J. Numer. Anal.*, **11**, 321–331.

Enright, W. H. (1974b) Optimal second derivative methods for stiff systems: in *Stiff Differential Systems*, ed. R. A. Willoughby, Plenum Press, New York, pp. 95–109.

Enright, W. H. and Hull, T. E. (1976). Test results on initial value methods for non-stiff ordinary differential equations, *SIAM J. Numer. Anal.*, **13**, 944–961.

Enright, W. H., Hull, T. E. and Lindberg, B. (1975) Comparing numerical methods for stiff systems of ODEs, *BIT*, **15**, 10–48.

Fehlberg, E. (1968) Classical fifth sixth, seventh and eighth order Runge–Kutta formulas with stepsize control, NASA TR R-287.

Fehlberg, E. (1969) Low order classical Runge–Kutta formulas with stepsize control and their application to some heat transfer problems, *NASA TR R-315*.

Frank, R., Schneid, J. and Ueberhuber, C. W. (1981) The concept of B-convergence, *SIAM J. Numer. Anal.*, **18**, 753–780.

Frank, R., Schneid, J. and Ueberhuber, C. W. (1985a) Stability properties of implicit Runge–Kutta methods, *SIAM J. Numer. Anal.*, **22**, 497–514.

Frank, R., Schneid, J. and Ueberhuber, C. W. (1985b) Order results for implicit Runge–Kutta methods applied to stiff systems, *SIAM J. Numer. Anal.*, **22**, 515–534.

Gear, C. W. (1965) Hybrid methods for initial value problems in ordinary differential equations, *SIAM J. Numer. Anal.*, **2**, 69–86.

Gear, C. W. (1967) The numerical integration of ordinary differential equations, *Math. Comp.*, **21**, 146–156.

Gear, C. W. (1969) The automatic integration of stiff ordinary differential equations: in *Information Processing 68: Proc. IFIP Congress, Edinburgh, 1968*, ed. A. J. H. Morrell, North-Holland, Amsterdam, pp. 187–193.

Gear, C. W. (1971a) *Numerical Initial Value Problems in Ordinary Differential Equations*, Prentice-Hall, Englewood Cliffs, New Jersey.

Gear, C. W. (1971b). Algorithm 407, DIFSUB, for solution of ordinary differential equations, *Comm. ACM*, **14**, 185–190.

Gear, C. W. and Tu, K. W. (1974) The effect of variable mesh size on the stability of multistep methods, *SIAM J. Numer. Anal.*, **11**, 1025–1043.

Gear, C. W. and Watanabe, D. S. (1974) Stability and convergence of variable order multistep methods, *SIAM J. Numer. Anal.*, **11**, 1044–1058.

Gourlay, A. R. and Watson, G. A. (1973) *Computational Methods for Matrix Eigenproblems*, Wiley, London.

Hahn, W. (1967) *Stability of Motion*, translated by A. P. Baartz, Springer-Verlag, Berlin.

Hairer, E., Nørsett, S. P. and Wanner, G. (1987) *Solving Ordinary Differential Equations I: Nonstiff Problems*, Springer-Verlag, Berlin.

Hall, G. (1976) Implementation of linear multistep methods: in *Modern Numerical Methods for Ordinary Differential Equations*, eds G. Hall and J. M. Watt, Clarendon Press, Oxford, pp. 86–104.

Henrici, P. (1962) *Discrete Variable Methods in Ordinary Differential Equations*, Wiley, New York.

Henrici, P. (1963) *Error Propagation for Difference Methods*, Wiley, New York.

Hindmarsh, A. C. (1974) GEAR: Ordinary differential equation system solver, *Lawrence Livermore Laboratory Rept. UCID-30001, Rev. 3*.

Hindmarsh, A. C. (1975) GEARB: Solution of ordinary differential equations having banded Jacobian, *Lawrence Livermore Laboratory Rept. UCID, Rev. 1*.

Hundsdorfer, W. H. and Spijker, M. N. (1981a) A note on *B*-stability of Runge–Kutta methods, *Numer. Math.*, **36**, 319–331.

Hundsdorfer, W. H. and Spijker, M. N. (1981b) On the existence of solutions to the algebraic equations in implicit Runge–Kutta methods, *Report No 81/49*, Inst. of Appl. Math. and Comp. Sci., University of Leiden.

Hull, T. E., Enright, W. H. and Jackson, K. R. (1976) User's guide for DVERK—a subroutine for solving non-stiff ODEs, *Dept. of Comp. Sci. Rept. 100*, University of Toronto.

Isaacson, E. and Keller, H. B. (1966) *Analysis of Numerical Methods*, Wiley, New York.

Kaps, P. and Rentrop, P. (1979) Generalized Runge–Kutta methods of order four with stepsize control for stiff ordinary differential equations, *Numer. Math.*, **33**, 55–68.

Kaps, P. and Wanner, G. (1981) A study of Rosenbrock-type methods of high order, *Numer. Math.*, **38**, 279–298.

Krogh, F. T. (1969) A variable step variable order multistep method for the numerical solution of ordinary differential equations: in *Information Processing 68: Proc. IFIP Congress, Edinburgh, 1968*, ed. A. J. H. Morrell, North-Holland, Amsterdam, pp. 194–199.

Krogh, F. T. (1973) Algorithms for changing the step size, *SIAM J. Numer. Anal.*, **10**, 949–965.

Krogh, F. T. (1974) Changing stepsize in the integration of differential equations using modified divided differences: in *Proc. Conf. on the Numerical Solution of Ordinary Differential Equations, University of Texas at Austin, 1972*, ed. D. G. Bettis, Lecture Notes in Mathematics No. 362, Springer-Verlag, Berlin, pp. 22–71.

Lambert, J. D. (1971) Predictor–corrector algorithms with identical regions of stability, *SIAM J. Numer. Anal.*, **8**, 337–344.

Lambert, J. D. (1973) *Computational Methods in Ordinary Differential Equations*, Wiley, London.

Lambert, J. D. (1975) Variable coefficient multistep methods for ordinary differential equations applied to parabolic partial differential equations: in *Topics in Numerical Analysis II, Proc. of Royal Irish Academy Conf. On Numerical Analysis*, ed. J. J. H. Miller, Academic Press, London, pp. 77–87.

Lambert, J. D. (1990) On the local error and the local truncation error of linear multistep methods, *BIT*, **30**, 673–681.

Lambert, J. D. and Sigurdsson, S. T. (1972) Multistep methods with variable matrix coefficients, *SIAM J. Numer. Anal.*, **9**, 715–733.

Lancaster, P. (1969) *Theory of Matrices*, Academic Press, New York.

Lindberg, B. (1971) On smoothing and extrapolation for the Trapezoidal Rule, *BIT*, **11**, 29–52.

Liniger, W. and Willoughby, R. A. (1970) Efficient integration methods for stiff systems of ordinary differential equations, *SIAM J. Numer. Anal.*, **7**, 47–66.

Mäkela, M., Nevanlinna, O. and Sipilä, A. H. (1974) On the concepts of convergence, consistency and stability in connexion with some numerical methods, *Numer. Math.*, **22**, 261–274.

Merson, R. H. (1957) An operational method for the study of integration processes: in *Proc. Symp. Data Processing*, Weapons Research Establishment, Salisbury, S. Australia.

Mitchell, A. R. and Griffiths, D. F. (1980) *The Finite Difference Method in Partial Differential Equations*, Wiley, Chichester.

Moulton, F. R. (1926) *New Methods in Exterior Ballistics*, University of Chicago.

Nevanlinna, O. and Liniger, W. (1978) Contractive methods for stiff differential equations. Part 1, *BIT*, **18**, 457–474.

Nevanlinna, O. and Liniger, W. (1979) Contractive methods for stiff differential equations. Part 2, *BIT*, **19**, 53–72.

Nordsieck, A. (1962) On numerical integration of ordinary differential equations, *Math. Comp.*, **16**, 22–49.

Nørsett, S. P. (1974) Semi-explicit Runge–Kutta methods, *Report Mathematics & Computation No. 6/74 Dept of Mathematics, University of Trondheim.*

Nørsett, S. P. (1976) Runge–Kutta methods with a multiple real eigenvalue only, *BIT*, **16**, 388–393.

Oliver, J. (1975) A curiosity of low order explicit Runge–Kutta methods, *Math. Comp.*, **29**, 1032–1036.

Prince, P. J. and Dormand, J. R. (1981) High order embedded Runge–Kutta formulae, *J. Comp. and Appl. Math.*, **7**, 67–75.

Prothero, A. (1976) Multistep methods for stiff problems: in *Modern Numerical Methods for Ordinary Differential Equations*, eds G. Hall and J. M. Watt, Clarendon Press, Oxford, pp. 152–163.

Prothero, A. and Robinson, A. (1974) On the stability and accuracy of one-step methods for solving stiff systems of ordinary differential equations, *Math. Comp.*, **28**, 145–162.

Richtmyer, R. D. and Morton, K. W. (1967) *Difference Methods for Initial Value Problems*, 2nd ed, Wiley, New York.

Robertson, H. H. and Williams, J. (1975) Some properties of algorithms for stiff differential equations, *J. Inst. Math. Appl.*, **16**, 23–24.

Rosenbrock, H. H. (1963) Some general implicit processes for the numerical solution of differential equations, *Comput. J.*, **5**, 329–330.

Sanz-Serna, J. M. (1981) Linearly implicit variable coefficient methods of Lambert-Sigurdsson type, *IMA J. Numer. Anal*, **1**, 39–45.

Seinfeld, J. K., Lapidus, L. and Hwang, M. (1970) Review of numerical integration techniques for stiff ordinary differential equations, *Ind. Eng. Chem. Fundam.*, **9**, 266–275.

Shampine, L. F. (1980) What everyone solving differential equations numerically should know: in *Computational Techniques for Ordinary Differential Equations*, eds I. Gladwell and D. K. Sayers, Academic Press, London, pp. 1–17.

Shampine, L. F. (1982) Type-insensitive ODE codes based on implicit $A(\alpha)$-stable formulas, *Math. Comp.*, **39**, 1109–123.

Shampine, L. F. and Baca, L. S. (1986), Fixed vs variable order Runge–Kutta, *ACM Trans. Math. Software*, **12**, 1–23.

Shampine, L. F. and Gordon, M. K. (1975) *Computer Solution of Ordinary Differential Equations: The Initial Value Problem*, Freeman, San Francisco.

Shampine, L. F. and Wisniewski, J. A. (1978) The variable order Runge–Kutta code RKSW and its performance, *Sandia Report SAND78-1347*, Sandia National Laboratories, Albuquerque, New Mexico.

Sharp, P. W. (1989) New low order explicit Runge–Kutta pairs, *University of Toronto Dept. Computer Science Tech. Rep. 222/89*.

Sigurdsson, S. T. (1973) Multistep methods with variable matrix coefficients for systems of ordinary differential equations, *Dept. of Comp. Sci. Rept. No. 1973. 04*, Chalmers Institute of Technology, Göteborg.

Skeel, R. D. and Kong, A. K. (1977) Blended linear multistep methods, *ACM Trans. Math. Software*, **3**, 326–343.

Spellman, J. W. and Hindmarsh, A. C. (1975) GEARS: solution of ordinary differential equations having a sparse Jacobian *Lawrence Livermore Rept. UCID-30116*.

Spijker, M. N. (1966) Convergence and stability of step-by-step methods for the numerical solution of initial value problems, *Numer. Math.*, **8**, 161–177.

Stetter, H. J. (1971) Stability of discretization on infinite intervals: in *Conf on Applications of Numerical Analysis, Dundee, 1971*, ed. J. Ll. Morris, Lecture Notes in Mathematics No. 228, Springer-Verlag, Berlin, pp. 207–222.

Stetter, H. J. (1973) *Analysis of Discretization Methods for Ordinary Differential Equations*, Springer-Verlag, Berlin.

Varga, R. S. (1961) On higher order stable implicit methods for solving parabolic partial differential equations, *J. Math. and Phys.* **40**, 220–231.

Verner, J. H. (1978) Explicit Runge–Kutta methods with estimates of the local truncation error, *SIAM J. Numer. Anal.*, **15**, 772–790.

Verwer, J. G. (1982) An analysis of Rosenbrock methods for nonlinear stiff initial value problems, *SIAM J. Numer. Anal.*, **19**, 155–170.

Vinograd, R. E. (1952) On a criterion of instability in the sense of Lyapunov of the solutions of a linear system of ordinary differential equations, *Dokl. Akad. Nauk. SSSR* **84**, pp. 201–204. (Russian.)

Wanner, G. (1987) Order stars and stability: in *The State of the Art in Numerical Analysis*, eds. A. Iserles and M. J. D. Powell, Clarendon Press, Oxford, pp. 451–471.

Wanner, G., Hairer, E. and Nørsett, S. P. (1978) Order stars and stability theorems, *BIT*, **18**, 475–489.

Widlund, O. B. (1967) A note on unconditionally stable linear multistep methods, *BIT*, **7**, 65–70.

Wright, K. (1970) Some relationship between implicit Runge–Kutta, Collocation and Lanczos τ-methods, and their stability properties, *BIT*, **10**, 217–227.

Index

The abbreviations LM, PC and RK are used to denote linear multistep, predictor–corrector and Runge–Kutta respectively. Major text references are in **bold type**.